Mathematical Control Theory of Coupled PDEs

CBMS-NSF REGIONAL CONFERENCE SERIES IN APPLIED MATHEMATICS

A series of lectures on topics of current research interest in applied mathematics under the direction of the Conference Board of the Mathematical Sciences, supported by the National Science Foundation and published by SIAM.

GARRETT BIRKHOFF, *The Numerical Solution of Elliptic Equations*
D. V. LINDLEY, *Bayesian Statistics, A Review*
R. S. VARGA, *Functional Analysis and Approximation Theory in Numerical Analysis*
R. R. BAHADUR, *Some Limit Theorems in Statistics*
PATRICK BILLINGSLEY, *Weak Convergence of Measures: Applications in Probability*
J. L. LIONS, *Some Aspects of the Optimal Control of Distributed Parameter Systems*
ROGER PENROSE, *Techniques of Differential Topology in Relativity*
HERMAN CHERNOFF, *Sequential Analysis and Optimal Design*
J. DURBIN, *Distribution Theory for Tests Based on the Sample Distribution Function*
SOL I. RUBINOW, *Mathematical Problems in the Biological Sciences*
P. D. LAX, *Hyperbolic Systems of Conservation Laws and the Mathematical Theory of Shock Waves*
I. J. SCHOENBERG, *Cardinal Spline Interpolation*
IVAN SINGER, *The Theory of Best Approximation and Functional Analysis*
WERNER C. RHEINBOLDT, *Methods of Solving Systems of Nonlinear Equations*
HANS F. WEINBERGER, *Variational Methods for Eigenvalue Approximation*
R. TYRRELL ROCKAFELLAR, *Conjugate Duality and Optimization*
SIR JAMES LIGHTHILL, *Mathematical Biofluiddynamics*
GERARD SALTON, *Theory of Indexing*
CATHLEEN S. MORAWETZ, *Notes on Time Decay and Scattering for Some Hyperbolic Problems*
F. HOPPENSTEADT, *Mathematical Theories of Populations: Demographics, Genetics and Epidemics*
RICHARD ASKEY, *Orthogonal Polynomials and Special Functions*
L. E. PAYNE, *Improperly Posed Problems in Partial Differential Equations*
S. ROSEN, *Lectures on the Measurement and Evaluation of the Performance of Computing Systems*
HERBERT B. KELLER, *Numerical Solution of Two Point Boundary Value Problems*
J. P. LASALLE, *The Stability of Dynamical Systems* - Z. ARTSTEIN, *Appendix A: Limiting Equations and Stability of Nonautonomous Ordinary Differential Equations*
D. GOTTLIEB AND S. A. ORSZAG, *Numerical Analysis of Spectral Methods: Theory and Applications*
PETER J. HUBER, *Robust Statistical Procedures*
HERBERT SOLOMON, *Geometric Probability*
FRED S. ROBERTS, *Graph Theory and Its Applications to Problems of Society*
JURIS HARTMANIS, *Feasible Computations and Provable Complexity Properties*
ZOHAR MANNA, *Lectures on the Logic of Computer Programming*
ELLIS L. JOHNSON, *Integer Programming: Facets, Subadditivity, and Duality for Group and Semi-Group Problems*
SHMUEL WINOGRAD, *Arithmetic Complexity of Computations*
J. F. C. KINGMAN, *Mathematics of Genetic Diversity*
MORTON E. GURTIN, *Topics in Finite Elasticity*
THOMAS G. KURTZ, *Approximation of Population Processes*

Jerrold E. Marsden, *Lectures on Geometric Methods in Mathematical Physics*
Bradley Efron, *The Jackknife, the Bootstrap, and Other Resampling Plans*
M. Woodroofe, *Nonlinear Renewal Theory in Sequential Analysis*
D. H. Sattinger, *Branching in the Presence of Symmetry*
R. Témam, *Navier-Stokes Equations and Nonlinear Functional Analysis*
Miklós Csörgo, *Quantile Processes with Statistical Applications*
J. D. Buckmaster and G. S. S. Ludford, *Lectures on Mathematical Combustion*
R. E. Tarjan, *Data Structures and Network Algorithms*
Paul Waltman, *Competition Models in Population Biology*
S. R. S. Varadhan, *Large Deviations and Applications*
Kiyosi Itô, *Foundations of Stochastic Differential Equations in Infinite Dimensional Spaces*
Alan C. Newell, *Solitons in Mathematics and Physics*
Pranab Kumar Sen, *Theory and Applications of Sequential Nonparametrics*
László Lovász, *An Algorithmic Theory of Numbers, Graphs and Convexity*
E. W. Cheney, *Multivariate Approximation Theory: Selected Topics*
Joel Spencer, *Ten Lectures on the Probabilistic Method*
Paul C. Fife, *Dynamics of Internal Layers and Diffusive Interfaces*
Charles K. Chui, *Multivariate Splines*
Herbert S. Wilf, *Combinatorial Algorithms: An Update*
Henry C. Tuckwell, *Stochastic Processes in the Neurosciences*
Frank H. Clarke, *Methods of Dynamic and Nonsmooth Optimization*
Robert B. Gardner, *The Method of Equivalence and Its Applications*
Grace Wahba, *Spline Models for Observational Data*
Richard S. Varga, *Scientific Computation on Mathematical Problems and Conjectures*
Ingrid Daubechies, *Ten Lectures on Wavelets*
Stephen F. McCormick, *Multilevel Projection Methods for Partial Differential Equations*
Harald Niederreiter, *Random Number Generation and Quasi-Monte Carlo Methods*
Joel Spencer, *Ten Lectures on the Probabilistic Method, Second Edition*
Charles A. Micchelli, *Mathematical Aspects of Geometric Modeling*
Roger Témam, *Navier–Stokes Equations and Nonlinear Functional Analysis, Second Edition*
Glenn Shafer, *Probabilistic Expert Systems*
Peter J. Huber, *Robust Statistical Procedures, Second Edition*
J. Michael Steele, *Probability Theory and Combinatorial Optimization*
Werner C. Rheinboldt, *Methods for Solving Systems of Nonlinear Equations, Second Edition*
J. M. Cushing, *An Introduction to Structured Population Dynamics*
Tai-Ping Liu, *Hyperbolic and Viscous Conservation Laws*
Michael Renardy, *Mathematical Analysis of Viscoelastic Flows*
Gérard Cornuéjols, *Combinatorial Optimization: Packing and Covering*
Irena Lasiecka, *Mathematical Control Theory of Coupled PDEs*

Irena Lasiecka
University of Virginia
Charlottesville, Virginia

Mathematical Control Theory of Coupled PDEs

siam
SOCIETY FOR INDUSTRIAL AND APPLIED MATHEMATICS
PHILADELPHIA

10 9 8 7 6 5 4 3 2 1

Library of Congress Cataloging-in-Publication Data

Lasiecka, I. (Irena), 1948-
Mathematical control theory of coupled PDEs / Irena Lasiecka.
p. cm. — (CBMS-NSF regional conference series in applied mathematics ; 75)
Includes bibliographical references and index.
ISBN 0-89871-486-9 (pbk.)
1. Control theory. 2. Differential equations, Hyperbolic. 3. Differential equations, Parabolic. 4. Coupled mode theory. I. Title. II. Series.

QA402.3 .L333 2002
629.8'312--dc21

2001042994

Contents

Preface

These lectures are an outgrowth of recent research in the area of control theory for systems governed by coupled partial differential equations (PDEs). While mathematical control theory for a single PDE has long occupied a central place within a broader discipline often referred to as control of infinite-dimensional systems, more complex and challenging issues have arisen recently in the context of coupled or interconnected PDE systems. This is in response to modern technological demands where the interaction (coupling) between various components of the model is a central mechanism in controlling the system. On the other hand, mathematical tools developed in the context of single equations more often than not are inadequate for dealing with and accounting for the more complex interactive nature of the structures. This has propelled a rapidly growing interest within the PDE community, which has posed an array of new questions and problems, including, How can we take advantage of the coupling within the model in order to improve system's performance? This question is clearly tightly connected to the issue of (controlled) propagation of certain properties from one component of a system into another, more generally, creating of new phenomena via coupling.

The purpose of these lectures is to describe several classes of coupled PDE models that display the above properties and to present mathematical tools and a cogent framework by which to successfully analyze the associated control problems.

We concentrate on systems of coupled PDEs of hyperbolic and parabolic type, which, moreover, are nonlinear and may include thermal effects. (The latter further contribute to the mixing of various dynamical components of the overall dynamics.) The controls are rough and as such naturally arise in the context of smart material technology. Particular emphasis is paid to boundary and point controls whose mathematical treatment is more involved and leads to the theory of semigroups with unbounded inputs.

For the above model, the ultimate goal of these lectures is to provide a mathematical theory for the solution of the following three main problems:

- well-posedness and regularity,
- stabilization and stability,
- optimal control for finite or infinite horizon problems and existence and uniqueness of associated Riccati equations.

When stabilization is sought, emphasis is placed on boundary feedback controls (or damping terms on the boundary). This, inevitably, leads to the fundamental question of solvability of associated Riccati equations and the regularity of the corresponding gain operators—a topic occupying a central place in these lectures.

Moreover, in the case of any of these three problems, the preliminary task is to provide the appropriate functional analytic or operator-theoretic setting, which includes suitable unbounded operators to model actuators and boundary feedback controls.

Throughout the development of our study of systems of coupled hyperbolic and parabolic PDEs, we endeavor to point out the key role played by two interweaving features of the respective dynamical components. These are (i) exceptional sharp regularity of the traces of the solutions of the hyperbolic component (variable), and (ii) analyticity of the solutions to the parabolic component, its propagation, and related analytic semigroup estimates.

The mechanism through which these two distinctly different phenomena, generated by the distinctly different dynamical components, propagate and interact through the overall structure to yield the desired mathematical and control-theoretic results is the main theme of these lectures. One of the payoffs obtained through this type of analysis is the *boundedness property* of feedback operators arising in an optimal control problem with highly unbounded control operators. Indeed, while it is known that optimal feedbacks arising in unbounded input control problems governed by hyperbolic dynamics are unbounded, it is precisely interaction with the parabolic component that provides local smoothing and allows one to show that feedback control operators are in fact much more regular than one may expect from the formulation of the problem. These results are critical for existence and uniqueness of the corresponding Riccati equations and well-posedness of feedback synthesis. The latter property is fundamental for construction of viable control algorithms.

Throughout these lectures we use the structural acoustic model as the canonical example illustrating various coupling phenomena arising in interconnected systems. This particular model consists of a coupled wave equation (acoustic medium) and a plate or shell equation (elastic structure) connected by an interface that is a portion of the boundary for the acoustic medium. This model, although rather specific, is nonetheless quite rich in its variety of phenomena of interest. On the other hand, the theory presented is more general and is applicable to other structures as well.

This book is intended for applied mathematicians and theoretical engineers with an interest in the mathematical quantitative analysis of control paradigms. The expected background includes exposure to modern PDE theory, semigroups, functional analysis, and basic infinite-dimensional control theory.

Acknowledgments

This book is an outgrowth of the lectures given by the author at the CMBS-NSF Regional Conference held at the University of Nebraska, Lincoln, August 4–9, 1999. The author sincerely thanks Professor Richard Rebarber for organizing the lectures and the Department of Mathematics and Statistics for hosting the conference.

The author also thanks Professor George Avalos for his help with the writing of Chapters 5–7, which deal with structural acoustic interactions. In fact, many of the results presented in these chapters are based on research papers authored or coauthored by Avalos.

This research was partially supported by NSF grant DMS-9804056 and Army Research Office grant DAAH04-96-0099.

Chapter 1

Introduction

1.1 Control Theory of Dynamical PDEs

What is control theory of a dynamical system? Traditionally, the classical viewpoint taken in the study of differential equations consisted of the (passive) analysis of the evolution properties displayed by a specific equation, or a class of equations, in response to given data. Control theory, however, injects an active mode of synthesis in the study of differential equations: it seeks to influence their dynamical evolution by selecting and synthesizing suitable data (input functions or control functions) from within a preassigned class, to achieve a predetermined desired outcome or performance. Within the deterministic (as opposed to stochastic) setting, canonical examples of sought-after goals include the following:

- The optimal control problem. This consists of the minimization, over a preassigned time horizon, of a preassigned performance cost functional, which penalizes both the control function and its corresponding solution, within the preassigned class of admissible controls. Such problems include reaching a target in minimal time or with minimal control and solution fuel (energy).
- The problem of controllability. This consists, depending on the dynamical properties of the differential equation at hand, of selecting a suitable control function, perhaps of minimal norm within a preassigned class, that steers a preassigned initial state exactly to (or to an arbitrarily small neighborhood of) a preassigned target state over a preassigned time interval, preferably the smallest one. Thus stated, this is an example of an open-loop control problem.
- Stabilization problem. This consists of forcing an original free (i.e., with no control) nonstable system to become asymptotically stable as time goes to infinity, by selecting, introducing, and synthesizing a suitable damping or dissipative term. The latter may be viewed as a control in feedback form, and thus the resulting problem is an example of closed-loop control.

These are just a few canonical illustrations of the target-oriented philosophy of synthesis, which is intrinsic to control theory. Moreover, the three sample control problems mentioned are typically interdependent. For instance, the controllability or the stabilization problem generally is a necessary prerequisite for the study of some optimal control problems over an infinite time horizon. Since the outgrowth, in the late 1950s, of control theory as a discipline in itself, within the fields of differential equations and optimization theory, the above problems have been recognized as being at the heart of control theory. They, in turn, produce a wealth of ramifications and variations.

1.1.1 Finite- versus infinite-dimensional control theory

Depending on the type of differential (or discrete) equations, the synthesis or implementation via actuators of the sought-after control function—and, more generally, the resulting control theory—takes on diverse, dynamics-dependent features. Models described by ordinary differential equations (ODEs) are usually referred to as being part of finite-dimensional control theory, because the state of such a model and the control function are finite-dimensional. By contrast, models described by (linear and nonlinear) PDEs may be viewed as (the most significant) part of infinite-dimensional control theory [152, 14, 60, 20, 28, 140].

Naturally, the specific and precise formulation of a control problem, such as the three examples above, need to be tuned to the particular class of dynamical systems at hand. These will then determine the techniques to be used and the nature of the results to be obtained. Within infinite-dimensional control theory, say, in the linear case, it is possible to invoke the unifying formulation provided by semigroup theory. This, however, works best in the case of nice classes of controls such as distributed controls, whose action on the abstract differential equation is exercised by virtue of a bounded control operator. In this case, regularity theory provided by semigroup theory is adequate to formulate and solve, with the help of operatory theory, a number of control-theoretic questions, or to ascertain their limitations in the context of infinite-dimensional spaces, where new pathological phenomena arise, of which there is no counterpart in the finite-dimensional case. All this, by and large, became available by the 1970s; see [152, 14].

1.1.2 Boundary/point control problems for single PDEs

By contrast, the situation is drastically different when the control action is rough, such as occurs when the control acts either as a boundary control on the boundary (or part thereof) of the spatial domain or as a point control through a Dirac measure (or its derivative) supported at, say, an interior point of the spatial domain. In these latter cases—which are mathematically much more challenging and physically much more appealing and relevant—the control operator has its range in a space definitively larger than the state space, and with weaker topology.

Naturally, to extract the best possible results and tune the technical tools to the problem at hand for a single PDE (as opposed to a system of PDEs), it is necessary to distinguish at the outset between different PDE classes: primarily, parabolic-like dynamics versus hyperbolic-like dynamics, with further subdistinctions in the latter class. This is due to well-known, intrinsically different dynamical properties between these two classes. As a consequence, they lead to two drastically different basic abstract models whose characterizing features set them apart. Accordingly, these two abstract models, therefore, need to be investigated by corresponding different technical strategies and tools. As a consequence, different types of distinctive results are achieved to characterize the two classes. All this dictates that, when dealing with a single equation, the abstract theory needs to bifurcate at the very outset into a parabolic-like model and hyperbolic-like basic models. Moreover, in the latter class, a further distinction into finite and infinite time horizons is called for when studying optimal control problems, to account for different, critical properties between these two cases.

In the case of classes of single PDEs—parabolic-like versus hyperbolic-like—such as arise in boundary or point control, a comprehensive, up-to-date, abstract treatment of quadratic optimal control problems over a finite or infinite time horizon, and related differential or algebraic Riccati equations, is given in [140]; see also [28]. Both continuous and approximation theories are included. These volumes present a technical extension of the deterministic optimal control problem, aimed at accommodating and encompassing multidimensional PDEs with boundary or point control or observation, in a natural way. Thus, throughout these volumes, emphasis is placed on unbounded control operators and, possibly, on unbounded observation operators as well, as they arise in the context of various abstract frameworks that are motivated by, and ultimately directed to, PDEs with boundary or point control and observation. A key feature of [140] is a wealth of concrete, multidimensional PDE illustrations, which naturally fit into the abstract theory, with no artificial assumptions imposed, at both the continuous and numerical level.

1.1.3 Boundary/point control problems for systems of coupled PDEs

The present lectures deal with boundary or point control problems for systems of highly coupled PDEs, possibly of different type, such as they arise in modern technological applications. A motivating driving force is the so-called structural acoustic problem, which, accordingly, receives much attention in these lectures.

In one canonical setting, the system to be studied couples a hyperbolic (wave) equation (within the acoustic chamber) with an elastic parabolic-like equation (because of the so-called structural damping) defined on a flat elastic wall of the chamber, with coupling taking place at the boundary. Thus, the overall system is more complex than just a single PDE; indeed, the overall system mixes a hyperbolic dynamical component with a parabolic dynamical component. The control action is rough. Then, for the quadratic optimal control problem, it is clear that the results on single PDE equations in [140] cannot be applied directly. While the single PDE treatment [140, 28] serves as a foundational basis for both results to be expected and techniques to be employed, some critical features of the problem are unpredictable, and new phenomena arise that require additional technical treatment. Would the parabolic component succeed in propagating its smoothing properties across, and onto, the whole coupled system? Would the resulting system behave globally more akin to a parabolic or to a hyperbolic structure? The theory included in these lectures will give a precise answer to these, and similar, questions.

The topic of systems of coupled PDEs and their control-theoretic properties, like many classes defined by exclusion, is simply too vast to be treated in a systematic way. Here, we narrow the focus to classes of systems of coupled PDEs which are driven by recent technological applications. Accordingly, we largely limit ourselves to optimal control problems defined for coupled hyperbolic and parabolic components, such as waves, plates, and heat equations. For these, we then attempt to bring forward several approaches, techniques, and methods that have proved successful in recent years in the resulting extension of the single PDE case of [140, 28]. As in the structural acoustic prototype model, we consider those situations where coupling takes place on the interface of two regions. At any rate, while we refer to established engineering literature for the modeling aspects, their range of validity, and related practical aspects of implementation, our focus here is on the mathematical analysis of the problem. In particular, we seek to unveil new mathematical questions, or difficulties, inherent to the more complex coupled model and then present suitable approaches for their solution.

1.2 Goal of the Lectures

We concentrate on systems of coupled PDEs of hyperbolic and parabolic type, which, moreover, are nonlinear and may include thermoelastic effects. (The latter further contributes to the mixing of various dynamical components of the overall dynamics.) The controls are rough as they naturally arise in the context of smart material technology. For the above model, the ultimate goal of these lectures is to provide a mathematical theory for the solution of the three main problems: well-posedness and regularity; stabilization and stability; and optimal control for finite or infinite horizon problems along with the associated Riccati equations.

When stabilization is sought, emphasis is placed on boundary feedback controls (or damping terms on the boundary). As noted, a positive solution of the stabilization problem serves as a necessary prerequisite for the optimal control problem over an infinite time horizon. Moreover, a preliminary task for any of the three problems consists of providing the appropriate functional analytic or operator-theoretic setting, which includes suitable unbounded operators to model actuators and boundary feedback controls.

To study a control problem (which, in some sense, can be seen as an inverse problem), it is necessary to understand well the properties of the forward problem. This entails an analysis of the well-posedness and regularity of the underlying PDE system. Thus, Chapter 2 provides the rudiments of such an analysis as specialized to the systems of interest, which are predominantly of second order in time and with rough forcing terms. Particular models are nonlinear wave equations, nonlinear plate equations (with and without thermoelasticity), and full von Karman systems, which couple elastodynamic waves with nonlinear plates. The analysis for waves and plates is a preliminary step toward the study of a well-posedness theory in structural acoustic interactions. Particular emphasis is given to the effects of nonlinear boundary conditions. These play a major role in the analysis to follow.

The study of asymptotic behavior and stabilizability is taken up in Chapters 3 and 4. Chapter 3 starts with a presentation of some preliminary abstract stability inequalities used throughout the lectures. Detailed analysis of the stabilization problem formulated for a benchmark case modeled by a semilinear wave equation is given in section 3.2. Here, particular emphasis is given to the issue of low regularity of the solutions, the nonlinearity of boundary conditions, and the potential nonuniqueness of the solutions. Section 3.4 provides a collection of relevant stability results pertinent to asymptotic stability of nonlinear plate dynamics. Building on the stability results for waves and plates, Chapter 4 deals entirely with stability issues pertinent to the nonlinear structural acoustic interactions. All three major damping mechanisms affecting flexible walls are discussed in detail: structural, mechanical, and thermal damping.

The remaining three chapters deal with optimal control problems formulated in the context of structural acoustic models, where the control actuation is modeled by highly unbounded operators (as they arise in smart material technology). Here, the dynamics considered are predominantly linear. Moreover, as noted before, coupling on the interface and propagation of analyticity and hyperbolicity play major role in the arguments. In Chapter 5 we provide a detailed description of various models which includes PDEs as well as abstract (semigroup) formulations for various structures considered.

Chapter 6 provides a complete solution to the optimal control problem defined on a finite time horizon. Here the main result is a derivation of an underlying Riccati equation, which culminates with a regularity analysis of the gain feedbacks. This, in turn, is critical for proper justification and formulation of the quadratic term in the differential Riccati equation. The subtlety of the issue is due to the unboundedness of the control operator, which enters quadratically into the Riccati equation. Thus, a priori, such an equation may not be properly formulated and the meaning of its solution is not clear in advance. On the other hand, in the case of dynamics generated by analytic semigroups, the theory of such Riccati equations is complete and well understood [140, 28]. This is due to regularizing effects of the analytic dynamics which permits a proper definition of the gain operators. The main contribution of Chapters 6 and 7 is in showing that for interactive structures that arise in structural acoustic problems (which are not analytic), we still have complete solvability of the Riccati equations. The corresponding Riccati equations do have well-defined solutions with meaningful gain operators that lead to a feedback solution of the optimization problem. Chapter 6 deals with the infinite horizon problem where issues already addressed in the finite horizon problem are compiled with an intrinsic stability and stabilization property for the overall system. It is here where we use the stability results developed in Chapters 3 and 4. The ultimate goal of this chapter is to provide a rigorous analysis of the existence and uniqueness of the resulting nonstandard algebraic Riccati equations.

As a final note we mention that little space is devoted in these lectures to the issue of exact controllability. The main reason for this is that point controls (as they typically arise in smart materials applications) offer very little in the context of exact controllability in the desired spaces of finite energy. In fact, most of the results here are of a negative nature. An additional difficulty of the problem is the hybrid connection of the system which is well known, even for much simpler models, to be not exactly controllable in such spaces [156]. Thus, the main difficulties with controlling exactly structural acoustic models are due to the hybrid nature of connections and the well-known and documented limited effectiveness of point controls in the context of infinite-dimensional (particularly, hyperbolic) systems.

Chapter 2

Well-Posedness of Second-Order Nonlinear Equations with Boundary Damping

2.1 Orientation

This chapter deals with an abstract second-order system written as a nonlinear second-order (in time) equation defined on a suitable Hilbert space. Since the aim of these lectures is directed toward the study of boundary control systems, particular emphasis is given to evolutionary problems with nonlinear terms on the boundary. These lead to nonlinear operators that are not defined on the basic state space. Our aim is to establish, under rather general hypotheses, existence, uniqueness, and regularity theory for such equations. This, besides being a topic of interest on its own, is a critical preliminary step in studying questions related to stabilizability, controllability, and optimization of such systems. While one does not need to defend a need for existence and uniqueness theory, it will become clear that the regularity of solutions is an indispensable part of control analysis. Indeed, many PDE calculations, which are necessary to establish control theoretic properties of the system, are formal only in the absence of sufficient regularity of solutions. Therefore, it is important to know when and under which conditions imposed on the data of the problem the desired regularity of solutions is available.

The main aim of this chapter is to establish well-posedness of several nonlinear dynamical systems that are relevant to the study of control systems dealt with in the subsequent lectures. Thus, here we formulate a fairly general second-order nonlinear evolutionary system [59] that is a benchmark model for several interactive control systems described by coupled PDEs. We present several well-posedness results pertaining to both local and, when available, global existence of weak solutions. Depending on the assumptions imposed, these solutions may be unique or may not. In sections 2.4–2.6, we provide several examples of dynamical models arising in nonlinear elasticity, which were meant to be prototypes for the abstract theory introduced in section 2. These include nonlinear von Karman plate equations, nonlinear waves, and nonlinear structural acoustic models that combine features of nonlinear waves and plates.

While the abstract setup of section 2 is fairly general, it does not encompass several dynamical models of physical interest. One example is a full von Karman system of dynamic elasticity in two dimensions. This particular system plays an important role in control theory of nonlinear elastic structures. For this reason, in section 2.7, we present an independent treatment of well-posedness theory that is tailored to this type of system.

2.2 Abstract Model

We shall treat the second-order abstract model

$$\begin{cases} Mu_{tt}(t) + Au(t) + AG\partial\Phi(G^*Au_t(t)) + AGf(u(t)) + Du_t(t) \ni F(u(t)); \\ u(t=0) = u_0, \quad u_t(t=0) = u_1 \end{cases} \tag{2.2.1}$$

under the following set of assumptions.

Assumption 2.2.1.

1. *A is a realization of a closed, linear positive self-adjoint operator $\tilde{A}$, acting as a Hilbert space H with $D(\tilde{A}) \subset H$, as an operator $D(\tilde{A}^{1/2}) \to D(\tilde{A}^{1/2})'$. We denote by $|\ |$ and $\|\ \|$ the norm of H and $D(\tilde{A}^{1/2})$, respectively. $(\ ,\ )$ denotes a scalar product in H. We use the same symbol to denote the duality pairing between $D(\tilde{A}^{1/2})$ and $D(\tilde{A}^{1/2})'$.*
2. *Let V be another Hilbert space such that $D(\tilde{A}^{1/2}) \subset V \subset V' \subset D(\tilde{A}^{1/2})'$, all injections being continuous and dense, $M \in L(V, V')$ and $(Mu, u) \geq \alpha_0|u|_V^2$, where $\alpha_0 > 0$ and $(,)$ is understood as a duality pairing between V and V'. Hence, $M^{-1} \in L(V', V)$. Setting $\bar{M} = M|_H$ with $D(\bar{M}) = u \in V, Mu \in H$, we have $D(\bar{M})^{1/2} = V$.*
3. *Let U and U_0 be other Hilbert spaces such that $U_0 \subseteq U \subseteq U_0'$. We denote by $\langle \cdot, \cdot \rangle$ the scalar product on U and the duality pairing between U_0 and U_0'. $\Phi : U \to R^1$ is a proper, convex, lower semicontinuous function with subgradient $\partial\Phi \in U_0 \times U_0'$; see [19, 36]. We assume that $0 \in \partial\Phi(0)$.*
4. *The linear operator $G : U_0' \to H$ satisfies $\tilde{A}^{1/2}G : U_0' \to H$ is bounded or, equivalently, $G^*A : D(\tilde{A}^{1/2}) \to U_0$ is bounded, where $\langle G^*u, v\rangle = (u, Gv)$. We assume that $G^*A : D(\tilde{A}^{1/2}) \to U_0$ is surjective or else that $\partial\Phi(u) = g(u)$, where g is a monotone, continuous vector valued function: $U_0 \to U_0'$.*
5. *The nonlinear operator $F : D(\tilde{A}^{1/2}) \to V'$ is locally Lipschitz, i.e., $|F(u_1) - F(u_2)|_{V'} \leq C(\| u_1 \|, \| u_2 \|) \| u_1 - u_2 \|$. Here $C(\| u \|, \| u_2 \|)$ denotes a function of $\| u_i \|$ which is bounded for bounded values of the argument.*
6. *The nonlinear operator $f : D(\tilde{A}^{1/2}) \to U$ is locally Lipschitz, $|f(u_1) - f(u_2)|_U \leq C(\| u_1 \|, \| u_2 \|) \| u_1 - u_2 \|$.*
7. *The operator $D : D(\tilde{A}^{1/2}) \to [D(\tilde{A}^{1/2})]'$ is assumed bounded, linear, and positive, i.e., $\mathrm{Re}\,(Du, u) \geq 0$; $u \in D(\tilde{A}^{1/2})$.*

Remark 2.2.1. *We note that the presence of the damping operator D is redundant in the model (2.2.1). Indeed, this term can be modeled by $AG\partial\Phi G^*A$, where we take $G \equiv A^{-1/2}$, $\partial\Phi \equiv A^{-1/2}DA^{-1/2}$. However, we prefer to single out this term,* *mostly for the convenience of the reader. In several applications (e.g., structural acoustic problems), this term plays a special role describing the coupling in the structure.*

In what follows we consider the inclusion (2.2.1) subject to Assumption 2.2.1 with the initial data $(u_0, u_1) \in D(\tilde{A}^{1/2}) \times V$. As we will see later, this model represents a large variety of nonlinear systems arising in nonlinear elasticity. Some of these are semilinear wave equations with nonlinear damping, semilinear Euler–Bernoulli or Kirchhoff plate models with nonlinear damping, van Karman plate equations, nonlinear models of shell equations, and nonlinear structural acoustic models.

We note that the composition operator, A G, defined from $D(\tilde{A}^{1/2}) \to D(\tilde{A}^{1/2})'$, represents various boundary operators; see [129]. Thus, the term $AGf(u)$ represents semilinear boundary conditions. Instead, the term $AG\partial\Phi G^* Au_t(t)$ models nonlinear boundary dissipation. We also note that the problem described by (2.2.1), under Assumption 2.2.1, is neither monotone nor (locally) Lipschitz. Indeed, the presence of operators $F(u)$ and $f(u)$ prevents monotonicity, while the presence of dissipation $AG\partial\Phi(G^* Au_t)$ and boundary semilinearity $AGf(u)$ makes the problem non-Lipschitz (on a natural state space which is $D(\tilde{A}^{1/2}) \times V$).

In what follows we discuss questions related to the existence, uniqueness, and regularity of solutions to the system in (2.2.1). Our results cover the following setups:

- The so-called regular case, when the boundary nonlinearity is monotone (i.e., $f(u)$ is affine) and the nonlinearity $F(u)$ complies with Assumption 2.2.1. In this case, the results obtained include local existence and uniqueness of weak solutions. Subject to appropriate structural Assumption 2.3.2, these solutions can be extended to global.
- The case when the boundary nonlinearity is not necessarily monotone, i.e., the nonlinear term $f(u)$ is explicitly included in the model. In this case some coercivity assumption imposed on the graph $\partial\Phi$ is required. Indeed, under the strong coercivity Assumption 2.3.1, both existence and uniqueness of weak solutions is proved in Theorem 2.3.1. A weaker-asymptotic-coercivity assumption in Assumption 2.3.3 provides, instead, only an existence result without the uniqueness; see Theorem 2.3.3.

2.3 Existence and Uniqueness: Statement of Main Results

To state our results, we need to give a precise definition of solutions. While in the linear theory the meaning of (weak) solutions is well understood, this is not the case for nonlinear equations. In fact, one needs to be very careful about what is meant by solution and how such a solution can be interpreted in terms of the quantities entering the formal equation. Our approach relies on two basic concepts, one related to weak solutions introduced via maximal monotone operator theory and another one, mild solutions, related to PDE variational framework. Often there is no relation between these two different concepts of solutions.

To accomplish our goal, we consider the following m-monotone nonhomogeneous problem:

$$u_{tt} + M^{-1} A[(u + G\partial\Phi G^* u_t + Gf_0)] + M^{-1}Du_t \ni M^{-1} F_1,$$
$$u(0) = u_0; u_t(0) = u_1, \qquad (2.3.2)$$

where $F_1 \in L_1(0,T;V')$, $f_0 \in L_2(0,T;U)$ are given elements.

Definition 2.3.1. *We say that the function $\tilde{u}(t) \equiv (u(t), u_t(t))$ is a* strong *solution to* (2.3.2) *on the interval* $(0,T)$ *iff*

$$\tilde{u} \in C([0,T]; D(\tilde{A}^{1/2}) \times V); \tilde{u}_t \in L_\infty([0,T]; D(\tilde{A}^{1/2}) \times V)),$$
$$A(u + G\partial\Phi G^* u_t + Gf_0) + Du_t \in L_\infty([0,T]; V'),$$

and inclusion (2.3.2) *holds almost everywhere (a.e.) on* $(0,T)$ *with the values in* V.

Weak solutions are defined as the appropriate limits of strong solutions; see [36, 19].

Definition 2.3.2. *We say that function $\tilde{u}(t) \equiv (u(t), u_t(t) \in C([0,T]; D(\tilde{A}^{1/2}) \times V)$ is a* weak *solution to* (2.3.2) *with prescribed forcing terms F_1, f_0 iff there exists a sequence of functions $F_{1,n}, f_{0,n}$ and of the corresponding strong solutions $\tilde{u}_n(t)$ of* (2.3.2) *such that $F_{1,n} \to F_1$ in $L_1(0,T,V')$, $f_{0,n} \to f_0$ in $L_2(0,T;U)$, and $\tilde{u}_n \to \tilde{u}$ in $C([0,T]; D(\tilde{A}^{1/2}) \times V)$.*

Definition 2.3.2 is a basis for defining weak solutions to the nonlinear problem (2.2.1).

Definition 2.3.3. *We say that function $\bar{u}(t) \equiv (u(t), u_t(t) \in C([0,T]; D(\tilde{A}^{1/2}) \times V)$ is a weak solution to* (2.2.1) *iff $\tilde{u}(t)$ is a weak solution to* (2.3.2) *with $F_1 \equiv F(u)$, $f_0 \equiv f(u)$*

Our first result deals with the case when the subgradient $\partial\Phi$ is assumed to be strictly coercive.

Assumption 2.3.1. *We assume that either $|G^*Au|_U \leq C|u|_V$ or $\partial\Phi$ is coercive on U. This is to say, there exists $\alpha > 0$ such that for all $(w,\eta) \in \partial\Phi \subset U_0 \times U_0'$ and $(z,\xi) \in \partial\Phi \in U_0 \times U_0'$, we have $\langle \xi - \eta, z - w\rangle \geq \alpha|w - z|_U^2$.*

Theorem 2.3.1 (local existence). *In addition to the standing Assumptions* 2.2.1, *we assume the coercivity Assumption* 2.3.1. *Then*

(i) *for all $(u_0, u_1) \in D(\tilde{A}^{1/2}) \times V$, there exists $T_0 > 0$ such that there exists a unique weak solution of problem* (2.2.1) *defined on the maximal interval of existence $[0, T_0)$; moreover,*
(ii) *the following regularity property holds:*

$$\int_0^{T_0} |G^*Au_t|_U^2 dt \leq C_{T_0}(\| u_0 \|, |u_1|_V). \qquad (2.3.3)$$

The proof of Theorem 2.3.1, which is based on the theory of maximal monotone operators and the contraction mapping principle, is given in [105].

Remark 2.3.1. *Notice that the estimate* (2.3.3) *does not follow from the regularity of weak solutions. It is an independent regularity result.*

In a special case when f is an affine function, the result of Theorem 2.3.1 holds without the additional coercivity Assumption 2.3.1. In fact, we have the following corollary.

Corollary 2.3.1. *Under Assumption* 2.2.1 *and*

$$f(u) = c_0 + c_1 G^* A u, \qquad c_0 \subset U, c_1 \in R, \tag{2.3.4}$$

the assertions of Theorem 2.3.1, *part* (i), *hold true.*

Global solutions to the problem (2.2.1) can be obtained under the additional structural assumptions imposed on the functions F and f. These are formulated below.

Assumption 2.3.2. *For all weak solutions to problem* (2.2.1), $\tilde{u} = (u, u_t) \in C([0, T_0]; D(\tilde{A}^{1/2}) \times V)$, *the following inequalities hold for* $0 \leq t \leq T_0$*:*

(F) $$\int_0^t (-F(u(\tau), u_t(\tau))d\tau + C_1 \int_0^t \|u(\tau)\|^2 + |u_t(\tau)|_V^2 d\tau + C_2(\|u_0\|, |u_1|_V) \geq 0,$$

(f) $$\int_0^t \langle f(u(\tau)), G^* A u_t(\tau)\rangle d\tau + C_1 \int_0^t \|u(\tau)\| |G^* A u_t(\tau)|_U d\tau + C_2(\|u_0\|, |u_1|_V) \geq 0,$$

where C_1 *is a suitable constant and* C_2 *denotes a function that is bounded for bounded values of the arguments.*

Theorem 2.3.2 (global existence). *In addition to the assumptions of Theorem* 2.3.1, *we assume structural Assumption* 2.3.2. *Then the weak solution* ($\tilde{u}$) *to problem* (2.2.1) *is global on* $[0, T]$ *for an arbitrary* $T < \infty$.

Proof of Theorem 2.3.2 is given in [105].

Corollary 2.3.2. *Under condition* 2.3.4 *of Corollary* 2.3.1 *and inequality* (F) *in Assumption* 2.3.2, *weak solutions to* (2.2.1) *are global on* $[0, T]$ *with* $T < \infty$ *arbitrary.*

Remark 2.3.2. *In Theorem* 2.3.2, *one may replace Assumption* 2.3.2 *by an existence of an a priori bound for the corresponding solutions. In fact, in several situations (see examples in sections* 2.5–2.7*) structural Assumption* 2.3.2 *may not hold, but a priori bounds in the right function spaces do hold and imply global existence of solutions.*

For the convenience of the reader, we formulate below a special version of Theorem 2.3.2 which deals with single-valued nonlinearities $\partial\Phi$.

Corollary 2.3.3. *With reference to the model* (2.2.1), *we assume that the operators* $\tilde{A}, M, D$ *satisfy parts* 1, 2, 5, *and* 7 *of Assumption* 2.2.1. *In addition, we assume*

- *$A^{1/2}G : U \to H$ is bounded;*
- *$\partial\Phi(u) = g(u), u \in U$, where a vector valued function g is monotone increasing, continuous $U \to U$, and $g(0) = 0$; and*
- *function* (F) *is subject to structural Assumption* 2.3.2 *and* (f) *satisfies condition* (2.3.4).

Then, there exists a unique weak, global solution $\tilde{u} \in C([0,T]; D(A^{1/2}) \times V)$ for all $T < \infty$.

We complete this section by providing a different framework for studying the problem without assuming strong coercivity Assumption 2.3.1. This, however, is at the price of imposing some bounds on the nonlinear operator $\partial\Phi$ and also certain good approximation properties of the nonlinear functions together with the a priori bounds. Situations of this type frequently arise in the context of Nemytski (substitution) operators in problems where the a priori bounds are secured by applications of appropriate energy methods. A good example for this model is a semilinear wave equation with boundary dissipation and boundary nonlinearity studied in section 2.5. In what follows we take, for simplicity, $D = 0$ (this is not essential) and we study the problem under the following hypotheses.

Assumption 2.3.3. (A1) (i) *There exists a sequence of Lipschitz continuous functions $f_n : D(\tilde{A}^{1/2}) \to U$ and $F_n : D(\tilde{A}^{1/2}) \to V'$ such that whenever $u_n \to u$ weakly in $D(\tilde{A}^{1/2})$,*

(if) $f_n(u_n) \to f(u)$ in U,
(iF) $F_n(u_n) \to F(u)$ *in* V'.

(ii) *For any function $u \in C([0,t]; D(\tilde{A}^{1/2}))$ and such that $G^*Au_t \in L_2(0,t,U)$, there exist positive constants $\rho_i, i = 1, 2$, such that*

$$\text{(iif)} \quad \int_0^t \langle f_n(u(\tau)), G^*Au_t(\tau)\rangle d\tau + \rho_1 \int_0^t \|u(\tau)\||G^*Au_t(\tau)|_U d\tau \\ +C(\|u_0\|, |v_0|_V) \geq 0;$$

$$\text{(iiF)} \quad \int_0^t -F_n(u(\tau), u_t(\tau))d\tau + \rho_2 \int_0^t \|u(\tau)\|^2 + |u_t(\tau)|_V^2 d\tau \\ +C(\|u_0\|, |v_0|_V) \geq 0.$$

(A2) $0 \in \partial\Phi(0)$, $\partial\Phi$: *$U \to U$ is bounded and for $\xi_n \in \partial\Phi(z_n)$*

$$\lim_{n\to\infty} \frac{\langle \xi_n, z_n\rangle}{|z_n|_U^2} \geq \alpha > 0 \text{ when } |z_n|_U \to \infty \text{ as } n \to \infty.$$

We note that assumption (A2) requires only certain asymptotic bound from below on the nonlinear term $\partial\Phi$, rather than the much stronger strict monotonicity and coercivity condition required in Assumption 2.3.1.

To state our main results we need another definition of the solution.

Definition 2.3.4. *We say that the function $\tilde{u} = (u, u_t) \in C[0,T; D(\tilde{A}^{1/2}) \times V]$ is a mild solution of* (2.2.1) *iff*

$$u(0) = u_0, \; u_t(0) = u_1, \qquad G^*Au_t \in L_2(0,T;U), \tag{2.3.5}$$

$$\frac{d}{dt}(\tilde{A}^{-1/2}Mu_t) + \tilde{A}^{1/2}[u + Gf(u) + G\partial\Phi(G^*Au_t)] \ni \tilde{A}^{-1/2}F(u) \text{ in } H^{-1}(0,T;H). \tag{2.3.6}$$

Remark 2.3.3. *Note that in general, there may be no relation between a weak and a mild (variational) solution. However, in many cases, under some additional information on the regularity of $\partial\Phi$ one could show that a weak solution is de facto a mild solution. This requires passage with the limit on a variational form of the equation.*

Theorem 2.3.3. (i) *In addition to Assumption* 2.2.1, *we assume Assumption* 2.3.3. *Then, for all $(u_0, v_0) \in D(\tilde{A}^{1/2}) \times V$, there exists a mild solution to* (2.2.1) *defined on the interval $[0, T)$, where T is arbitrary.*

(ii) *If, in addition, we assume that either Assumption* 2.3.1 *holds or else f satisfies* (2.3.4), *then the mild solution of part* (i) *is unique.*

Proof of Theorem 2.3.3 is given in [120].

Remark 2.3.4. *Notice that, in general, one does not have the uniqueness statement without the strong coercivity Assumption* 2.3.1.

Remark 2.3.5. *The results stated above provide an existence and uniqueness of weak or mild solutions only. Under more regularity assumptions imposed on the data of the problem, one obtains more regular solutions. We do not pursue this topic here, referring the reader instead to* [61, 64] *for general technique, which allows one to deduce the additional regularity of weak solutions.*

In sections 2.4–2.7 we describe several PDE examples illustrating the theory presented in this section. Our goal is to show how various assumptions imposed on the nonlinear setup fit naturally in describing the nature of nonlinearities arising in many physical models of dynamic elasticity. Theorems 2.3.1 and 2.3.2 are applied in sections 2.4 and 2.6, respectively, to deduce the well-posedness of nonlinear plate equations and structural acoustic models, respectively. Semilinear wave equations with boundary dissipation and boundary nonlinearities are dealt with in section 2.5 by applying Theorem 2.3.3.

2.4 Nonlinear Plates: von Karman Equations

Here we deal with a nonlinear model of elasticity describing vibrations of a thin plate defined on a two-dimensional bounded domain $\Omega \subset R^2$ with a smooth boundary Γ. It is assumed that the displacements may be large, while the strains are still assumed to be small. This kind of phenomenon is described by the so-called von Karman system [99, 150], given by the set of equations consisting of (2.4.7) and (2.4.8):

$$w_{tt} - \gamma\Delta w_{tt} + \Delta^2 w = [w, \mathcal{F}(w)] \text{ in } \Omega \times (0, \infty), \tag{2.4.7}$$

where $\gamma > 0$ is proportional to the square of the thickness of the plate and hence is assumed small. From the point of view of physical interpretation, the presence of the term $\gamma \Delta w_{tt}$ represents rotational inertia. Some of the von Karman models account for the limit case $\gamma = 0$, and we do likewise. Thus, in what follows we consider the two cases $\gamma > 0$ and $\gamma = 0$. The von Karman bracket $[u, v]$ is given by

$$[u, v] \equiv u_{xx}v_{yy} + u_{yy}v_{xx} - 2u_{xy}v_{xy}.$$

The Airy stress function $\mathcal{F}$ satisfies the elliptic equation

$$\Delta^2 \mathcal{F} = -\frac{Eh}{2D}[w, w] \text{ in } \Omega, \tag{2.4.8}$$

$$\mathcal{F} = \frac{\partial}{\partial \nu}\mathcal{F} = 0 \text{ on } \Gamma, \tag{2.4.9}$$

where E is the Young's modulus and D denotes the modulus of elasticity and ν denotes a positively oriented unit normal to the boundary.

With (2.4.7) we associate the *free-type* boundary conditions given by

$$\Delta w + (1-\mu)B_1 w = -g_1\left(\frac{\partial}{\partial \nu}w_t\right) - f_1\left(\frac{\partial}{\partial \nu}w\right) \quad \text{on } \Gamma \times (0, \infty), \tag{2.4.10}$$

$$\frac{\partial}{\partial \nu}\Delta w + (1-\mu)B_2 w - \gamma \frac{\partial}{\partial \nu}w_{tt} = g_2(w_t) + f_2(w) \text{ on } \Gamma \times (0, \infty), \tag{2.4.11}$$

where the boundary operators [99] are given by

$$B_1 \equiv 2\nu_1\nu_2 \frac{\partial^2}{\partial x \partial y} - \nu_1^2 \frac{\partial^2}{\partial y \partial y} - \nu_2^2 \frac{\partial^2}{\partial x \partial x},$$

$$B_2 \equiv \frac{\partial}{\partial \tau}\left[(\nu_1^2 - \nu_2^2)\frac{\partial^2}{\partial x \partial y} + \nu_1\nu_2\left(\frac{\partial^2}{\partial y \partial y} - \frac{\partial^2}{\partial x \partial x}\right)\right] + lI, \qquad l \geq 0.$$

The functions g_i, f_i are subject to the following hypotheses.

Assumption 2.4.1. (g) *Functions $g_i \in C(R)$ are monotone increasing and such that $g_i(0) = 0$. In addition*

$$(g_i(s) - g_i(v))(s - v) \geq \alpha_i(s - v)^2, \qquad s, v \in R,$$

for some constants $\alpha_1 > 0$ and $\alpha_2 \geq 0$. If f_1 satisfies (2.3.4), *we can also have $\alpha_1 = 0$.*

(f) *Functions $f_i \in C(R)$ are locally Lipschitz. Moreover, the derivative of function f_1 is of the polynomial growth.*

The initial conditions associated with the model (2.4.7) are given by

$$w(0) = w_0, \qquad w_t(0) = w_1 \text{ in } \Omega.$$

We note that the value of the parameter γ in (2.4.7) distinguishes between two critical states: hyperbolic and nonhyperbolic, which correspond, respectively, to the cases $\gamma > 0$ and $\gamma = 0$. In the latter case, the von Karman system is often referred to as a modified von Karman system. Our analysis for the well-posedness of this system is separate for these two cases.

2.4.1 Case $\gamma > 0$

Theorem 2.4.1. *Under Assumption* 2.4.1, *for any initial data* $w_0 \in H^2(\Omega)$, $w_1 \in H^1(\Omega)$, *we have the following:*

- *There exists a unique weak solution that is local and defined on the maximal interval of existence,* $[0, T_0)$, *with the values in* $H^2(\Omega) \times H^1(\Omega)$.
- *If, in addition, the functions* f_i *satisfy* $f_i(s)s \geq 0$, $i = 1, 2$, *then such solution is global.*

Remark 2.4.1. *We note that in the case when the function* f_1 *is an affine function, i.e.,* $f_1(\frac{\partial}{\partial\nu}w) = a\frac{\partial}{\partial\nu}w + b, a \in R, b \in L_2(\Gamma)$, *there is no need for strong monotonicity of function* g_1. *In fact, we can take both* α_i *equal to zero. The reason for strong coercivity of* g_1 *is to offset the unboundedness introduced by the boundary nonlinearity* f_1.

Remark 2.4.2. *The result stated in Theorem* 2.4.1 *requires minimal assumptions on regularity of the dissipation* g_i. *Indeed, typical assumptions imposed in the literature* [99, 97], *even in the case of regular plates (i.e.,* $\mathcal{F} = 0$*), require that the dissipation be linearly bounded at the infinity.*

Remark 2.4.3. *Theorem* 2.4.1 *remains valid with other choices of boundary conditions (clamped or hinged). The assumptions involving boundary dissipation and boundary nonlinearities need to be appropriately modified.*

Proof of Theorem 2.4.1. The proof relies on an application of Theorems 2.3.1 and 2.3.2. To accomplish this we put the problem into an abstract form. We define the following spaces and operators:

$$H = L_2(\Omega), U = L_2(\Gamma), \qquad Aw \equiv \Delta^2 w; w \in D(A),$$

$$D(A) = \left\{ w \in H^4(\Omega); \Delta w + (1-\mu)B_1 w = 0; \frac{\partial}{\partial\nu}\Delta w + (1-\mu)B_2 w = 0 \text{ on } \Gamma \right\}.$$

It is well known that $D(A^{1/2}) \sim H^2(\Omega)$.

$$A_N u \equiv -\Delta u; \; D(A_N) \equiv \left\{ u \in H^2(\Omega); \frac{\partial}{\partial\nu} u = 0 \text{ on } \Gamma \right\}.$$

We define the Neumann map $N : L_2(\Gamma) \to L_2(\Omega)$ by

$$(\Delta - I)Ng = 0 \text{ in } \Omega, \qquad \frac{\partial}{\partial\nu} Ng = g \text{ on } \Gamma.$$

The remaining two Green's maps are defined analogously: $G_i : L_2(\Gamma) \to L_2(\Omega)$, $i = 1, 2$, are given by

$$\Delta^2 G_i g = 0 \text{ in } \Omega$$

and

$$\Delta G_1 g + (1-\mu)B_1 G_1 g = g, \quad \frac{\partial}{\partial\nu}\Delta G_1 g + (1-\mu)B_2 G_1 g = 0 \text{ on } \Gamma, \tag{2.4.12}$$

$$\Delta G_2 g + (1-\mu)B_1 G_2 g = 0, \quad \frac{\partial}{\partial\nu}\Delta G_2 g + (1-\mu)B_2 G_2 g = g \text{ on } \Gamma. \tag{2.4.13}$$

Finally, the nonlinear functions g_i can be identified with subgradients of appropriate convex functions (antiderivatives of g_i), so we have

$$\partial\Phi_1 = g_1, \qquad \partial\Phi_2 = g_2.$$

This is done, as usual, by defining

$$\Phi_i(u) \equiv \begin{cases} \int_\Gamma g_i(u(x))dx \text{ if } g_i(u) \in L_1(\Gamma), \\ \infty \text{ otherwise.} \end{cases}$$

With the above notation, the original PDE equation (2.4.7) with the boundary conditions (2.4.11) can be rewritten in an abstract form as [34, 140]

$$\begin{aligned}(I+\gamma A_N)w_{tt} + Aw + AG_1 g_1\left(\frac{\partial}{\partial\nu} w_t|_\Gamma\right) + AG_2 g_2(w_t|_\Gamma) \\ + AG_1 f_1\left(\frac{\partial}{\partial\nu} w\right) + AG_2 f_2(w|_\Gamma) = [\mathcal{F}(w), w]. \qquad (2.4.14)\end{aligned}$$

Noting that by the virtue of Green's formula we have (see [34])

$$G_1^* A = \frac{\partial}{\partial\nu} u, \quad G_2^* Au = u|_\Gamma \text{ for } u \in D(A^{1/2}),$$

and denoting

$$M \equiv I + \gamma A_N$$

so $D(M^{1/2}) \equiv V \sim H^1(\Omega)$, $\mathcal{H} = H^2(\Omega) \times H^1(\Omega)$, we obtain exactly the same structure as the one postulated in abstract equation (2.2.1). Moreover, requirements 1–4 of Assumption 2.2.1 are satisfied. To see this, we notice that with the choices

$$U_{01} \equiv H^{1/2}(\Gamma), U_{02} \equiv H^{3/2}(\Gamma),$$

the operators $G_1^* A : H^2(\Omega) \to U_{01}$ and $G_2^* A : H^2(\Omega) \to U_{02}$ are bounded and surjective. (This follows from the trace theorem.)

Parts 5 and 6 of Assumption 2.2.1 are satisfied as well, as we see below.

Verification of part 5 *in Assumption* 2.2.1. We introduce the following notation:

$$\mathcal{G}f \equiv u \text{ iff } \Delta^2 u = -\frac{Eh}{2D} f \text{ in } \Omega \text{ and } u = \frac{\partial}{\partial\nu} u = 0 \text{ on } \Gamma.$$

We write the Airy stress function as

$$\mathcal{F}(w) = \mathcal{G}[w, w]$$

and $F(w) = [\mathcal{G}[w,w], w]$. Moreover,

$$\begin{aligned} F(u_1) - F(u_2) &= [\mathcal{G}[u_1,u_1], u_1 - u_2] + [\mathcal{G}[u_1,u_1] - \mathcal{G}[u_2,u_2], u_2] \\ &= \mathcal{G}[u_1,u_1], u_1 - u_2] + [\mathcal{G}[u_1 - u_2, u_1 + u_2], u_2]. \qquad (2.4.15)\end{aligned}$$

Since [62]

$$|[u,v]|_{-1-\epsilon,\Omega} \leq C|u|_{2,\Omega}|v|_{2,\Omega},$$

elliptic regularity gives

$$|\mathcal{G}[u,v]|_{3-\epsilon} \le C|u|_{2,\Omega}|v|_{2,\Omega}.$$

On the other hand, from Sobolev's embeddings we also have that

$$|[u,v]|_{-\epsilon,\Omega} \le C|u|_{3-\epsilon,\Omega}|v|_{2,\Omega}.$$

Combining these inequalities gives

$$\begin{aligned}|[\mathcal{G}[u_1,u_1],u_1-u_2]|_{-\epsilon,\Omega} &\le c|\mathcal{G}[u_1,u_1]|_{3-\epsilon,\Omega}|u_1-u_2|_{2,\Omega} \\ &\le C|u_1|^2_{2,\Omega}|u_1-u_2|_{2,\Omega} \le C \parallel u_1 \parallel^2 \parallel u_1-u_2 \parallel,\end{aligned} \tag{2.4.16}$$

$$\begin{aligned}&|[\mathcal{G}[u_1-u_2,u_1+u_2],u_2]|_{-\epsilon,\Omega} \le c|\mathcal{G}[u_1-u_2,u_1+u_2]|_{3-\epsilon,\Omega}|u_2|_{2,\Omega} \\ &\le C|u_1-u_2|_{2,\Omega}|u_1+u_2|_{2,\Omega}|u_2|_{2,\Omega} \le C \parallel u_2 \parallel\parallel u_1-u_2 \parallel\parallel u_1+u_2 \parallel .\end{aligned} \tag{2.4.17}$$

Combining (2.4.15)–(2.4.17) gives

$$|F(u_1)-F(u_2)|_{-\epsilon,\Omega} \le (\parallel u_1 \parallel^2 + \parallel u_2 \parallel^2 + \parallel u_1 \parallel) \parallel u_2 \parallel\parallel u_1-u_2 \parallel, \tag{2.4.18}$$

which in view of the fact that $H^{-\epsilon}(\Omega) \subset V' \sim H^1(\Omega)'$ [67] implies the validity of part 5 of Assumption 2.2.1.

Verification of part 6 *in Assumption* 2.2.1. By using the assumption that the derivative of f_1 is of the polynomial growth, integral version of the mean value theorem combined with the Hölder inequality gives the estimate

$$\left|f_1\left(\frac{\partial}{\partial\nu}u_1\right)-f_1\left(\frac{\partial}{\partial\nu}u_2\right)\right|_{0,\Gamma} \le C\left(\left|\frac{\partial}{\partial\nu}u_1\right|_{L_p(\Gamma)},\left|\frac{\partial}{\partial\nu}u_2\right|_{L_p(\Gamma)}\right)\left|\frac{\partial}{\partial\nu}u_1-\frac{\partial}{\partial\nu}u_2\right|_{L_r(\Gamma)}$$

for some, possibly large (depending on the growth of the nonlinearity), values of p, r. Thus, by Sobolev's embeddings, $H^{1/2}(\Gamma) \subset L_p(\Gamma)$, for any large value of p,

$$\left|f_1\left(\frac{\partial}{\partial\nu}u_1\right)-f_1\left(\frac{\partial}{\partial\nu}u_2\right)\right|_{0,\Gamma} \le C\left(\left|\frac{\partial}{\partial\nu}u_1\right|_{1/2,\Gamma},\left|\frac{\partial}{\partial\nu}u_2\right|_{1/2\Gamma}\right)\left|\frac{\partial}{\partial\nu}u_1-\frac{\partial}{\partial\nu}u_2\right|_{1/2,\Gamma},$$

and by the trace theorem,

$$\left|f_1\left(\frac{\partial}{\partial\nu}u_1\right)-f_1\left(\frac{\partial}{\partial\nu}u_2\right)\right|_{0,\Gamma} \le C(\parallel u_1 \parallel, \parallel u_2 \parallel) \parallel u_1-u_2 \parallel .$$

The above estimate implies condition 6 in Assumption 2.2.1.

The proof in the case of f_2 is even simpler, due to more regularity of the boundary elements $w|_\Gamma$.

Thus, Assumption 2.2.1 is satisfied and the coercivity Assumption 2.3.1 holds true due to strict monotonicity of the function g_1. For function g_2, we do not need a coercivity assumption, because $|G_2^*Au_t|_U = |u_t|_{0,\Gamma} \le C|u_t|_{1,\Omega} \le C|u_t|_V$.

The first statement in Theorem 2.4.1 follows now as a direct application of Theorem 2.3.1.

As for the second statement in the theorem, by virtue of Theorem 2.3.2, we need to verify the structural Assumption 2.3.2.

Verification of part (F) *in Assumption* 2.3.2. We compute

$$\int_0^t (F(u(\tau), u_t(\tau))d\tau = \int_0^t ([\mathcal{F}(u), u], u_{\tau)}d\tau$$

by using commutativity of the bracket

$$= \int_0^t (\mathcal{F}(u), [u, u_\tau])d\tau = 1/2 \int_0^t (\mathcal{F}(u), \frac{d}{d\tau}[u, u]d\tau$$

by using (2.4.9)

$$= -1/2c_0 \int_0^t \left(\mathcal{F}(u), \frac{d}{d\tau}\Delta^2 \mathcal{F}(u)\right) d\tau$$

by using clamped boundary conditions satisfied by $\mathcal{F}$

$$= -1/2c_0 \int_0^t \left(\Delta\mathcal{F}(u), \frac{d}{d\tau}\Delta\mathcal{F}(u)\right) d\tau = -1/4c_0 \int_0^t \frac{d}{d\tau}|\Delta\mathcal{F}(u)|^2 d\tau$$
$$= -1/4c_0|\Delta\mathcal{F}(u(t))|^2 + 1/4|\Delta\mathcal{F}(u(0))|^2, \qquad (2.4.19)$$

where $c_0 = \frac{Eh}{2D}$. Since

$$|\Delta\mathcal{F}(u)| \leq |\mathcal{F}(u)|_{2,\Omega} \leq C|u|^2_{2,\Omega},$$

we obtain from (2.4.19)

$$\int_0^t (F(u(\tau), u_t(\tau))d\tau \leq C|u(0)|^4_{2,\Omega} \leq C \parallel u_0 \parallel^4, \qquad (2.4.20)$$

which is the desired inequality.

Verification of part (f) *in Assumption* 2.3.2. This follows from the dissipativity condition imposed on the functions f_1. Indeed, we have

$$\int_0^t \left\langle f_1\left(\frac{\partial}{\partial\nu}u(\tau)\right), G_1^* A u_t(\tau)\right\rangle d\tau = \int_0^t \left(f_1\left(\frac{\partial}{\partial\nu}u(\tau), \frac{\partial}{\partial\nu}u_t(\tau)\right)_\Gamma d\tau\right.$$
$$= \int_0^t \frac{d}{d\tau}\int_\Gamma \tilde{f}_1\left(\frac{\partial}{\partial\nu}u\right) d\Gamma d\tau,$$

where $\tilde{f}_1$ denotes the antiderivative of f_1

$$= \int_\Gamma \tilde{f}_1\left(\frac{\partial}{\partial\nu}u(t)\right) d\Gamma - \int_\Gamma \tilde{f}_1\left(\frac{\partial}{\partial\nu}u(0)\right) d\Gamma. \qquad (2.4.21)$$

Since $f_1(s)s \geq 0$, the antiderivative $\tilde{f}_1$ is a positive function and (2.4.21) implies

$$\int_0^t \left\langle f_1\left(\frac{\partial}{\partial\nu}u(\tau)\right), G_1^* A u_t(\tau)\right\rangle d\tau \geq -C\int_\Gamma \tilde{f}_1\left(\frac{\partial}{\partial\nu}u(0)\right) d\Gamma;$$

since $\tilde{f}_1$ is of the polynomial growth,
Sobolev's embeddings and the trace theorem give

$$\geq -C\left(\left|\frac{\partial}{\partial_\nu}(0)\right|_{1/2,\Gamma}\right) \geq -C(|u(0)|_{2,\Omega}). \qquad (2.4.22)$$

This proves the assertion in part (f) of Assumption 2.3.2 for the function f_1. The analysis in the case of function f_2 is even simpler and hence is omitted. The proof of global existence, stated in Theorem 2.4.1, is thus completed. □

2.4.2 Case $\gamma = 0$

We are led to consider the following problem:

$$w_{tt} + \Delta^2 w = [w, \mathcal{F}(w)] \text{ in } \Omega \times (0, \infty), \tag{2.4.23}$$

$$\begin{aligned} \Delta^2 \mathcal{F} &= -\frac{Eh}{2D}[w, w] \text{ in } \Omega, \\ \mathcal{F} &= \frac{\partial}{\partial \nu}\mathcal{F} = 0 \text{ on } \Gamma, \end{aligned} \tag{2.4.24}$$

$$\begin{aligned} \Delta w + (1-\mu)B_1 w &= -g_1\left(\frac{\partial}{\partial \nu} w_t\right) - f_1\left(\frac{\partial}{\partial \nu} w\right) \text{ on } \Gamma \times (0, \infty), \\ \frac{\partial}{\partial \nu}\Delta w + (1-\mu)B_2 w &= g_2(w_t) + f_2(w) \text{ on } \Gamma \times (0, \infty). \end{aligned} \tag{2.4.25}$$

This model is mathematically more challenging. This is due to insufficient regularity of the nonlinear term $[\mathcal{F}(w), w]$ with respect to a new state space $H^2(\Omega) \times L_2(\Omega)$. Indeed, the standard regularity of Airy stress function $\mathcal{F}(w)$ does not allow us to conclude that this nonlinear term is bounded on $L_2(\Omega)$. To cope with the problem, "sharp" regularity of Airy's stress function will be necessary.

The main result in this case is given by the following theorem.

Theorem 2.4.2. *Under Assumption* 2.4.1 *we consider the equations given by* (2.4.23), (2.4.24), *and* (2.4.25). *In addition, we assume that either f_i satisfy condition* (2.3.4) *or else the functions g_i are strongly monotone, i.e., the constants α_i in Assumption* 2.4.1 *are both positive.*

Then, for any initial data $w_0 \in H^2(\Omega), w_1 \in L_2(\Omega)$, we have the following:

- *There exists a unique weak solution that is local and defined on the maximal interval of existence, $[0, T_0)$, with the values in $H^2(\Omega) \times L_2(\Omega)$.*
- *If, in addition, the functions f_i satisfy $f_i(s)s \geq 0$, then such solution is global.*

Remark 2.4.4. *In the case that one of the f_i is affine, there is no need for strict monotonicity of the corresponding g_i. This follows from Corollaries* 2.3.1 *and* 2.3.2.

Proof of Theorem 2.4.2. As before, the proof relies on the application of Theorems 2.3.1 and 2.3.2. The spaces H, U, U_0 and the operators A, G, Φ are the same as in the previous subsection. With the above notation, the original PDE model can be rewritten in an abstract form as

$$\begin{aligned} w_{tt} + Aw + AG_1 g_1 &\left(\frac{\partial}{\partial \nu} w_t|_\Gamma\right) + AG_2 g_2(w_t|_\Gamma) \\ + AG_1 f_1 &\left(\frac{\partial}{\partial \nu} w\right) + AG_2 f_2(w|_\Gamma) = [\mathcal{F}(w), w]. \end{aligned}$$

Recalling that by the virtue of Green's formula we have [129]

$$G_1^* A = \frac{\partial}{\partial \nu} u, \quad G_2^* A u = u|_\Gamma \text{ for } u \in D(A^{1/2}),$$

and taking $M \equiv I$, we obtain the exact same structure as the one postulated by abstract equation (2.2.1). Note that in this case we have $V = H$. $D(M^{1/2}) \equiv H = L_2(\Omega)$ and the corresponding state space becomes $H^2(\Omega) \times L_2(\Omega)$. Moreover, the requirements 1–4 of Assumption 2.2.1 are satisfied with

$$U_{01} \equiv H^{1/2}(\Gamma), U_{02} \equiv H^{3/2}(\Gamma).$$

The main task is to verify condition 5 of Assumption 2.2.1. Here the arguments are very different from those in the previous case as they rely on a newly developed regularity of the Airy stress function.

Verification of 5 *in Assumption* 2.2.1. Here, a critical role is played by the sharp regularity of the Airy stress function. This is given in the following.

Lemma 2.4.1 (see [62, 63]). *With the notation introduced in the preceding subsection, we have*

$$|\mathcal{G}[u, v]|_{W^{2,\infty}(\Omega)} \leq C|u|_{2,\Omega}|v|_{2,\Omega}. \tag{2.4.26}$$

Hence, in particular,

$$|\mathcal{F}(u)|_{W^{2,\infty}(\Omega)} \leq C|u|^2_{2,\Omega}. \tag{2.4.27}$$

Remark 2.4.5. *We note that this result does not follow from standard PDE arguments producing only the regularity* [150]

$$|\mathcal{F}(u)|_{3-\epsilon,\Omega} \leq C|u|^2_{2,\Omega},$$

which is obviously insufficient.

Armed with the result of Lemma 2.4.1, we proceed with the verification of our hypothesis:

$$\begin{aligned}
|F(u_1) - F(u_2)|_H &= |[\mathcal{G}[u_1, u_1], u_1 - u_2] \\
&\quad + [\mathcal{G}[u_1 - u_2, u_1 + u_2], u_2]|_{L_2(\Omega)} \\
&\quad \text{by direct estimates} \\
&\leq C|[\mathcal{G}[u_1, u_1]|_{W^{2,\infty}(\Omega)}|u_1 - u_2|_{2,\Omega} \\
&\quad + C|\mathcal{G}[u_1 - u_2, u_1 + u_2]|_{W^{2,\infty}(\Omega)}|u_2|_{2,\Omega} \\
&\quad \text{by sharp regularity} \\
&\leq C|u_1|^2_{2,\Omega}|u_1 - u_2|_{2,\Omega} + C|u_1 - u_2|_{2,\Omega}|u_1 + u_2|_{2,\Omega}|u_2|_{2,\Omega} \\
&\leq C|u_1 - u_2|_{2,\Omega}[|u_1|^2_{2,\Omega} + |u_1 + u_2|_{2,\Omega}|u_2|_{2,\Omega}] \\
&\leq C \parallel u_1 - u_2 \parallel [\parallel u_1 \parallel^2 + \parallel u_2 \parallel^2 + \parallel u_1 \parallel \parallel u_2 \parallel].
\end{aligned} \tag{2.4.28}$$

The verification of part 6 of Assumption 2.2.1 is the same as before. The coercivity Assumption 2.3.1 is now required from both components of function g. The reason for this is that this time we do not have a regularizing effect of the operator M^{-1} and $G^* A u_t$ is not bounded on the state space H.

The remaining parts of the proof are identical to the case $\gamma > 0$ and hence are not repeated. □

Remark 2.4.6. *We note that the issue of the uniqueness of weak solutions to von Karman models with any boundary conditions (clamped, hinged, or free), in the case of $\gamma = 0$, has been until recently an open problem* [150, 99]. *In fact, this was the case even in the absence of nonlinear boundary dissipation. Thus, the result of Theorem* 2.4.2 *settles this issue not only for the basic von Karman equation but for even more general von Karman models with nonlinear boundary dissipation.*

Remark 2.4.7. *The results presented in this section do not require any constraints imposed on the growth of nonlinear functions g_i. This is in contrast with most of the literature related to this topic* [99, 97].

Remark 2.4.8. *All the results presented in this section apply to a von Karman system equipped with either clamped or hinged boundary conditions. That is to say that instead of boundary conditions* (2.4.11), *one may consider clamped boundary conditions $w = \frac{\partial}{\partial\nu} w = 0$ on $\Gamma \times (0, \infty)$ or hinged boundary conditions*

$$w = 0, \ \Delta w = -g_1\left(\frac{\partial}{\partial\nu} w_t\right) - f_1\left(\frac{\partial}{\partial\nu} w\right) \quad on\ \Gamma \times (0, \infty)$$

with the same hypotheses imposed on g_1, f_1.

2.5 Semilinear Wave Equation

We consider an example of a semilinear wave equation with boundary dissipation and boundary nonlinearity, a model which arises in the context of boundary stabilization. This particular example, discussed in detail in [120], gives rise to the situation where, in general, we may not have the uniqueness of solutions but we will have global existence. This is precisely the situation dealt with in Theorem 2.3.3.

Let $\Omega \in R^n$ be a bounded domain with a smooth boundary Γ consisting of two subportions denoted by Γ_1 and Γ_0. We assume that $\overline{\Gamma_1} \cap \overline{\Gamma_0} = \emptyset$. In what follows we denote

$$H^1_{\Gamma_0}(\Omega) = \{u \in H^1_{\Gamma_0}(\Omega); u = 0 \text{ on } \Gamma_0\}.$$

We consider the following semilinear wave equation:

$$\begin{cases} u_{tt} = \Delta u - F(u) \ \text{ in } \ \Omega \times (0, \infty), \\ \dfrac{\partial u}{\partial\nu} = -g(u_t) - f(u) \ \text{ on } \ \Gamma_1 \times (0, \infty), \\ u = 0 \text{ on } \Gamma_0 \times (0, \infty), \\ u(0) = u_0 \in H^1_{\Gamma_0}(\Omega), \quad u_t(0) = u_1 \in L_2(\Omega). \end{cases} \tag{2.5.29}$$

The nonlinear functions F, g, f (Nemytski's operators) are subject to the following hypotheses.

Assumption 2.5.1.
(g) 1. $g \in C(R)$ *is monotone increasing,* $g(0) = 0$,
2. $g(s)s \geq ms^2, \qquad |s| > 1.$

(F) 1. $F \in W^{1,\infty}_{loc}(R)$,
2. $F(s)s \geq 0,\ s \in R$,
3. $F'(s) \leq C[1 + |s|^{k-1}]$ *a.e. in* s, $1 < k < \frac{n}{n-2}, n \geq 3$, *and* $k < \infty$ *for* $n = 2$.

(f) 1. $f \in W^{1,\infty}_{loc}(R)$,
2. $f(s)s \geq 0,\ s \in R$,
3. $|f(s)| < C[|s| + |s|^r]$ *a.e. in* s, $r < \frac{n-1}{n-2}$.

Our main result is as follows.

Theorem 2.5.1. *We assume*[1] *Assumption* 2.5.1:

- *Assume that either* $f(u) = c_0 + c_1 u, c_0 \in L_2(\Gamma), c_1 \in R$, *or else* $g(s)$ *is strictly coercive, i.e.,*
$$(g(s) - g(t))(s - t) \geq m|s - t|^2.$$
Then, there exists a unique, weak, and global solution to problem (2.5.29) *such that with an arbitrary* $T < \infty$ *we have*
$$u \in C([0, T]; H^1_{\Gamma_0}(\Omega)) \cap C^1([0, T]; L_2(\Omega)).$$

- *If, instead,* $g(s)$ *satisfies the following growth condition at infinity* $sg(s) \leq Ms^2$, *for* $|s| \geq 1$, *and the constant* k *in Assumption* 2.5.1 *is subcritical, i.e.,* $k < \frac{n}{n-2}$, *then there exists a mild global solution to problem* (2.5.29) *such that with an arbitrary* $T < \infty$,
$$u \in C([0, T]; H^1(\Omega)) \cap C^1([0, T]; L_2(\Omega)).$$

Remark 2.5.1. *Note that if there is no semilinear term* $f(u)$ *on the boundary, the nonlinear dissipation* $g(s)$ *does not need to satisfy any growth conditions. However, in the presence of nonlinear boundary term* $f(u)$, *the result of Theorem* 2.5.1 *provides a trade-off in terms of the hypotheses imposed on the nonlinear dissipation* $g(s)$. *It needs to be either strictly coercive or subject to linear growth at infinity.*

Remark 2.5.2. *Proof of Theorem* 2.5.1 *is much simpler, and it follows from Theorem* 2.3.1, *in the case when the boundary term* $f(u)$ *is affine or the dissipation* $g(u)$ *is strictly coercive. To handle nonlinear* $f(u)$ *and noncoercive* $g(u)$, *we need the more complicated result stated in Theorem* 2.3.3.

Proof. Proof of the first statement is an easy application of Theorems 2.3.1 and 2.3.2. The details are similar to those in subsection 2.4 and hence are omitted. Proof of the second part relies on the application of Theorem 2.3.3

Step 1. *Abstract setup.* To put problem (2.5.29) into an abstract form, we introduce the following spaces and operators:

$$H \equiv L_2(\Omega), \quad U \equiv L_2(\Gamma_1), \qquad A : L_2(\Omega) \to L_2(\Omega),$$

$$A(u) = -\Delta u : u \in D(A) = \left\{ u \in H^2(\Omega); \frac{\partial}{\partial \nu} u = 0 \text{ on } \Gamma_1, u = 0 \text{ on } \Gamma_0 \right\},$$

[1] If f is linear or g is coercive, one can take $k \leq \frac{n}{n-2}$ and Assumption 2.5.1 (g)(2) is not needed.

$$G \equiv N : L_2(\Gamma_1) \to L_2(\Omega)$$

$$Ng = v \text{ iff } \Delta v = 0 \text{ in } \Omega, \quad \frac{\partial}{\partial \nu} v = g \text{ on } \Gamma_1, \ v = 0 \text{ on } \Gamma_0.$$

By Green's formula we have

$$G^* Au = u|_{\Gamma_1}, \qquad u \in H^1_{\Gamma_0}(\Omega),$$

where $H^1_{\Gamma_0}(\Omega) \equiv \{u \in H^1(\Omega); \ u = 0 \text{ on } \Gamma_0\}$.

Taking $M = I, V = L_2(\Omega)$, and $\partial\Phi = g$ we obtain an abstract form of (2.5.29):

$$u_{tt} + Au + AG\partial\Phi(G^* Au_t) + AGf(u) = F(u).$$

Step 2. *Verification of Assumption* 2.2.1. Statements 1–3 in Assumption 2.2.1 are obvious. As for the statement 4, we first note that $D(A^{1/2}) = H^1_{\Gamma_0}(\Omega)$. With $U_0 = H^{1/2}(\Gamma_1)$, we have that G^*A is bounded and surjective $G^*A : D(A^{1/2}) \to U_0$.

Statements 5 and 6 follow from Sobolev's embeddings and parts $(f), (F)$ in Assumption 2.5.1. Computations here are routine based on the integral form of the mean value theorem [120].

Step 3. *Verification of Assumption* 2.3.3. For simplicity of the exposition we consider case $n = 2$. We define

$$F_n(s) \equiv \begin{cases} F(s), & |s| \le n, \\ F(n), & s > n, \\ F(-n), & s < -n, \end{cases} \tag{2.5.30}$$

$$f_n(s) \equiv \begin{cases} f(s), & |s| \le n, \\ f(n), & s > n, \\ f(-n), & s < -n, \end{cases} \tag{2.5.31}$$

and we set

$$F_n(u)(s) \equiv F_n(u(s)), \qquad f_n(u)(s) \equiv f_n(u(s)).$$

Verification of (A1)(i) *in Assumption* 2.3.3. This follows from the following Lemma.

Lemma 2.5.1. *Under Assumption* 2.5.1, *the following convergence holds for any weakly convergent sequence* $u_n \xrightarrow{\omega} u$ *in* $H^1(\Omega)$*:*

(if) $f_n(u_n|_{\Gamma_1}) \to f(u|_{\Gamma_1})$ *in* $L_2(\Gamma_1)$,

(iF) $F_n(u_n) \to F(u)$ *in* $L_2(\Omega)$.

Proof. We write

$$\int_{\Gamma_1} |f_n(u_n) - f(u_n)|^2 d\Gamma_1 \le 2 \int_{\Gamma_n} |f(u_n)|^2 + f(n)^2 + f(-n)^2 d\Gamma_n, \tag{2.5.32}$$

where

$$\Gamma_n \equiv \{x \in \Gamma_1; |u_n(x)| \ge n\}.$$

Since $L_p(\Gamma) \subset H^{1/2}(\Gamma)$, for any value $p \ge 1$

$$\int_{\Gamma_n} n^p \le \int_{\Gamma_n} |u_n(x)|^p dx \le C|u_n|^p_{1/2,\Gamma_1} \le C|u_n|^p_{1,\Omega}. \tag{2.5.33}$$

Hence

$$meas\Gamma_n \leq Cn^{-p}|u_n|_{1,\Omega}^p \text{ for any } p > 0. \tag{2.5.34}$$

Recalling Assumption 2.5.1,

$$\int_{\Gamma_n} |f(u_n)|^2 d\Gamma_n \leq C \int_{\Gamma_n} |u_n(x)|^{2r} dx \leq C \left[\int_{\Gamma_n} |u_n(x)|^{4r} dx\right]^{1/2} [meas\Gamma_n]^{1/2}$$
$$\leq C|u_n|_{1,\Omega}^{2r}[meas\Gamma_n]^{1/2} \leq C[meas\Gamma_n]^{1/2} \to 0 \text{ as } n \to \infty, \tag{2.5.35}$$

where we used in the last step inequality in (2.5.34). Similarly,

$$\int_{\Gamma_n} |f(n)|^2 d\Gamma_n \leq C \int_{\Gamma_n} |n|^{2r} d\Gamma_n \leq Cn^{2r}[meas\Gamma_n]$$
$$\text{by (2.5.34)}$$
$$\leq C|u_n|_{1,\Omega}^p n^{-p} n^{2r} \to 0 \text{ as } n \to \infty, \tag{2.5.36}$$

where we took $p > 2r$.

This way we have proved that

$$f(u_n) - f_n(u_n) \to 0 \text{ in } L_2(\Gamma_1) \text{ when } n \to \infty.$$

To conclude the proof of part (if), we need to show that

$$f(u_n) - f(u) \to 0 \text{ in } L_2(\Gamma_1) \text{ when } u_n \overset{\omega}{\to} u \text{ in } H^{1/2}(\Gamma_1).$$

But this follows from strong convergence of $u_n \to u$ in $H^{1/2-\epsilon}(\Gamma_1)$ together with continuity of f on $H^{1/2-\epsilon}(\Gamma_1)$. This last property is due to the polynomial growth and continuity assumed on f.

Proof of part (iF) is the same and hence is omitted. □

Verification of (A1)(ii) *in Assumption* 2.3.3. This follows immediately from the dissipativity conditions assumed on f, F. Indeed,

$$\int_0^t \int_{\Gamma_1} f_n(u(\tau))u_t(\tau)d\Gamma_1 d\tau = \int_0^t \int_{\Gamma_1} \frac{d}{d\tau}\tilde{f}_n(u(\tau))d\Gamma_1 d\tau,$$

where $\tilde{f}_n$ is an antiderivative of f, and, by virtue of the Assumption 2.5.1 and the definition of f_n, we have that $f_n(s)s \geq 0 \Rightarrow \tilde{f}_n \geq 0$. Hence

$$\int_0^t \int_{\Gamma_1} f_n(u(\tau))u_t(\tau)d\Gamma_1 d\tau = \int_{\Gamma_1} [\tilde{f}_n(u(t)) - \tilde{f}_n(u(0))]d\Gamma_1 \geq -\int_{\Gamma_1} \tilde{f}_n(u(0))d\Gamma_1$$
$$\text{by Sobolev's embeddings and growth condition imposed on } f$$
$$\geq -C(|u(0)|_{L_{r+1}(\Gamma_1)}) \geq -C(|u(0)|_{H^{1/2}(\Gamma_1)}) \geq -C(|u(0)|_{1,\Omega}), \tag{2.5.37}$$

which implies the desired inequality (with $\rho_1 = 0$) in Assumption 2.3.3, part (iif). A similar argument applies to part (iiF).

Verification of (A2) *in Assumption* 2.3.3. This follows from the hypothesis imposed on the nonlinear function g. Indeed, we have

$$\int_{\Gamma_1} g(z_n)z_n d\Gamma_1 \geq m \int_{x\in\Gamma_1; z_n(x)\geq 1} {z_n}^2(x)dx + \int_{x\in\Gamma_1; z_n(x)<1} g(z_n(x))z_n(x)dx$$
$$\geq m \int_{\Gamma_1} {z_n}^2(x)dx - [g(1)+m]meas\Gamma_1 = m|z_n|^2_{0,\Gamma_1} - [g(1)+m]meas\Gamma_1. \tag{2.5.38}$$

Thus,

$$\frac{\int_{\Gamma_1} g(z_n)z_n d\Gamma_1}{|z_n|^2_{0,\Gamma_1}} \geq m/2 \text{ for } |z_n|^2_{0,\Gamma_1} \to \infty,$$

which gives the desired coercivity of the subgradient $\partial\Phi$ with a constant $\alpha = m/2$.

We have verified Assumption 2.3.3. Theorem 2.3.3 yields the result of Theorem 2.5.1. □

2.6 Nonlinear Structural Acoustic Model

A structural acoustic interaction can be represented by a coupled system of equations that describes (i) an acoustic medium in a given two- or three-dimensional chamber (wave equation) and (ii) a structure (beam, plate, or shell equation) representing a flexible wall of the chamber (see [24, 15, 16, 155, 166]).

The interaction takes place on the boundary between the acoustic medium and the structure (flexible wall). This leads to a mathematical model of coupled hyperbolic (wave) and parabolic/hyperbolic (wall) equations. In some cases, when the structure (wall) is modeled by higher-order approximations such as a Kirchhoff model without accounting for structural damping, the coupling is between two hyperbolic systems.

In this section we provide a model that is fairly general and that encompasses a variety of more specific models to be discussed later. We refer to [108] for more detailed discussion of various setups.

Let $\Omega \subset R^n$, $n = 2, 3$, be an open bounded domain with boundary Γ. We shall assume that Ω is either smooth or convex. We assume that $\Gamma = \bar{\Gamma_0} \cup \bar{\Gamma_1}$, where $\Gamma_i, i = 0, 1$, are two disjoint sets that are open and simply connected. Thus, Γ_0 is a simply connected, bounded manifold in R^{n-1}. The acoustic medium is described by the wave equation in the variable z, where the quantity ρz_t describes the acoustic pressure, ρ being the density of the fluid. The structural vibrations of the (structure) described by the variable $\mathbf{v}$ will then satisfy a suitable beam, plate, or shell equation defined on the manifold Γ_0. The last component of the vector $\mathbf{v}$, denoted by v, represents vertical vibrations of the wall. The nonlinear boundary dissipation is described by nonlinear, monotone, continuous functions $g_1 \in C(R), g_2 \in C(R^s)$, where $s = 1, 2$, or 3. We also assume that $g_i(0) = 0$. In the case of beams or plates we have $s = 1$, and in the case of shells we have $s = 3$. (The first two components of the vector $\mathbf{v}$ denote plane displacements.)

The following PDE model describes the problem.

Acoustic medium.

$$\begin{aligned} z_{tt} &= c^2\Delta z + F_1(z) \text{ in } \Omega \times (0,\infty), \\ \frac{\partial}{\partial \nu} z &= 0 \text{ on } \Gamma_1 \times (0,\infty), \\ \frac{\partial}{\partial \nu} z + g_1(z_t) &= v_t \text{ on } \Gamma_0 \times (0,\infty), \\ z(0) = z_0 \in H^1(\Omega), &\quad z_t(0) = z_1 \in L_2(\Omega). \end{aligned} \tag{2.6.39}$$

We assume that the nonlinear function $F_1 : H^1(\Omega) \to L_2(\Omega)$ is locally Lipschitz.

Structure-elastic wall. To describe a flexible wall, we introduce the following operators:

- Self-adjoint, positive operators

$$\mathcal{A} : [L_2(\Gamma_0)]^s \to [L_2(\Gamma_0)]^s \text{ and } \mathcal{M} : [L_2(\Gamma_0)]^s \to [L_2(\Gamma_0)]^s,$$

  where $s = 1, 2$, or 3.
- A Green's map $\mathcal{G}$ which is bounded $\mathcal{G} : [L_2(\partial\Gamma_0)]^s \to [L_2(\Gamma_0)]^s$. Moreover, $\mathcal{G}^*\mathcal{A}^{1/2}$ is bounded and surjective from $D(\mathcal{A}^{1/2})$ to $H^\beta(\partial\Gamma_0)$ for an appropriate value of $\beta \geq 1/2$.
- A nonlinear operator F_2 which is assumed locally Lipschitz from $D(\mathcal{A}^{1/2}) \to D(\mathcal{M}^{1/2})'$.
- A linear, positive operator $\mathcal{D}$ on $L_2(\Gamma_0)$ such that $\mathcal{D} \in \mathcal{L}(\mathcal{D}(\mathcal{A}^{1/2}) \to \mathcal{D}(\mathcal{A}^{1/2})')$.

A general form of the equation describing the wall can be taken as

$$\begin{aligned} \mathcal{M}\, \mathbf{v}_{tt} + \mathcal{A}\mathbf{v} + \alpha\mathcal{A}\mathbf{v}_t + \mathcal{C}\rho z_t|_{\Gamma_0} + \mathcal{A}\mathcal{G}g_2(\mathcal{G}^*\mathcal{A}\mathbf{v}_t) + d\mathcal{D}(\mathbf{v}_t) &= F_2(\mathbf{v}), \\ \mathbf{v}(0) = \mathbf{v}_0 \in D(\mathcal{A}^{1/2}), \qquad \mathbf{v}_t(0) = \mathbf{v}_1 &\in D(\mathcal{M}^{1/2}). \end{aligned} \tag{2.6.40}$$

The operator $\mathcal{C}$ simply denotes a projection of the vector $\mathbf{v}$ to the third coordinate of v (relevant only when $s > 1$; when $s = 1$ we have $\mathcal{C}\rho z_t|_{\Gamma_0} = \rho z_t|_{\Gamma_0}$).

The constants $\alpha, d \geq 0$, represent a potential structural (viscous) damping. If $\alpha > 0$, then the coupling on the interface describes hyperbolic-parabolic coupling of the structure [48]. If, instead, $\alpha = 0$, then the corresponding coupling is hyperbolic-hyperbolic. In practical applications, the operator $\mathcal{A}$ is a biharmonic operator with appropriate boundary conditions (clamped, hinged, or free). The operator $\mathcal{M}$ corresponds typically to $I - \gamma\Delta$, where the constant $\gamma > 0$ is small (proportional to the square of the thickness of the plate) and describes moments of inertia. For concrete PDE representations of these operators, we refer the reader to Chapter 5 (see also [108]).

Using the notation introduced in the previous section, we rewrite the system consisting of (2.6.39), (2.6.40) as

$$z_{tt} + c^2 A_N z + A_N N g_1(N^* A_N z_t) - A_N N v_t = F_1(z), \tag{2.6.41}$$

$$\begin{aligned} \mathcal{M}\, \mathbf{v}_{tt} + \mathcal{A}\mathbf{v} + \alpha\mathcal{A}\mathbf{v}_t + \rho\mathcal{C}N^*A_N z_t + \mathcal{A}\mathcal{G}g_2(\mathcal{G}^*\mathcal{A}\mathbf{v}_t) + d\mathcal{D}\mathbf{v}_t &= F_2(\mathbf{v}), \\ z_0, z_1, \mathbf{v}_0, \mathbf{v}_1 \in H^1(\Omega) \times L_2(\Omega) \times D(\mathcal{A}^{1/2}) \times D(\mathcal{M}^{1/2}).& \end{aligned} \tag{2.6.42}$$

We rewrite the above system as an equation in the form given in (2.2.1). To simplify the notation we take $\rho = 1$. (Otherwise a rescaling of the inner product is necessary.)

We introduce the following, natural, notation:

$$H \equiv L_2(\Omega) \times [L_2(\Gamma_0)]^s, \quad U \equiv L_2(\Gamma_0) \times [L_2(\partial\Gamma_0)]^3, \quad V \equiv L_2(\Omega) \times D(\mathcal{M}^{\ 1/2}),$$

$$U_0 \equiv H^{1/2}(\Gamma_0) \times H^{\beta}(\partial\Gamma_0),$$

$$A \equiv \begin{pmatrix} c^2 A_N & 0 \\ 0 & \mathcal{A} \end{pmatrix}, \quad M \equiv \begin{pmatrix} I & 0 \\ 0 & \mathcal{M} \end{pmatrix},$$

$$G \equiv \begin{pmatrix} N & 0 \\ 0 & \mathcal{G} \end{pmatrix}, \quad D \equiv \begin{pmatrix} 0 & -A_N N\mathcal{C} \\ \mathcal{C}N^* A_N & \alpha\mathcal{A} + d\mathcal{D} \end{pmatrix},$$

$$\partial\Phi \equiv \begin{pmatrix} g_1 & 0 \\ 0 & g_2 \end{pmatrix}, \quad F \equiv \begin{pmatrix} F_1 & 0 \\ 0 & F_2 \end{pmatrix}.$$

With the above notation and setting $u \equiv (z, \mathbf{v})$, the system consisting of (2.6.39), (2.6.40) can be written in the form as in (2.2.1). It can be verified immediately that hypotheses 1–6 of Assumption 2.2.1 are satisfied. Hypothesis 7 also can be easily verified. Indeed, since $D(A^{1/2}) = D(A_N^{1/2}) \times D(\mathcal{A}^{1/2})$, and the following operators are bounded

$$A_N N = A_N^{1/2} A_N^{1/2} N : L_2(\Gamma_0) \to D(A_N^{1/2})',$$

$$N^* A_N : D(A_N^{1/2}) \sim H^1(\Omega) \to L_2(\Gamma_0),$$

operator D is bounded from $D(A^{1/2}) \to D(A^{1/2})'$. On the other hand, simple computations with $u \in D(A^{1/2})$ give

$$\begin{aligned}(Du, u)_H &= (-A_N N v, z)_{L_2(\Omega)} + (N^* A_N z, v)_{L_2(\Gamma_0)} \\ &\quad + \alpha(\mathcal{A}\mathbf{v}, \mathbf{v})_{L_2(\Gamma_0)} + d(\mathcal{D}\mathbf{v}, \mathbf{v})_{L_2(\Gamma_0)} \geq 0,\end{aligned}$$

since all the hypotheses in Assumption 2.2.1 have been verified. Noting that $f = 0$ in (2.2.1), we are in a position to apply the result of Corollary 2.3.2. This gives the following theorem.

Theorem 2.6.1. *For all initial data* $z_0, z_1, \mathbf{v}_0, \mathbf{v}_1 \in H^1(\Omega) \times L_2(\Omega) \times D(\mathcal{A}^{1/2}) \times D(\mathcal{M}^{\ 1/2})$*, there exists a unique, weak, and local solution to* (2.6.39), (2.6.40) *such that*

$$(z, z_t, \mathbf{v}, \mathbf{v}_t) \in C([0,T]; H^1(\Omega) \times L_2(\Omega) \times D(\mathcal{A}^{1/2}) \times D(\mathcal{M}^{\ 1/2})),$$

where $T < T_0$*, which may be finite.*

To obtain global solutions, we need to impose some structural hypotheses on nonlinear terms F_1, F_2. In fact, we assume that these operators satisfy structural hypotheses in Assumption 2.3.2. Under this condition, Theorem 2.3.2 provides global existence of the corresponding solutions.

A practical example of this situation is the following.

Assumption 2.6.1. F_1 *is a Nemytski operator of a polynomial growth if* $\dim\Omega = 2$ *(or of cubic growth at most if* $\dim\Omega = 3$*) and such that* $F_1(s)s \leq 0$.

F_2 *is a nonlocal operator representing the von Karman nonlinearity. That is to say,* $F_2(v) = [v, \mathcal{F}v]$*, where* $[,]$ *is the von Karman bracket and* $\mathcal{F}$ *is the Airy's stress function introduced in* (2.4.8).

By applying the arguments that are identical to those given in sections 2.4 and 2.5, we verify that the structural Assumption 2.3.2 holds. Thus, Theorem 2.3.2 yields the following.

Theorem 2.6.2. *In addition to the hypotheses in Theorem* 2.6.1, *we assume that Assumption* 2.6.1 *holds. Then, the local solution of Theorem* 2.6.1 *becomes global and one can take* $T_0 = \infty$.

Remark 2.6.1. *We note that the results of Theorem* 2.6.1 *and Theorem* 2.6.2 *are valid for any* $\alpha \geq 0$. *In particular, one can take* $\alpha = 0$, *which means that there is no structural damping in the model. If* $\alpha \geq 1/2$, *then it is shown in* [48] *that the linear semigroup corresponding to the plate is analytic. In such a case, the corresponding solutions are more regular than predicted by the theorems above. We return to this point later.*

Remark 2.6.2. *One can also study a structural acoustic model with the effects of thermoelasticity. The qualitative statement for the well-posedness result is the same as provided in the theorems above after accounting for appropriate spaces representing thermal effects* [112, 114, 113].

2.7 Full von Karman Systems

Although the framework introduced in section 2.2 is fairly general and covers a variety of PDE examples arising in dynamic elasticity, there are systems of great practical and physical interest that do not fit this setup. Indeed, the main limiting factor in the abstract theory presented in sections 2.2 and 2.3 is that the nonlinear forcing term $F(u)$ is required to satisfy the local Lipschitz condition. On the other hand, several systems of interest to control theory do not comply with this requirement. Therefore, different methods have been developed to encompass this class of problem. A particular model we have in mind is the full von Karman system which couples the dynamic system of elasticity with the scalar plate equation. In contrast to the scalar version of von Karman equations considered in section 2.4, the full von Karman system, which accounts as well for in-plane accelerations, is a system of three equations (two wave equations and a plate equation) that are strongly coupled. The coupling is nonlinear and is described by functions $F(u)$, which are not even bounded on the energy space. In what follows, we focus on the full von Karman system, which also accounts for thermal effects. This will give us a richer class of problems to discuss. We also mention that nonlinear shell equations, which have a mathematical structure very similar to the full von Karman system, can be treated by methods presented in this section (see [186, 119]).

Depending on the values of certain physical parameters in the model (moments of inertia), we witness different dynamical behavior governed either by hyperbolic-hyperbolic coupling or by hyperbolic-parabolic coupling.

2.7.1 Model

We consider a model describing nonlinear oscillations of a plate subjected to thermal effects and accounting for in-plane accelerations. The resulting system is referred to as the thermoelastic full von Karman system. References related to general von Karman systems are [103, 101, 100, 99, 90, 51, 57]. This includes a model of the von Karman system accounting for mechanical variables with free boundary conditions that was studied in [99]. Thermoelastic von Karman systems with Dirichlet boundary conditions were considered in [27]. The derivation of thermoelastic von Karman models [27, 25] parallels the arguments presented in [102] for linear thermoelastic equations after incorporating the nonlinear strain-displacement relations.

Here, the variables w and $u = (u_1, u_2)$ represent, respectively, the vertical and in-plane displacement of a thin plate occupying a two-dimensional domain Ω with sufficiently smooth boundary $\Gamma = \overline{\Gamma}_0 \cup \overline{\Gamma}_1$. We assume that $\overline{\Gamma}_0 \cap \overline{\Gamma}_1 = \emptyset$. The variables θ and ϕ describe the thermal strain resultants affecting the vertical displacement and in-plane displacements, respectively. More specifically, θ is the average (across the thickness of the plate) of thermal moment while ϕ is the average of thermal stress (see formulas (6.8) and (6.9) in [102]). The governing equations, defined on $\Omega \times (0, \infty)$, are given by

$$\begin{aligned} u_{tt} - \operatorname{div}[\mathcal{C}[\epsilon(u) + f(\nabla w)]] + \nabla\phi = 0, \\ w_{tt} - \gamma\Delta w_{tt} + D\Delta^2 w - \operatorname{div}[\mathcal{C}[\epsilon(u) + f(\nabla w)]\nabla w + \phi\nabla w] + \Delta\theta = 0 \end{aligned} \tag{2.7.43}$$

with Dirichlet boundary conditions on the portion of the boundary Γ_0

$$u = w = \nabla w = 0 \quad \text{on } \Gamma_0 \times (0, \infty).$$

Boundary conditions prescribed on $\Gamma_1 \times (0, \infty)$ are of the free type and are given by

$$\begin{aligned} \mathcal{C}[\epsilon(u) + f(\nabla w)]\nu + ku - \phi\nu = -g(u_t), \\ D[\Delta w + (1-\mu)B_1 w] + \theta = 0, \\ D\left[\frac{\partial}{\partial\nu}\Delta w + (1-\mu)B_2 w\right] - \gamma\frac{\partial}{\partial\nu}w_{tt} - \mathcal{C}[\epsilon(u) + f(\nabla w)]\nabla w\nu + \frac{\partial}{\partial\nu}\theta - \phi\frac{\partial}{\partial\nu}w = 0. \end{aligned} \tag{2.7.44}$$

Here, the vector ν denotes the outward normal direction to the boundary. We denote by $\frac{\partial}{\partial\tau}$ the tangential direction on Γ. The constant $\gamma \geq 0$ is proportional to the square of the thickness of the plate and hence is assumed small. The term $\gamma\Delta w_{tt}$ represents moments of inertia, and $k \geq 0$.

The temperature is described by the system of equations

$$\begin{aligned} \phi_t = \Delta\phi - \operatorname{div} u_t + \nabla w \nabla w_t \quad \text{in } \Omega \times (0, \infty), \\ \theta_t = \Delta\theta + \Delta w_t \quad \text{in } \Omega \times (0, \infty) \end{aligned} \tag{2.7.45}$$

with the boundary conditions imposed on thermal variables ϕ, θ,

$$\frac{\partial}{\partial\nu}\theta + \lambda_1\theta = 0, \quad \frac{\partial}{\partial\nu}\phi + \lambda_2\phi = 0 \quad \text{on } \Gamma \times (0, \infty),$$

where the constants $\lambda_i \geq 0$, $i = 1, 2$.

With (2.7.43) and (2.7.45) we associate the initial conditions

$$\begin{aligned} u(0) = u_0, \quad u_t(0) = u_1, \quad w(0) = w_0, \\ w_t(0) = w_1, \quad \theta(0) = \theta_0, \quad \phi(0) = \phi_0 \text{ in } \Omega. \end{aligned} \tag{2.7.46}$$

D represents the flexural rigidity, and the constant $0 < \mu < 1/2$ is Poisson's modulus. The fourth-order tensor $\mathcal{C}$ is defined by

$$\mathcal{C}\epsilon \equiv \frac{E}{(1-2\mu)(1+\mu)}[\mu \; trace \; \epsilon \; I + (1-2\mu)\epsilon],$$

where the strain tensor is given by $\epsilon(u) \equiv 1/2(\nabla u + \nabla^T u)$. It can be easily verified that the tensor $\mathcal{C}$ is symmetric and strictly positive. The function f is given by

$$f(s) \equiv (1/2)s \otimes s, \quad s \in R^2,$$

and the boundary operators B_i are defined by

$$B_1 \equiv 2\nu_1\nu_2 D^2_{x,y} - \nu_1^2 D^2_{y,y} - \nu_2^2 D^2_{x,x},$$

$$B_2 \equiv \frac{\partial}{\partial\tau}[(\nu_1^2 - \nu_2^2)D^2_{x,y} + \nu_1\nu_2(D^2_{y,y} - D^2_{x,x})] + lI, \qquad l \geq 0.$$

The dissipation in the system is represented by a nonlinear vector function g which is subject to the following hypothesis.

Assumption 2.7.1. *$g(u = [u_1, u_2]) = [g_1(u_1), g_2(u_2)]$, where $g_i \in C(R)$ are assumed monotone increasing, zero at the origin, and of linear growth at infinity.*

A natural energy function associated with this model is given by

$$E_\gamma(t) = E_{k,\gamma}(t) + E_p(t), \quad E_{k,\gamma}(t) = \int_\Omega [|u_t|^2 + w_t^2 + \gamma|\nabla w_t|^2]d\Omega \tag{2.7.47}$$

and the potential energy

$$E_p(t) = \int_\Omega \left[\mathcal{C}N(u,w)N(u,w) + |\theta|^2 + |\phi|^2\right] d\Omega + a(w,w) + k\int_{\Gamma_1} |u|^2 d\Gamma_1, \tag{2.7.48}$$

where the stress resultant $N(u,w)$ is given by $N(u,w) \equiv \epsilon(u) + f(\nabla w)$ and

$$a(w,z) \equiv D\int_\Omega [\, w_{x,x}z_{x,x} + w_{y,y}z_{y,y} + \mu w_{x,x}z_{y,y} + \mu w_{y,y}z_{x,x}$$
$$+ 2(1-\mu)w_{x,y}z_{x,y}]d\Omega + l\int_{\Gamma_1} wz d\Gamma_1.$$

It is well known that $E_p(t)$ is topologically equivalent to $H^2_{\Gamma_0}(\Omega) \times [H^1_{\Gamma_0}(\Omega)]^2 \times [L_2(\Omega)]^2$ topology. Here the spaces $H^1_{\Gamma_0}(\Omega)$ and $H^2_{\Gamma_0}(\Omega)$ denote the usual Sobolev spaces which incorporate zero boundary conditions on Γ_0

In particular, the following inequalities resulting from Korn's inequality and Sobolev's embeddings are used frequently:

$$|u|^2_{H^1(\Omega)} \leq C\left[|N(u,w)|^2_{L_2(\Omega)} + |\nabla w|^4_{L_4(\Omega)} + k\int_{\Gamma_1} |u|^2 d\Gamma_1\right]$$
$$\leq C\left[|N(u,w)|^2_{L_2(\Omega)} + |w|^4_{H^2(\Omega)} + k\int_{\Gamma_1} |u|^2 d\Gamma_1\right], \tag{2.7.49}$$

where $k > 0$ whenever $\Gamma_0 = \emptyset$. The kinetic portion of the energy is topologically equivalent to

$$[L_2(\Omega)]^2 \times H^1_\gamma(\Omega),$$

where $H^1_\gamma(\Omega)$ denotes the $H^1_{\Gamma_0}(\Omega)$ space when $\gamma > 0$ and the $L_2(\Omega)$ space when $\gamma = 0$.

The model described by (2.7.43) and (2.7.45) can be easily put in the framework of (2.2.1) in section 2.2, after setting

$$H \equiv [L_2(\Omega)]^5, \qquad V \equiv [L_2(\Omega)]^2 \times H^1_\gamma(\Omega) \times [L_2(\Omega)]^2$$

and after giving an appropriate definition to all operators involved (note that $D = 0$ and $f(u)$ is linear).

However, the nonlinear term $F(u)$ coupling all five equations is strongly nonlinear and it is not locally Lipschitz. To see this, it suffices to analyze just one term: $\operatorname{div}[\mathcal{C}f(\nabla w)]$ with w in the energy space, i.e., $w \in H^2(\Omega)$. Simple calculations based on Sobolev's embeddings imply that $\operatorname{div}[\mathcal{C}f(\nabla w)] \in H^{-\epsilon}(\Omega)$ and not in $L_2(\Omega)$, as required by Assumption 2.2.1. There is a lack of ϵ derivative. In the case $\gamma = 0$, the gap is even larger and the $1 + \epsilon$ derivative is missing. For this reason we will not attempt to put this problem into an abstract framework, which is not of much use in the present context. Instead, we deal directly with variational formulation of (2.7.43) and (2.7.45). Multiplying these equations by the appropriate test functions and integrating by parts leads to the following definition of *weak*, often referred as the variational or finite energy solution.

Definition 2.7.1. *We say that the function*

$$u, w, \phi, \theta \in C([0,T]; [H^1_{\Gamma_0}(\Omega)]^2 \times H^2_{\Gamma_0}(\Omega) \times [L_2(\Omega)]^2)$$

and such that

$$(u_t, w_t, \phi, \theta) \in C([0,T]; [L_2(\Omega)]^2 \times H^1_\gamma(\Omega)) \times [L_2(0,T; H^1(\Omega))]^2$$

is a finite energy *solution to* (2.7.43), (2.7.45) *iff* u, w, ϕ, θ *satisfy the variational form*

$$
\begin{aligned}
(u_{tt}, \xi)_\Omega + (\mathcal{C}N(u,w), \epsilon(\xi))_\Omega + \langle g(u_t) + ku - \phi\nu, \xi\rangle_{\Gamma_1} + (\nabla\phi, \xi)_\Omega = 0,\\
(w_{tt}, \psi)_\Omega + \gamma(\nabla w_t, \nabla\psi)_\Omega + a(w,\psi) + (\mathcal{C}N(u,w), \nabla\psi \otimes \nabla w)_\Omega\\
+ (\phi\nabla w, \nabla\psi)_\Omega + \left\langle \theta, \frac{\partial}{\partial\nu}\psi \right\rangle_{\Gamma_1} - (\nabla\theta, \nabla\psi)_\Omega = 0,\\
(\phi_t, \delta)_\Omega + (\nabla\phi, \nabla\delta)_\Omega - (u_t, \nabla\delta)_\Omega + \langle u_t, \nu\delta\rangle_{\Gamma_1}\\
+\lambda_2\langle\phi, \delta\rangle_{\Gamma_1} = 1/2\left(\frac{d}{dt}|\nabla w|^2, \delta\right)_\Omega,\\
(\theta_t, \eta)_\Omega + (\nabla\theta, \nabla\eta)_\Omega - \left\langle \frac{\partial}{\partial\nu} w_t, \eta \right\rangle_{\Gamma_1} + \lambda_1\langle\theta, \eta\rangle_{\Gamma_1} + (\nabla w_t, \nabla\eta)_\Omega = 0 \qquad (2.7.50)
\end{aligned}
$$

for all test functions $\xi \in [H^1_{\Gamma_0}(\Omega)]^2$, $\psi \in H^2_{\Gamma_0}(\Omega)$; $\delta, \eta \in H^1(\Omega)$. *The above equalities are understood in the sense of* $H^{-1}(0,T)$ *topology when* $\gamma = 0$ *and* $L_2(0,T)$ *when* $\gamma > 0$.

Remark 2.7.1. *One could also add appropriate nonlinear damping terms in the boundary conditions of the plate equation. Since this will not add new mathematical difficulties, we prefer to keep a simpler model.*

Well-posedness of regular solutions, in the context of nonlinear dynamic elasticity, is typically approached by some sort of fixed-point theorem or nonlinear Galerkin method. This rather standard procedure is often applicable because the smoothness properties of regular solutions are sufficient to engage in a fixed-point argument or to pass with the limit (on the nonlinear terms), in the case of the Galerkin method. However, the situation is much more subtle when it comes to the well-posedness of finite energy solutions. Indeed, the main difficulty in proving the well-posedness of weak solutions (particularly the uniqueness, which is critical for stability results presented later) is related to the fact that the nonlinear terms in the equation are not bounded on the energy space (this is due to the lack of appropriate Sobolev's embeddings in the two-dimensional case). This difficulty, typical to dynamic von Karman systems, is well recognized in the literature, where uniqueness of weak or finite energy solutions to von Karman equations is quoted as an open problem [150, 57, 99, 96, 177, 178]. Some progress in solving this open problem has been achieved, and the issue of the uniqueness of weak solutions to some von Karman systems has been settled in the affirmative. In fact, for the modified (scalar) von Karman equations, the uniqueness of weak solutions is the result (see section 2.4) of establishing sharp regularity of Airy's stress function and compensated compactness methods [62, 63]. Since there is no Airy stress function in the full von Karman system, the arguments used in [62] are not relevant to the analysis of these equations. The uniqueness question of finite energy solutions for Vlasov–Maguerre equations with homogeneous clamped boundary conditions was dealt with in [186], where a rather special technique based on duality was proposed. This technique has been extended and applied successfully to von Karman systems with a variety of boundary conditions, including nonhomogeneous and boundary feedback terms [107, 106, 110, 26]. We see below that the results and the methods strongly differ depending on whether $\gamma = 0$ or $\gamma > 0$ and whether thermoelastic effects are included. For this reason, we distinguish between these cases.

2.7.2 Formulation of the results: Case $\gamma = 0$

Theorem 2.7.1 (see [110]). 1. Finite energy solutions. *With reference to the system in* (2.7.43)–(2.7.45) *with* $\gamma = 0$,

(i) *there exists a unique, global solution of finite energy. That is, for any initial data*

$$u_0, u_1, \theta_0, \phi_0 \in [H^1_{\Gamma_0}(\Omega)]^2 \times [L_2(\Omega)]^2 \times L_2(\Omega) \times L_2(\Omega),$$
$$w_0, w_1 \in H^2_{\Gamma_0}(\Omega) \times L_2(\Omega),$$

there exists a unique solution

$$\begin{aligned}(u, w, \theta, \phi) &\in C([0,T]; [H^1(\Omega)]^2 \times H^2(\Omega) \times L_2(\Omega) \times L_2(\Omega),\\ (u_t, w_t) &\in C([0,T]; [L_2(\Omega)]^3),\\ \theta, \phi &\in L_2(0,T; H^1(\Omega)),\\ (w, w_t) &\in L_2(0,T; H^3(\Omega) \times H^1(\Omega)),\end{aligned}$$

which depends continuously on the initial data.

(ii) *If, in addition,*

$$(g(s), s)_{R^2} \geq m|s|^2, \qquad m > 0, |s| \geq 1, \tag{2.7.51}$$

then the additional boundary regularity is valid: $u_t, \nabla u \in L_2((0,T) \times \Gamma_1)$.

2. Regular solutions. *Assume, in addition, that* $g \in C^1$. *For any initial data subject to the regularity in part* (1) *and, in addition,*

$$u_0 \in [H^2(\Omega)]^2, \; u_1 \in [H^1_{\Gamma_0}(\Omega)]^2, \; w_0 \in H^4(\Omega), \; w_1 \in H^2(\Omega),$$

$$\theta_0, \phi_0 \in H^2(\Omega),$$

and subject to compatibility on the boundary, there exists a unique, global solution

$$(u, w, \theta, \phi) \in C([0,T]; [H^2(\Omega)]^2 \times H^4(\Omega) \times H^2(\Omega) \times H^2(\Omega)),$$

$$(u_t, w_t, \phi_t, \theta_t) \in C([0,T]; [H^1(\Omega)]^2 \times H^2(\Omega) \times L_2(\Omega) \times L_2(\Omega)),$$

where $T > 0$ *is arbitrary.*

Moreover, regular solutions depend continuously on the initial data in the topology of regular solutions (as defined above).

Remark 2.7.2. *The result as stated in Theorem* 2.7.1 *holds for von Karman systems with different boundary conditions, such as clamped or hinged. This is to say, one may consider (see* [26]*)* $w = \frac{\partial}{\partial \nu} w = 0$ *on* Γ_1 *(clamped) or* $w = \Delta w + (1 - \nu) B_1 w + \theta = 0$ *on* Γ_1 *(hinged). Since the free boundary conditions are mathematically most demanding (Lopatinski condition is not satisfied here), we concentrate on this particular case.*

Comments about the proof. The proof of Theorem 2.7.1 is given in [110]. Without getting into lengthy details of the proofs, we wish to point out that the key behind the proof is newly discovered analyticity of the semigroup generated by a subcomponent of the system. We use properties of partial analyticity of thermoelastic semigroups with free boundary conditions which are associated with the linearization of the equation describing the transversal displacements of the plate [136, 137]. These regularity properties are propagated onto the hyperbolic component of the system, which allows us to obtain appropriate estimates for the nonlinear terms.

To accomplish this, the following lemma is critical.

Lemma 2.7.1 (see [137]). *We consider the following linear system of equations:*

$$w_{tt} + D\Delta^2 w + \Delta\theta = 0, \quad \theta_t = \Delta\theta + \Delta w_t \quad \textit{in } \Omega \times (0, \infty), \tag{2.7.52}$$

with Dirichlet boundary conditions on the portion of the boundary Γ_0

$$w = \nabla w = 0 \quad \textit{on } \Gamma_0 \times (0, \infty)$$

and boundary conditions on $\Gamma_1 \times (0, \infty)$ *are free and given by*

$$D[\Delta w + (1 - \mu) B_1 w] + \theta = 0,$$

$$D\left[\frac{\partial}{\partial \nu} \Delta w + (1 - \mu) B_2 w\right] + \frac{\partial}{\partial \nu} \theta = 0. \tag{2.7.53}$$

Thermal variable θ satisfies the usual Robin boundary conditions on Γ.

$$\frac{\partial}{\partial \nu}\theta + \lambda_1 \theta = 0. \tag{2.7.54}$$

The linear semigroup generated by this system written in the variables (w, w_t, θ) is analytic on the space $H^2_{\Gamma_0}(\Omega) \times L_2(\Omega) \times L_2(\Omega)$.

We note that while the analyticity results are known for thermoelastic plates with Dirichlet boundary conditions [160], the situation in the case of free boundary conditions is very different. This is due to boundary coupling occurring in the model, which prevents applicability of methods known in the literature. Nevertheless, it was shown in [137] that free boundary conditions give rise to an analytic semigroup.

Once equipped with Lemma 2.7.1, the strategy in proving Theorem 2.7.1 is to apply a fixed-point theorem. However, one needs to be careful about the choice of the space on which fixed-point property can be shown. In fact, a natural choice would be to take the energy space. However, this space is too large. It turns out that the right choice is a more restrictive subspace given by

$$X \equiv \{(u, w, \theta, \phi) \in C([0,T]; [H^1(\Omega)]^2 \times H^2(\Omega) \times L_2(\Omega) \times L_2(\Omega));$$
$$(u_t, w_t) \in C([0,T]; [L_2(\Omega)]^3; (w, w_t, \theta, \phi) \in L_2(0,T; H^3(\Omega) \times [H^1(\Omega)]^3)\}. \tag{2.7.55}$$

2.7.3 Formulation of the results: Case $\gamma > 0$

In the case $\gamma > 0$ the situation is very different. The main difference is that the dynamics corresponding to the w equation is no longer analytic [136]. Thus, there is no mechanism to cover the ϵ gap in the differentiability of the nonlinear term. In fact, this difference is reflected quantitatively in the statement of the main result.

Theorem 2.7.2 (see [107, 26]). *With reference to the system* (2.7.43)–(2.7.45) *with $\gamma > 0$, the following apply.*

1. Finite energy solutions.

   (i) *There exists a global solution of finite energy. This is to say that for any initial data*

   $$u_0, u_1, \theta_0, \phi_0 \in [H^1_{\Gamma_0}(\Omega)]^2 \times [L_2(\Omega)]^2 \times [L_2(\Omega)]^2, \; w_0, w_1 \in H^2_{\Gamma_0}(\Omega) \times H^1_{\Gamma_0}(\Omega),$$

   there exists a weak solution such that

   $$(u, w, \theta, \phi) \in C([0,T]; [H^1(\Omega)]^2 \times H^2(\Omega) \times [L_2(\Omega)]^2),$$
   $$(u_t, w_t) \in C([0,T]; [L_2(\Omega)]^2 \times H^1(\Omega)), \quad \theta, \phi \in L_2(0,T; H^1(\Omega)).$$

   (ii) *If, in addition,*

   $$(g(s), s)_{R^2} \geq m|s|^2, \qquad m > 0, |s| \geq 1, \tag{2.7.56}$$

   then the additional boundary regularity is valid: $u_t, \nabla u \in L_2((0,T) \times \Gamma_1)$.

(iii) *Assuming, additionally, that the vector function g is linear, the solution in part* (i) *is unique and it depends continuously on the initial data with respect to finite energy norms* [95].

2. Regular solutions. *Assume, in addition, that $g \in C^1(R^2)$. For any initial data subject to the regularity in part* (1) *and*

$$u_0 \in [H^2(\Omega)]^2, \; u_1 \in [H^1_{\Gamma_0}(\Omega)]^2, \; w_0 \in H^3(\Omega), \; w_1 \in H^2(\Omega),$$

$$\theta_0, \phi_0 \in H^2(\Omega),$$

and subject to compatibility on the boundary, there exists a unique, global solution

$$(u, w, \theta, \phi) \in C([0,T]; [H^2(\Omega)]^2 \times H^3(\Omega) \times [H^2(\Omega)]^2),$$

$$(u_t, w_t, \phi_t, \theta_t) \in C([0,T]; [H^1(\Omega)]^2 \times H^2(\Omega) \times [L_2(\Omega)]^2),$$

where $T > 0$ is arbitrary.

Moreover, regular solutions depend continuously on the initial data in the topology of regular solutions (as defined above).

Remark 2.7.3. *We note that while the existence of finite energy solutions is established for both $\gamma = 0$ and $\gamma > 0$, and the velocity component w_t is in $H^1(\Omega)$ for both cases, the mechanism by which this is achieved is very different. For $\gamma > 0$, the H^1 regularity of w_t is a natural inheritance of the regularity displayed by the energy norm. Instead, in the case $\gamma = 0$, this regularity is more subtle and it follows from the smoothing effect of the analyticity.*

More important, in the case $\gamma > 0$, there is no statement regarding the uniqueness of finite energy solutions. However, these are shown to be unique [107] *if the boundary dissipation is linear. Whether the uniqueness of weak solutions may be extended to nonlinear boundary dissipation is an open problem.*

Remark 2.7.4. *We note that the existence and uniqueness of solutions to this problem can be shown to hold in the intermediate spaces (between weak and regular solutions)* [106]. *In this case, the results hold under the assumption of C^1 regularity of the boundary damping.*

In the case $\gamma > 0$, thermoelastic effects do not seem to play any major role for the well-posedness. Exactly the same result holds for full von Karman model accounting for mechanical displacements only [106, 107] (set $\theta = \phi \equiv 0$ in the corresponding equations). In fact, the following purely mechanical full von Karman system is a special case of the model considered in [107]. In $\Omega \times (0, \infty)$, we consider

$$\begin{aligned} u_{tt} - \operatorname{div}[\mathcal{C}[\epsilon(u) + f(\nabla w)]] &= 0, \\ w_{tt} - \gamma \Delta w_{tt} + D\Delta^2 w - \operatorname{div}[\mathcal{C}[\epsilon(u) + f(\nabla w)]\nabla w] &= 0 \end{aligned} \tag{2.7.57}$$

with Dirichlet boundary conditions on the portion of the boundary Γ_0

$$u = w = \nabla w = 0 \quad \text{on } \Gamma_0 \times (0, \infty).$$

The boundary conditions on $\Gamma_1 \times (0, \infty)$ are free and given by

$$\begin{aligned} \mathcal{C}[\epsilon(u) + f(\nabla w)]\nu + ku &= -g(u_t), \\ D[\Delta w + (1-\mu)B_1 w] &= -h_1\left(\frac{\partial}{\partial \nu} w_t\right), \\ D\left[\frac{\partial}{\partial \nu}\Delta w + (1-\mu)B_2 w\right] - \gamma\frac{\partial}{\partial \nu}w_{tt} - \mathcal{C}[\epsilon(u) + f(\nabla w)]\nabla w \nu &= -\frac{\partial}{\partial \tau}h_2\left(\frac{\partial}{\partial \tau}w_t\right). \end{aligned} \tag{2.7.58}$$

With reference to the model given by (2.7.57) and (2.7.58) and $\gamma > 0$, the following result holds.

Theorem 2.7.3 (see [107]). *Assume that functions g, h_1, h_2 satisfy Assumption* 2.7.1.

1. Finite energy solutions.
   (i) *For any initial data*

   $$u_0, u_1 \in [H^1_{\Gamma_0}(\Omega)]^2 \times [L_2(\Omega)]^2, \;\; w_0, w_1 \in H^2_{\Gamma_0}(\Omega) \times H^1_{\Gamma_0}(\Omega),$$

   there exists a weak solution such that

   $$(u, w) \in C([0,T]; [H^1(\Omega)]^2 \times H^2(\Omega)),$$
   $$(u_t, w_t) \in C([0,T]; [L_2(\Omega)]^2 \times H^1(\Omega)).$$

   (ii) *Assuming, additionally, that the functions g, h_1, h_2 are linear, the solution in part* (i) *is unique and it depends continuously on the initial data of finite energy* [95].

2. Regular solutions. *Assume, in addition, that $g \in C^1(R^2), h_1, h_2 \in C^1(R)$. For any initial data subject to the regularity in part* (1) *and, in addition,*

   $$u_0 \in [H^2(\Omega)]^2, \;\; u_1 \in [H^1_{\Gamma_0}(\Omega)]^2, \;\; w_0 \in H^3(\Omega), \;\; w_1 \in H^2(\Omega),$$

   and subject to compatibility on the boundary, then there exists a unique, global solution

   $$(u, w) \in C([0,T]; [H^2(\Omega)]^2 \times H^3(\Omega)),$$
   $$(u_t, w_t) \in C([0,T]; [H^1(\Omega)]^2 \times H^2(\Omega)),$$

   where $T > 0$ is arbitrary.

Moreover, regular solutions depend continuously on the initial data in the topology of regular solutions (as defined above).

Comments about the proof of Theorems 2.7.2 *and* 2.7.3. As expected, the proof of Theorem 2.7.2 is based on a completely different mechanism than the one for the case $\gamma = 0$. While existence of finite energy solutions follows from application of a nonlinear Galerkin method, the uniqueness of finite energy solutions is much more involved, and here the argument is based on a device developed in [186]. The main idea relies on the following observation: To prove uniqueness of finite energy solutions, it suffices to show that the difference of these solutions is equal to zero when measured in a weaker topology. Thus, the main technical point is to find a right measure. In the case of linear boundary dissipation, it is possible to use certain duality arguments to construct a right dual space with appropriate topology. However, this argument breaks down in the case of nonlinear boundary dissipation, simply because the corresponding linearization around zero is not a good approximation due to the roughness of boundary operators. Whether this problem can be dealt with remains to be seen, and as far as we know this is an open question.

The well-posedness of a regular solution relies on more classical methods. However, the main complications here are due to the nonlinearity of boundary dissipation. The approach taken in [107] relies on showing that Galerkin approximations display higher a priori bounds for more regular initial data. Then, the passage to the limit on these higher norms yields the final result. The details are technical and are given in [107]. We also mention that the issue of well-posedness of regular solutions for a purely mechanical von Karman system with linear boundary damping was considered in [177]. However, the method employed in [177], based on application of a fixed-point argument, appears to break down when dealing with a nonlinear dissipation.

2.8 Comments and Open Problems

Several comments and open questions can be addressed in the context of the material presented in Chapter 2.

- The addition of thermoelastic effects does not have any substantial bearing on the well-posedness of the solutions to the modified (scalar) von Karman equations considered in section 2.4. Indeed, the arguments and the results are the same as for the mechanical plate. If we consider

$$\begin{aligned} w_{tt} - \gamma\Delta w_{tt} + \Delta^2 w + \Delta\theta = [\mathcal{F}(w), w] \text{ in } \Omega \times (0,T), \\ \Delta^2\mathcal{F}(w) = -[w,w], \\ \theta_t - \Delta\theta = \Delta w_t \text{ in } \Omega \times (0,T), \end{aligned} \qquad (2.8.59)$$

  then for all values of parameters $\gamma \geq 0$, and for all combinations of boundary conditions (clamped, hinged, or free), the system is well-posed in the finite energy space. By this, we mean that the solution exists, is unique, and depends continuously on the initial data with respect to the topology induced by finite energy, i.e., $H^2(\Omega) \times H^1_\gamma(\Omega) \times L_2(\Omega)$, equipped with appropriate boundary conditions (if relevant). In addition, the solutions decay exponentially to zero in finite energy norm [9, 13].
- Similarly, the addition of thermal effects to the structural acoustic model discussed in section 2.6 has no effect on the well-posedness properties of the overall system. The final conclusions are the same as in section 2.6, after accounting for the space representing the thermal variable. We replace the equation for the structure given in (2.6.40) by the thermoelastic plate equation in (2.8.59).

- One may also consider the structural acoustic problem studied in section 2.6 with the wall structure described by the full von Karman system. In this case, (2.6.40) should be replaced by (2.7.43). The use of nonlinear spherical shells was studied in [35].
- Uniqueness of weak (finite energy) solutions is stated in Theorem 2.7.3 for the case of the full von Karman system with $\gamma > 0$ and with linear boundary damping. Whether this result can be extended to the nonlinear, monotone damping is an open problem.
- In the case of the full von Karman system discussed in section 2.8, the presence of thermal effects is critical for the case $\gamma = 0$. Indeed, when $\gamma = 0$, the gap in the degree of the unboundedness of the nonlinear term with respect to the topology generated by finite energy is $1 + \epsilon$, rather than only ϵ, in the case $\gamma = 0$. While thermal effects have no substantial influence on the well-posedness for $\gamma > 0$, where the thermoelastic plate still retains a strong hyperbolic component, the situation changes drastically when $\gamma = 0$. Indeed, in the presence of thermal dissipation, the regularizing parabolic effect is strong enough to counteract the $1 + \epsilon$ gap mentioned above. Instead, if $\theta \equiv 0$ in (2.7.43) of section 2.7, there is no smoothing (parabolic) effect to close this gap. This leads to several major technical difficulties, and it is not clear at the present time to what extent the theory existing for $\gamma > 0$ remains valid for $\gamma = 0$ and $\theta = 0$. Thus the result stated in Theorem 2.7.3, particularly the uniqueness of finite energy solutions, is at present an open problem with $\gamma = 0$.
- Uniqueness of weak solutions to full von Karman system defined on full R^2 and with $\gamma > 0$ was also shown in [190] by using Strichratz estimates. However, an extension of this result to arbitrary bounded domains and arbitrary boundary conditions runs into difficulties due to the intrinsic limitations of applicability Strichratz estimates.

Chapter 3

Uniform Stabilizability of Nonlinear Waves and Plates

3.1 Orientation

In this chapter we consider several models of semilinear waves and plates that display conservative behavior, i.e., the energy of the system is conserved for all time. Our main goal is to construct stabilizing boundary feedbacks that cause the energy to dissipate at a uniform rate. Since we are interested in stabilizing the systems by means of a suitable boundary feedback, we consider classes of boundary feedbacks acting on the velocity of boundary displacement. The advantage of these feedbacks, from the implementation point of view, is their relative simplicity. Indeed, standard damping mechanisms encountered in engineering practice are realized in a form of boundary dashpoints or elements inducing friction on the boundary, etc. All these are represented by monotone (nonlinear) graphs passing through the origin. Thus the requirement of monotonicity of boundary feedbacks is prevalent in our analysis. In addition, we impose certain growth conditions at infinity.

As mentioned, we consider nonlinear wave and plate models. The existence and, in some cases, uniqueness of solutions follow from general abstract results presented in the previous chapter. The key to stabilization estimates is the observability estimate. These estimates are derived, initially, for smooth solutions. Since the solutions of nonlinear problems may not display enough regularity, we introduce a regularization scheme whose purpose is to smooth out the problem. Indeed, instead of the original nonlinear equation, we consider an appropriate linear problem displaying smooth solutions that are close to the given solution of the nonlinear equation. The original observability estimates will be applied to the regularized scheme. Passage with the limit will reconstruct the desired estimates for the original solutions. This described method has an additional advantage in the cases when there is no uniqueness of solutions. Indeed, in such cases an ordinary passage with the limit on smooth solutions (provided such can be constructed) will not guarantee that the decay rates obtained will hold for a given (weak) solution of the nonlinear problem (because of the lack of uniqueness). Instead, our technique guarantees that the regularization converges precisely to a given solution of the nonlinear equation.

In what follows we are concerned with the same general model (2.2.1) considered in the previous chapter. However, we slightly specialize the setup to focus on issues related to stability. More specifically, we deal with the abstract equation

$$\begin{cases} Mu_{tt}(t) + Au(t) + AG\partial\Phi(G^*Au_t(t)) + AGf(u(t)) + Du_t(t) = F(u(t)), \\ u(t=0) = u_0, \quad u_t(t=0) = u_1, \end{cases} \tag{3.1.1}$$

under Assumption 2.2.1 and assume additionally that $\partial\Phi$ is single valued and $\partial\Phi(0) = 0$. In what follows we need to impose on $\partial\Phi$ some growth conditions at infinity.

In view of the well-posedness results presented in the previous chapter, it makes sense to introduce the following energy function:

$$E(t) \equiv |M^{1/2}u_t(t)|^2 + |A^{1/2}u(t)|^2.$$

To have this energy globally defined, one would need to impose certain structural assumptions imposed on $f(u)$ and $F(u)$. Depending on the problem considered, these assumptions may take various forms. We will be more specific when discussing particular examples. However, in general, the following structural hypothesis is enforced:

$$\int_0^t (F(u), u_t) ds \leq 0 \text{ for } (u, u_t) \in C([0,t]; D(A^{1/2}) \times V),$$

$$\int_0^t (f(u), G^*Au_t)_U ds \geq 0 \text{ for } u \in L_2(0,t; D(A^{1/2})), \quad u_t \in L_2(0,t; D(G^*A)). \tag{3.1.2}$$

Under conditions (3.1.2), one can show that the energy is nonincreasing, i.e., any weak solution to (2.2.1) satisfies

$$E(t) \leq E(s), \qquad t > s. \tag{3.1.3}$$

Our main goal is to show that under appropriate conditions imposed on nonlinear terms, the energy of the system decays to zero at a uniform rate. That is, we aim to show that there exists a scalar function, say, $S(t)$, that typically depends on the initial energy $E(0)$ such that

$$S(t) \to 0 \text{ when } t \to \infty$$

and

$$E(t) \leq S(t), \qquad t > T_0.$$

It is important to emphasize that the function $S(t)$ depends not on a profile of the initial conditions but only on the norm $E(0)$.

To achieve this goal, we begin by presenting a very general method that specifies the estimates necessary to obtain uniform decays for the energy function. This is done in section 3.2. Since the topic of boundary stabilizability of waves and plates is rather extensive, we are not in a position to provide here a complete account of the existing results and techniques. For this reason we concentrate on a prototype problem, a wave equation with Neumann boundary conditions, which are treated in section 3.3 in fair generality and with complete details. The main concepts, ideas, and tools introduced for the wave equation are also used in the study of boundary stabilization of the plates. In section 3.4 we mention several results pertinent to plate problems. By necessity, this section is brief and provides only the statements of the main results and indicates appropriate references for various results. We believe that this will suffice to orient the reader to the results available in the literature.

3.2 Abstract Stabilization Inequalities

To derive estimates describing the decay rates for the solutions, we apply comparison theorems [120] which will allow us to compare the asymptotic behavior of the energy function associated with a given nonlinear model to an asymptotic behavior of a solution of an appropriate nonlinear ordinary differential equation (ODE) problem. The solution to this nonlinear ODE problem is precisely the function $S(t)$ introduced in the previous section.

Let T be a given positive constant. Let a function $p(s)$ be a given nonlinear function that is positive, strictly increasing, and zero at the origin. We assume that we are given a scalar, positive function $E(t)$ with the following properties.

Assumption 3.2.1. *$E(t)$ is monotone decreasing for $t \in R^+$. This is to say, $E(t_2) \leq E(t_1)$, $t_2 \geq t_1 \geq 0$.*

$$p(E((m+1)T)) + E((m+1)T) \leq E(mT)$$

for $m = 0, 1, 2 \ldots$ and $p(s)$ does not depend on m.

Our result states as follows.

Theorem 3.2.1. *Assume Assumption 3.2.1. Let us define a monotone increasing function $q(s)$ given by the formula*

$$q(s) \equiv s - [I + p]^{-1}(s).$$

Then

$$E(t) \leq S(t/T - 1), \ \ for\ t > T,$$

where the function $S(t)$ is a solution of the following ODE:

$$S_t(t) + q(S(t)) = 0, \qquad S(0) = E(0). \tag{3.2.4}$$

Moreover, if $p(s) > 0$ for $s > 0$, then

$$S(t) \to 0, \qquad t \to \infty.$$

Remark 3.2.1. *We note that our result is a very natural generalization of a linear stability criterion credited to Datko and Pazy [175]. Indeed, in the linear case, the function p is linear, i.e., $p(s) = as$, $a > 0$. Thus, Assumption 3.2.1 says that $aE(T)+E(T) \leq E(0)$. But this implies $E(T) \leq \frac{1}{1+a}E(0)$; hence exponential stability follows by well-known linear criteria.*

Remark 3.2.2. *In the truly nonlinear case, the function p is, of course, nonlinear. As we will see later, this function depends on the parameters in the equation and, in particular, on the behavior of nonlinear damping at the origin (as expected). If, for instance, function p is of polynomial growth (i.e., the damping is of polynomial growth at the origin), then the decay rates computed from our algorithm will be algebraic. If, instead, function p is rapidly approaching zero as an exponentially decaying function, then the corresponding decay rates are logarithmic.*

Proof of the Theorem. We begin with the following lemma.

Lemma 3.2.1. *Consider a sequence s_n of positive numbers such that*

$$s_{n+1} + p(s_{n+1}) \leq s_n. \tag{3.2.5}$$

Then we have

$$s_n \leq S(n), \tag{3.2.6}$$

where $S(t)$ is a solution of (3.2.4) *with the initial data $S(0) = s_0$.*

Proof. We use induction. Since $S(0) = s_0$, it suffices to prove the implication,

$$s_n \leq S(n) \Rightarrow s_{n+1} \leq S(n+1). \tag{3.2.7}$$

Inequality (3.2.5) is equivalent to

$$(I + p)s_{n+1} \leq s_n,$$

and since $(I + p)^{-1}$ is monotone increasing,

$$s_{n+1} \leq (I + p)^{-1} s_n,$$

hence $s_{n+1} - s_n \leq -q(s_n)$ or, equivalently,

$$s_{n+1} \leq s_n - q(s_n). \tag{3.2.8}$$

On the other hand, since q is increasing, the solution $S(t)$ of (3.2.4) describes a nonlinear semigroup of contractions. Hence

$$S(t) \leq S(\tau) \quad \text{for all} \quad t \geq \tau. \tag{3.2.9}$$

Integrating (3.2.4) from n to $n + 1$ gives

$$S(n+1) - S(n) + \int_n^{n+1} q(S(\tau))d\tau = 0. \tag{3.2.10}$$

Since q is increasing, (3.2.9) yields

$$\begin{aligned} S(n+1) \geq S(n) - q(S(n)) &= (I - q)S(n) = [I + p]^{-1} S(n) \\ &\geq [I + p]^{-1} s_n = s_n - q(s_n), \end{aligned} \tag{3.2.11}$$

where we have used the inductive assumption $s_n \leq S(n)$ and the increasing property of the function $[I + p]^{-1}$. Comparing (3.2.8) with (3.2.11) yields the desired inequality stated by (3.2.7), hence proving the lemma. □

Completion of the Proof of the Theorem. We apply the result of the Lemma 3.2.1 with

$$s_n \equiv E(nT), \qquad s_0 \equiv E(0). \tag{3.2.12}$$

By the result of the Lemma 3.2.1,

$$E(nT) \leq S(n), \qquad n = 0, 1, 2, 3 \ldots. \tag{3.2.13}$$

Setting $t = nT + \tau$ for $0 \leq \tau \leq T$ and evoking decreasing property of the energy function together with Assumption 3.2.1 gives

$$E(t) \leq E(nT) \leq S(n) \leq S\left(\frac{t-\tau}{T}\right) \leq S(t/T - 1), \qquad t > T, \tag{3.2.14}$$

which completes the proof of the theorem. □

It is well known that for linear dynamics, uniform decay rates are equivalent to exponential decay rates. This, of course, refers to the situation where the bounds in decay rate estimates are uniform with respect to the underlined topology for the problem (say, $E(t)$). In some (even linear) cases it may be difficult or impossible to obtain decay rate estimates where trajectories and initial conditions are measured with respect to the same topology. The constants describing inequalities may depend on higher topology of initial conditions. In such cases one does not expect exponential decay rates (also for linear problems), but some other type of decay rates, typically polynomial.

The result stated below pertains to this situation. To describe the result we introduce three energy levels with an increasing topological strength,

$$E_1(t) \leq AE(t) \leq BE_{-1}(t)$$

with some positive constants A, B. Typically, in application, the middle energy $E(t)$ is a standard energy function used to describe solutions. $E_1(t)$ is a higher energy level, often corresponding to the norms induced by the graph norm of the generator (or its fractional powers) of the corresponding semigroup. $E_{-1}(t)$ will be a dual energy to $E_1(t)$ with respect to $E(t)$. We postulate the following condition.

Assumption 3.2.2. *There exist $T > 0$ and the constants $C_i, i = 0, 1, 2,$ such that*

- observability, $E_{-1}((m-1)T) \leq C_0[E((m-1)T) - E(mT)], \quad m = 1, 2 \ldots,$
- interpolation, $E(t) \leq C_1 E_{-1}(t)^\theta E_1(t)^{1-\theta}, \quad 0 \leq \theta \leq 1,$
- dissipativity, $E(t) \leq E(s), t \geq s, \quad C_2 E_1(t) \leq E_1(0), t \geq 0.$

Theorem 3.2.2. *Under Assumption* 3.2.2 *the decay estimate*

$$E(t) \leq CE_1(0) t^{\frac{\theta}{\theta - 1}}, \qquad t \geq 0, \tag{3.2.15}$$

holds true for some positive constant C.

Remark 3.2.3. *We note that the decay estimate in Theorem* 3.2.2 *does not imply exponential decay rates, even in the linear case. Indeed, this is because the right-hand side of* (3.2.15) *depends on $E_1(0)$, which term represents higher norms for the initial conditions.*

Remark 3.2.4. *In typical applications of Theorem* 3.2.2 *we take for* $E_1(t)$ *the energy corresponding to the topology generated by the graph norm of a fractional power of the generator corresponding to the overall dynamics. Similarly, the* $E_{-1}(t)$ *will correspond to the topology dual to some fractional powers of this generator, and duality is taken with respect to pivot space corresponding to* $E(t)$. *In this case, the dissipativity condition in Assumption* 3.2.2 *follows from dissipativity of natural energy* $E(t)$. *Similarly, interpolation inequality in Assumption* 3.2.2 *follows from interpolation properties of domains of fractional powers of generators. The observability inequality in the assumption is a classical relation between the initial energy and the damping, expressed, however, with respect to two different norms.*

Proof of Theorem 3.2.2. From the interpolation property we obtain that

$$\frac{E((m-1)T)^{\frac{1}{\theta}}}{E_1((m-1)T)^{\frac{1-\theta}{\theta}}} \le C_1^{\frac{1}{\theta}} E_{-1}((m-1)T).$$

Combining the above inequality with the observability inequality in Assumption 3.2.2 we obtain

$$\frac{E((m-1)T)^{\frac{1}{\theta}}}{E_1((m-1)T)^{\frac{1-\theta}{\theta}}} \le C_0 C_1^{\frac{1}{\theta}} [E((m-1)T) - E(mT)].$$

Since the energy $E(t)$ is dissipative, $E(mT) \le E((m-1)T)$ and

$$\frac{E(mT)^{\frac{1}{\theta}}}{E_1((m-1)T)^{\frac{1-\theta}{\theta}}} \le C_0 C_1^{\frac{1}{\theta}} [E((m-1)T) - E(mT)]$$

and

$$E(mT) + \frac{1}{C_0 C_1^{\frac{1}{\theta}}} \frac{E(mT)^{\frac{1}{\theta}}}{E_1((m-1)T)^{\frac{1-\theta}{\theta}}} \le E((m-1)T).$$

Now we use the boundedness of $E_1(t) \le C_2^{-1} E_1(0)$, which gives

$$E(mT) + c \frac{1}{E_1(0)^{-1}} \frac{E(mT)^{\frac{1}{\theta}}}{E_1(0)^{\frac{1}{\theta}}} \le E((m-1)T),$$

where

$$c \equiv C_0 C_1^{\frac{1}{\theta}} C_2^{\frac{1-\theta}{\theta}}.$$

Dividing the above relation by $E_1(0)$ and denoting

$$\tilde{E}(t) \equiv \frac{E(t)}{E_1(0)},$$

we arrive at

$$\tilde{E}(mT) + p(\tilde{E}(mT)) \le \tilde{E}((m-1)T), \tag{3.2.16}$$

where

$$p(s) = cs^{\frac{1}{\theta}}.$$

Thus, asymptotically $q(s) \sim cs^{\frac{1}{\theta}}$, and by solving the nonlinear node problem (3.2.4) we conclude from the result of Theorem 3.2.1 the final conclusion. □

The method described above, along with appropriate observability estimates, provides an array of stability results valid for various PDEs describing waves, plates, shells, and combinations thereof [120, 83, 84, 88, 112]. However, to convey the main ideas, and at the same time to display the main features of the stabilization problem, we concentrate in the next section on a benchmark problem of a nonlinear wave equation. This particular equation (discussed in full detail in [120]) may serve as a prototype for general technique. Another reason to focus on the wave equation is that to analyze stability properties of structural acoustic models, the intrinsically unstable wave component is of prime interest. Although many papers dealt with a question of boundary stabilizability of wave equations [99, 97, 45] and references therein, these results are often too restrictive to be of sufficient use in the study of interactive systems. We need more general models, under less restrictive assumptions (regarding geometry and nonlinearity), which will accommodate additional interactions with other subsystems. The main goal of the next section is to provide stability analysis of a rather general nonlinear model of the wave equation though with constant coefficients and with boundary damping subject to minimal assumptions regarding the smoothness of the nonlinear dissipation as well as the geometry of the domain. To incorporate this extra degree of generality, in addition to standard multipliers, other methods and techniques (e.g., microlocal analysis) are necessary. We finally note that boundary stabilization of wave equations with *variable* principal part is more technical [22, 189, 143] and beyond the scope of these notes. One of the reasons is that conditions necessary for validity of observability estimates are hard to check (except the constant coefficient case) and therefore of limited applicability in the context of structure.

3.3 Semilinear Wave Equation with Nonlinear Boundary Damping

Let Ω be a bounded, open domain in R^n with sufficiently smooth boundary Γ. We assume that the boundary Γ consists of two disjoint parts: Γ_0 and Γ_1, $\Gamma = \Gamma_0 \cup \Gamma_1$. Γ_0 may be empty. We consider the following model of a semilinear wave equation with nonlinear boundary damping acting via Neumann boundary conditions:

$$\begin{aligned} w_{tt} &= \Delta w - f_0(w) \;\text{ in } \Omega \times (0,\infty), \\ \frac{\partial}{\partial \nu} w + f_1(w) &= u \;\text{ on }\; \Gamma_1 \times (0,\infty), \\ w &= 0 \;\text{ on }\; \Gamma_0 \times (0,\infty), \\ w(0) = w_0, \quad w_t(0) &= w_1 \;\text{ in }\; \Omega. \end{aligned} \tag{3.3.17}$$

The following assumptions are made on the nonlinear functions.

Assumption 3.3.1. 1. *$f_0(s) \in W^{1,\infty}_{loc}(R)$ is a piecewise $C^1(R)$ function, differentiable at $s = 0$ and such that there exists constant $M > 0$,*

$$f_0(s)s \geq 0, \quad |f_0'(s)| \leq M(1 + |s|^{k_0 - 1}), \quad |s| > N, \; N \textit{ large enough,}$$

where $1 \leq k_0 < n/(n-2)$ for $n \geq 2$, $k_0 < \infty$ for $n = 2$.[1]

[1]If f_1 is linear, we can take $k \leq \frac{n}{n-2}$.

2. *$f_1(s) \in C(R)$ is a differentiable function at $s = 0$ and such that*

$$f_1(s)s \geq 0, \quad |f_1(s)| \leq M[|s|^{k_1} + |s|],$$

where $1 \leq k_1 < (n-1)/(n-2)$ *for* $s \in R$, $n \geq 2$.

With the above model we associate the energy function given by

$$E_w(t) \equiv \int_\Omega [|\nabla w|^2 + |w_t|^2] \, dt. \tag{3.3.18}$$

Under these assumptions, it can be shown by PDE methods (see [130]) that for a given control function in, say, $L_2(0, T; H^{1/2}(\Gamma_1))$, there exists a solution of finite energy. However, our main goal is to study a closed-loop system where the control function is given in a feedback form. Therefore, we consider a problem of finding a control function u, in a feedback form, such that the resulting closed-loop system is uniformly stable. Accordingly, we consider a feedback control given in the form

$$u(t) = -g(w_t(t)), \tag{3.3.19}$$

where the nonlinear function g satisfies the following hypotheses.

Assumption 3.3.2. 1. *$g(s)$ is a continuous, monotone increasing function such that*

2. *$g(s)s > 0$ for $s \neq 0$,*
3. *there exist constants M_1, M_2 positive such that*

$$M_2|s|^2 \leq g(s)s \leq M_1|s|^2 \quad \textit{for} \quad |s| \geq 1.$$

We note that the system in (3.3.17) with a control given in (3.3.19) can be written in an abstract form (2.2.1) with the following definitions of abstract operators:

$$M \equiv I; \; A \equiv A_N, \quad G \equiv N; \quad \partial\Phi \equiv g, \quad f \equiv f_1, \; F \equiv -f_0.$$

Our main goal is to derive uniform decay rates for the energy of the system given by (3.3.17), (3.3.19).

Remark 3.3.1. *Unlike most of the literature (see [97, 99] and references therein), we do not assume any conditions on the growth of nonlinearities at the origin.*

Remark 3.3.2. *Instead of taking Neumann boundary conditions on a stabilized portion of the boundary, one could take the Dirichlet boundary conditions. This leads to the model*

$$\begin{aligned} w_{tt} &= \Delta w - f_0(w) \quad \textit{in } \Omega \times (0, \infty), \\ w &= g\left(\frac{\partial}{\partial \nu} A_D^{-1} w_t\right) \quad \textit{on} \;\; \Gamma_1 \times (0, \infty), \\ w &= 0 \quad \textit{on} \;\; \Gamma_0 \times (0, \infty), \\ w(0) &= w_0, \quad w_t(0) = w_1 \quad \textit{in } \Omega. \end{aligned} \tag{3.3.20}$$

The appropriate energy function associated with this model is

$$E_w(t) \equiv \int_\Omega [|w|^2 + |w_t|^2_{H^{-1}(\Omega)}]\, dt. \tag{3.3.21}$$

Exponential stability for the energy (3.3.21) *of this model, in the linear case, was first established in* [124], *in the case in which the domain Ω is convex. Subsequently, by using tools of microlocal analysis, this convexity assumption was altogether eliminated in* [131]. *By combining the techniques developed in this chapter with the analysis of* [131], *one can obtain uniform stability estimates for the model* (3.3.20) *under appropriate conditions imposed on the nonlinear terms f_i and g. In fact, the conditions imposed on function g are the same as in the Neumann case, but the conditions assumed on semilinear terms f_i need to be more restrictive.*

3.3.1 Formulation of the results

To proceed, we begin with the well-posedness result valid for the system (3.3.17) with the feedback control given by (3.3.19).

Theorem 3.3.1. *Assume Assumption* 3.3.1 *with $k_0 < \frac{n}{n-2}$*[2] *and Assumption* 3.3.2. *Then, for each initial data*

$$w_0, w_1 \in H^1_{\Gamma_0}(\Omega) \times L_2(\Omega),$$

there exists at least one finite energy (mild) solution, i.e.,

$$w \in C_{loc}(0,\infty; H^1(\Omega)), \quad w_t \in C_{loc}(0,\infty; L_2(\Omega)).$$

Moreover, the following boundary regularity takes place:

$$w_t \in L_2(0,\infty, L_2(\Gamma)), \quad \frac{\partial}{\partial\nu} w \in L_2(0,\infty, L_2(\Gamma)).$$

The result of Theorem 3.3.1 follows from the abstract result in Theorem 2.5.1 in Chapter 2 (see also [120]).

Remark 3.3.3. *We note that the regularity of solutions on the boundary does not follow from interior regularity. This is an independent regularity result.*

Note that the theorem above says nothing about the uniqueness of solutions. In fact, under the above generalities we may not have unique solutions. To guarantee the uniqueness, the additional hypotheses imposed on the feedback g or semilinear term f_1 are needed.

Corollary 3.3.1. *Along with Assumption* 3.3.1 *and Assumption* 3.3.2 *we assume that*

- *either f_1 is affine, i.e., satisfies* (2.3.4),
- *or else $[g(s_1) - g(s_2)][s_1 - s_2] \geq \alpha |s_1 - s_2|^2$.*

[2]If f_1 is linear or g is coercive, we can take $k_0 \leq \frac{n}{n-2}$.

Then, the solution (w, w_t) *with regularity stated in Theorem* 3.3.1 *exists and is unique in the energy space.*

The corollary follows from Theorem 2.5.1.

To state our stability results, we introduce some notation. Let $h(s)$ be a real valued function defined for $s \geq 0$; it is concave, strictly increasing, $h(0) = 0$, and it satisfies

$$h(sg(s)) \geq s^2 + g^2(s) \text{ for } |s| \leq N \text{ for some } N > 0. \tag{3.3.22}$$

Such a function h can always be constructed by virtue of Assumption 3.3.2. Indeed, define the increasing functions k_1, k_2 by

$$k_1(sg(s)) \geq s^2 + g^2(s) \text{ for } s \geq 0, \tag{3.3.23}$$
$$k_2(sg(s)) \geq s^2 + g^2(s) \text{ for } s \leq 0. \tag{3.3.24}$$

Then the function

$$h \equiv conc(\max\{k_1, k_2\})$$

has the desired properties.

Let $T > 0$ be a given constant, $\Sigma_1 \equiv \Gamma_1 \times (0, T)$. We define

$$\tilde{h}(x) \equiv h(x/meas\Sigma_1) \text{ for } x \geq 0.$$

Since $\tilde{h}$ is monotone increasing, $cI + \tilde{h}$ is invertible for any positive value of a constant c. Define

$$p(x) \equiv [cI + \tilde{h}]^{-1}(Kx), \tag{3.3.25}$$

where K is a positive constant. Then p is a positive, continuous, strictly increasing function with $p(0) = 0$. Let

$$q(x) \equiv x - (I + p)^{-1}(x), \qquad x > 0. \tag{3.3.26}$$

Since $p(x)$ is positive, increasing, so is $q(x)$.

To state our stability result, we need to introduce the following two additional assumptions.

Assumption 3.3.3.

$$(x - x_0) \cdot \nu \leq 0 \quad on \quad \Gamma_0. \tag{3.3.27}$$

Assumption 3.3.4. *If* $\Gamma_0 = \emptyset$, *then* $f_0(s)s + f_1(s)s \geq \epsilon s^2$.

We note that the geometric Assumption 3.3.3 needs to hold only on the uncontrolled portion of the boundary. This is in contrast to other results in the literature; see [99, 97] and references therein. Also, Assumption 3.3.4 will be needed only in the compactness-uniqueness part of the argument. It is, in fact, a necessary condition for stability.

We are ready to state the main result of this section.

Theorem 3.3.2. *Assume Assumptions* 3.3.1–3.3.4. *Let* w, w_t *be any finite energy solution corresponding to* (3.3.17) *with the feedback control* (3.3.19). *Then, for some* $T_0 > 0$,

$$E_w(t) \leq S(t/T_0 - 1), \qquad t > T_0, \tag{3.3.28}$$

where the real variable function $S(t)$ *is the solution of the following nonlinear differential equation:*

$$S_t(t) + q(S(t)) = 0, \quad S(0) = E_w(0), \tag{3.3.29}$$

with function $q(s)$ *given by* (3.3.26) *where the constant* K *in* (3.3.25) *depends generally on* $E_w(0)$, *and where the constant* $c = 1/meas\Sigma_1(M_1 + M_2^{-1})$.

A priori knowledge of the rate growth of the feedback g at the origin allows one to determine from the estimate (3.3.29) the explicit decay rates for the energy function. Indeed, we have the following corollary.

Corollary 3.3.2. *In addition to Assumptions* 3.3.1–3.3.4, *we assume that for some positive constants* a, b *the following conditions hold:*

$$g(s)s \leq bs^2, \tag{3.3.30}$$

$$g(s)s \geq as^{p+1}, \quad |s| \leq 1 \textit{ for some } p \geq 1. \tag{3.3.31}$$

Then

$$E(t) \leq Ce^{-\omega t} \textit{ if } p = 1, \tag{3.3.32}$$

$$E(t) \leq Ct^{2/(1-p)} \textit{ if } p > 1. \tag{3.3.33}$$

Here, the constants C, ω *typically depend on* $E(0)$ *(unless* $k = 1$*).*

Proof of Corollary 3.3.2. It is enough to construct a function h with the property (3.3.22). Indeed, we take

$$h(s) = a^{-2/p+1}(1 + b^2)s^m, \quad m = 2/(p+1) \leq 1.$$

Then,

$$p(s) = (cI + \tilde{h})^{-1}(Ks);$$

i.e., $cp + d(a,b)p^m = Ks$, where d is a suitable constant depending on a, b. Also, recall

$$q(s) = s - (I + p)^{-1}(s).$$

Since asymptotically (for s small) we have that for some constant ω depending in general (unless $k = 1$) on $E(0)$,

$$p(s) \sim \omega s^{1/m} \text{ and therefore } q(s) \sim \omega s^{1/m}.$$

Solving the nonlinear ODE (3.3.29) with the above choice of function q gives

$$S(t)x = \begin{cases} c_1(t + c_2 x^{(1-p)/2})^{2/1-p} & \text{if } p > 1, \\ e^{-\omega t}x & \text{if } p = 1, \end{cases}$$

where c_1, c_2 depend only on ω, p. The conclusion in the corollary follows now from the estimate (3.3.28). □

In a similar manner, we obtain optimal decay rates for nonlinear functions that decay to zero at the origin faster than any polynomial.

Corollary 3.3.3. *In addition to Assumptions* 3.3.1–3.3.4, *we assume that for some positive constants* a, b *the following conditions hold:*

$$g(s)s \le bs^2, \tag{3.3.34}$$

$$g(s)s \ge s^{-1}e^{\frac{-a}{s^2}}, \quad |s| \le 1, \ a > 0. \tag{3.3.35}$$

Then

$$E(t) \le C\frac{2a}{K}\left[ln\left(\frac{at}{K} + e^{\frac{2a}{KE(0)}}\right)\right]^{-1}. \tag{3.3.36}$$

Here, the constants K *typically depend on* $E(0)$ *(unless* $k = 1$*).*

Proof. Direct application of the algorithm presented above yields $h(s) = -\frac{(a+\epsilon)}{lns}$; $q(s) \sim e^{\frac{-2a}{Ks}}$. Solving the differential equation leads to the desired conclusion in the corollary. □

Remark 3.3.4. *A similar conclusion follows for more general nonlinear functions such that* $g(s) \ge s^p e^{\frac{-a}{s^2}}$, $p \ge -1$. *In this case, the decay rates are the same as in the corollary above, with* a *replaced by* $a - \epsilon$, *where* ϵ *is an arbitrary small constant.*

Remark 3.3.5. *One could also replace the Dirichlet boundary conditions on* Γ_0 *by the Neumann boundary conditions,*

$$\frac{\partial}{\partial\nu}w = 0, \ \ on \ \Gamma_0 \times (0, \infty),$$

under the additional assumption that Γ_0 *is "convex."*[3] *In fact, in this case we do not need the requirement that the intersection of the closures of* Γ_0 *and* Γ_1 *is empty. This feature is of particular interest in the study of structural acoustic models, when making a hole in the acoustic chamber is highly undesirable. To achieve this level of generalization, the techniques presented below need a substantial adjustment. The standard vector field* $(x - x_0)$ *is no longer sufficient, and more general vector fields need to be considered. The details are in* [112, 144].

Remark 3.3.6. *The same results as posted above can be obtained for the Dirichlet problem formulated in Remark* 3.3.2, *under appropriate conditions imposed on the nonlinear function* $f(w)$ *and with appropriate definition of the energy function that is given in Remark* 3.3.2. *The conditions imposed on the nonlinear function are the same. Moreover, there is no need to require that* $\bar{\Gamma}_0$ *and* $\bar{\Gamma}_1$ *be separated.*

Boundary stabilizability of the wave equation has been a subject of intense study and many papers are devoted to the topic. See [47, 45, 99, 97] and the lists of references therein. The major tool that has been utilized for proving the requisite energy decays is differential multipliers. This method forces, in the context of both linear and nonlinear problems, restrictive geometric and analytic constraints. The major ones, frequently encountered in the literature, are the following:

---

[3] We say that Γ_0 is "convex" if the level set function representing Γ_0 is convex in the neighborhood of Γ_0 on the side of Ω; see [144].

- Star-shaped geometric conditions are imposed on the boundary Γ_1.
- Semilinear terms $(f_0(u))$ in the equation, if considered, are subject to structural (superlinear) restrictions.
- Semilinear terms in the boundary conditions $(f_1(u))$ are not considered.
- The nonlinear dissipation is subject to polynomial decay restrictions at the origin.
- The obtained decay rates are valid for regular solutions only. This issue is particularly sensitive when the solutions may not be unique and the usual regularity-density argument is not applicable.

The main goal of Theorem 3.3.2 was to remove the above restrictions. To accomplish this, more complicated methods, which involve elements of microlocal analysis, need to be brought in. One may argue that the advantage of the multiplier method is that it leads to rather explicit estimates for the decay rates. Indeed, since the methods of estimates are direct, it is possible to track all the constants. This is not the case for more general and involved methods using, among other things, microlocal analysis techniques [22]. On the other hand, microlocal analysis leads to much more general results in terms of the geometry of the domains, which is particularly relevant in the context of interactive systems. This motivates our choice for presenting rather general and fairly complete treatment of stabilization problem, where a combination of multiplier method with microlocal analysis leads to explicit, optimal results applicable to a large class of geometries. We follow methodology introduced in [120] which precisely accomplishes this goal. In particular, it dispenses with all the restrictions listed above.

The remainder of this section is devoted to the proof of Theorem 3.3.2. Our program consists of the following four steps. We first prove (section 3.3.2) an approximation theorem (Theorem 3.3.3 and Corollary 3.3.4), which states that any weak solution of the original problem is approximated by smooth solutions of an appropriately constructed regularized problem. In step 2 (section 3.3.3), we prove PDE estimates valid for smooth solutions of the approximating problem. These PDE estimates involve microlocal calculus, whose application mandates sufficient smoothness of the solutions. Passage with the limit by using the approximating theorem from section 3.3.2 allows us to establish the observability estimate, stated in Lemma 3.3.3, which is valid for weak solutions of the original problem. This estimate is polluted by lower-order terms. To absorb these lower-order terms, we apply unique continuation results valid for overdetermined PDEs. This is part of step 3 culminating with Lemma 3.3.4. Combining the observability estimate in Lemma 3.3.3 with Lemma 3.3.4 we obtain the final form of a nonlinear observability estimate stated in Lemma 3.3.5. The last step in the proof relies on an application of abstract stability inequality in Theorem 3.2.1, which in turn yields the desired decay rates.

3.3.2 Regularization

The main purpose of this section is to construct smooth approximations for any given solution to the nonlinear problem (3.3.17), (3.3.19) which will converge in the energy norm to that solution and whose boundary traces will converge as well. We need this kind of regularization because in the process of proving stability estimates we need to perform various PDE calculations. To justify rigorously these computations, an appropriate amount of solution regularity is needed. The original solutions (typically weak) to the nonlinear problem will not display enough smoothness. Therefore, the idea is to consider the regularized problem with smooth solutions and to perform all the calculations on these solutions. In the final stage we pass to the limit—but only on the desired stability estimates that call for less regularity. In view of this, it is important to have a right regularization that will converge, in the energy norm, to a given predetermined solution of the original nonlinear problem and the appropriate traces will converge as well. In this context we mention that for the usual wave equation with monotone damping and taking $f_0 = f_1 = 0$ (which admits, of course, unique solutions), this goal can be easily achieved by applying classical regularization of the initial data and of the nonlinear monotone function g [97]. Then, the passage to the limit is straightforward and based on application of monotone operator theory. However, in a more general case, as considered in this section, this procedure is not applicable. Particular difficulties are brought about by the potential nonuniqueness of solutions and the fact that the entire equation does not correspond to the monotone problem (due to the presence of f_0, f_1). For this reason, a regularization procedure very different from that in [97], based on [120], is presented below. The remainder of this section describes the details of the regularization.

We consider the following linear wave equation:

$$
\begin{aligned}
w_{tt} - \Delta w = f \in L_1(0,T; L_2(\Omega)),&\\
w(0) = w_0 \in H^1(\Omega),\ w_t(0) = w_1 \in L_2(\Omega),&\\
w = 0 \ \text{ on } \ \Gamma_0.&
\end{aligned}
\tag{3.3.37}
$$

Note that we do not impose any boundary conditions on Γ_1. Our goal is to construct a suitable smooth approximation of (3.3.37). The main result in this direction is formulated in the following theorem.

Theorem 3.3.3. *Let w be a given solution to the linear problem* (3.3.37) *such that*

$$
w \in C([0,T]; H^1(\Omega)) \cap C^1([0,T]; L_2(\Omega)), \tag{3.3.38}
$$

$$
w_t, \frac{\partial}{\partial\nu} w \in L_2(\Sigma_1). \tag{3.3.39}
$$

Then, there exists a sequence of functions

$$
w_l \in C([0,T]; H^2(\Omega)) \cap C^1([0,T]; H^1(\Omega)), \tag{3.3.40}
$$

$$
f_l \in C([0,T]; H^1_{\Gamma_0}(\Omega)) \tag{3.3.41}
$$

such that

$$
w_{l,t,t} = \Delta w_l + f_l, \tag{3.3.42}
$$

$$
f_l \to f \ \text{ in } \ L_1([0,T]; L_2(\Omega)), \tag{3.3.43}
$$

$$
w_l \to w \ \text{ in } \ C([0,T]; H^1(\Omega)), \tag{3.3.44}
$$

$$
w_{l,t} \to w_t \ \text{ in } \ C([0,T]; L_2(\Omega)), \tag{3.3.45}
$$

$$
w_{l,t}|_{\Gamma_1} \to w_t|_{\Gamma_1} \ \text{ in } \ L_2(\Sigma_1), \tag{3.3.46}
$$

$$
\frac{\partial}{\partial\nu} w_l|_{\Gamma_1} \to \frac{\partial}{\partial\nu} w|_{\Gamma_1} \ \text{ in } \ L_2(\Sigma_1). \tag{3.3.47}
$$

Remark 3.3.7. *We note that the regularization w_l, constructed above, satisfies an approximated equation, which is linear but nonhomogeneous. Moreover, and this is very important for the development, the boundary traces of the regularization provide good $L_2(\Sigma_1)$ approximations of the traces of the original nonlinear solutions. The above property is not typically obtained with other forms of regularization. Indeed, a typical smooth approximation of a solution, obtained by taking smooth initial data, will not necessarily give a good approximation of the boundary traces.*

Remark 3.3.8. *The result remains valid if we replace zero Dirichlet boundary conditions on Γ_0 by* (i) $w|_{\Gamma_0} \in H^1(\Sigma_0)$ *or* (ii) $\frac{\partial}{\partial\nu} w \in H^{1/2}(\Sigma_0)$. *In this case, we obtain an appropriate convergence of the traces $w_l|_{\Gamma_0}$ in* (i) *and of $\frac{\partial}{\partial\nu} w_l|_{\Gamma_0}$ in* (ii). *Also, in case* (ii), *there is no need to assume that $\overline{\Gamma_0} \cap \overline{\Gamma_1}$ is an empty set.*

To prove Theorem 3.3.3, we need a result from linear theory that provides regularity of a linear wave equation with absorbing boundary conditions. We consider the problem

$$\begin{aligned} v_{tt} &= \Delta v + f \quad \text{in } \Omega \times (0,\infty), \\ \frac{\partial}{\partial\nu} v + v_t &= g \quad \text{on } \Gamma_1 \times (0,\infty), \\ v &= 0 \quad \text{on } \Gamma_0 \times (0,\infty), \\ v(0) = v_0, &\quad v_t(0) = v_1 \quad \text{in } \Omega. \end{aligned} \tag{3.3.48}$$

Lemma 3.3.1. *Let v be a solution to* (3.3.48) *with*

$$f \in C([0,T], H^1_{\Gamma_0}(\Omega)), \tag{3.3.49}$$

$$g \in H^1(0,T; L_2(\Gamma_1)) \cap C([0,T]; H^{1/2}(\Gamma_1)), \qquad g(t=0) = 0, \tag{3.3.50}$$

$$v_0 \in H^2(\Omega) \cap H^1_{\Gamma_0}(\Omega)), \quad v_1 \in H^1_{\Gamma_0}(\Omega)), \quad \textit{and} \quad \frac{\partial}{\partial\nu} v_0 + v_1 = 0 \quad \textit{on} \quad \Gamma_1. \tag{3.3.51}$$

Then,

$$v \in C([0,T]; H^2(\Omega) \cap H^1_{\Gamma_0}(\Omega)), \quad v_t \in C([0,T]; H^1_{\Gamma_0}(\Omega)). \tag{3.3.52}$$

Proof of Lemma 3.3.1. The result of Lemma 3.3.1 follows from general abstract Theorem 2.3.2 dealing with well-posedness and regularity of boundary feedback problems. However, we present here a simple self-contained proof.

Let A_F denote a feedback generator corresponding to absorbing boundary conditions, i.e.,

$$A_F y = \begin{bmatrix} v \\ \Delta w \end{bmatrix},$$

where $y \equiv (w,v) \in \mathcal{D}(A_F)$,

$$\mathcal{D}(A_F) \equiv \left\{ (w,v) \in H^2(\Omega) \cap H^1_{\Gamma_0}(\Omega)) \times H^1_{\Gamma_0}(\Omega)); \ \frac{\partial}{\partial\nu} w + v = 0 \text{ on } \Gamma_1 \right\}.$$

It is well known that the operator A_F generates a strongly continuous semigroup of contractions. It is also clear that the initial data $(v_0, v_1) \in \mathcal{D}(A_F)$. The solution to the linear problem (3.3.48) can be written via variation of the semigroup formula as

$$\begin{bmatrix} v(t) \\ v_t(t) \end{bmatrix} = e^{A_F t} \begin{bmatrix} v_0 \\ v_1 \end{bmatrix} + \int_0^t e^{A_F(t-s)} \left(\begin{bmatrix} 0 \\ f(s) \end{bmatrix} + \begin{bmatrix} 0 \\ A_N N g(s) \end{bmatrix} \right) ds,$$

where the operator A_N corresponds to the Laplacian with Neumann boundary conditions and the operator N is the Neumann map. Due to the presence of absorbing boundary conditions, it is known that the operator

$$(Lg)(t) \equiv \int_0^t e^{A_F(t-s)} \begin{bmatrix} 0 \\ A_N N g(s) \end{bmatrix}$$

is bounded from $L_2(\Sigma_1) \to C([0,T], E)$, where E denotes the energy space, i.e.,

$$E \equiv H^1_{\Gamma_0}(\Omega) \times L_2(\Omega).$$

We show

$$|Lg|_{C(0,T;H^2(\Omega)\times H^1(\Omega))} \leq C[|g|_{H^1(0,T;L_2(\Gamma_1))} + |g|_{C(0,T;H^{1/2}(\Gamma_1))}] \tag{3.3.53}$$

for all g such that $g(t=0)=0$. Indeed, we write

$$(Lg)(t) = A_F^{-1} \begin{bmatrix} 0 \\ A_N N g(t) \end{bmatrix} - e^{A_F t} A_F^{-1} \begin{bmatrix} 0 \\ A_N N g(0) \end{bmatrix}$$
$$+ \int_0^T e^{A_F(t-s)} A_F^{-1} \begin{bmatrix} 0 \\ A_N N g_t(s) \end{bmatrix} ds.$$

Hence

$$(Lg)(t) = A_F^{-1}(Lg_t)(t) + \begin{bmatrix} Ng(t) \\ 0 \end{bmatrix}.$$

Since $N \in \mathcal{L}(H^{1/2}(\Gamma_1) \to H^2(\Omega))$ and $A_F^{-1}L \in \mathcal{L}(L_2(\Sigma_1); C([0,T]; \mathcal{D}(A_F)))$, we obtain the desired conclusion in (3.3.53). Noting that the element

$$\begin{bmatrix} 0 \\ f(s) \end{bmatrix} \in \mathcal{D}(A_F),$$

the final result in Lemma 3.3.1 follows from a variation of the parameter formula together with the regularity in (3.3.53). □

Proof of Theorem 3.3.3. Let f_l be any sequence in $C([0,T]; H^1_{\Gamma_0}(\Omega))$ such that $f_l \to f$ in $L_1(0,T; L_2(\Omega))$. Let g_l denote a sequence of functions with the following properties:

$$g_l \in H^1(0,T; L_2(\Gamma_1)) \cap C([0,T]; H^{1/2}(\Gamma_1)), \qquad g_l(t=0) = 0, \tag{3.3.54}$$

$$g_l \to \frac{\partial}{\partial \nu} w + w_t|_{\Gamma_1} \text{ in } L_2(\Sigma_1) \text{ as } l \to \infty. \tag{3.3.55}$$

We define the regularization w_l as a solution to the following linear problem:

$$\begin{aligned} w_{l,tt} &= \Delta w_l + f_l \quad \text{in } \Omega \times (0,\infty), \\ \frac{\partial}{\partial \nu} w_l + w_{l,t} &= g_l \quad \text{on } \Gamma_1 \times (0,\infty), \\ w_l &= 0 \quad \text{on } \Gamma_0 \times (0,\infty), \\ w_l(0) = w_{l,0}, \quad & w_{l,t}(0) = w_{l,1} \quad \text{in } \Omega, \end{aligned} \tag{3.3.56}$$

where

$$(w_{l,o}, w_{l,1}) \in \mathcal{D}(A_F) \text{ and } (w_{l,o}, w_{l,1}) \to (w_0, w_1) \text{ in } H^1_{\Gamma_0}(\Omega)) \times L_2(\Omega).$$

From Lemma 3.3.1 and the characterization of $\mathcal{D}(A_F)$ we infer that

$$w_l \in C([0,T]; H^2(\Omega) \cap H^1_{\Gamma_0}(\Omega)), \quad w_{l,t} \in C([0,T]; H^1_{\Gamma_0}(\Omega)). \tag{3.3.57}$$

Therefore, we are in a position to apply the standard energy argument applied to the equation resulting from taking the difference of two solutions to (3.3.56). This gives

$$\begin{aligned} |\nabla(w_l - w_m)(t)|^2_{0,\Omega} + |(w_{l,t} - w_{m,t})(t)|^2_{0,\Omega} + 1/2 \int_0^t |(w_{l,t} - w_{m,t})(t)|^2_{0,\Gamma_1} ds \\ \leq \int_0^t |f_l - f_m|_{0,\Omega} |(w_{l,t} - w_{m,t})(t)|_{0,\Omega} ds + 1/2 \int_0^t |g_l - g_m|^2_{0,\Gamma_1} ds \\ + |\nabla(w_l - w_m)(0)|^2_{0,\Omega} + |(w_{l,t} - w_{m,t})(0)|^2_{0,\Omega}. \end{aligned} \tag{3.3.58}$$

Using Gronwall's inequality and passing with the limit on (3.3.58) yields

$$w_l \to w^* \text{ in } C(0,T; H^1_{\Gamma_0}(\Omega)) \cap C^1([0,T]; L_2(\Omega)), \tag{3.3.59}$$

$$w_{l,t}|_{\Gamma_1} \to w^*_t|_{\Gamma_1} \text{ in } L_2(\Sigma_1). \tag{3.3.60}$$

Reading off boundary conditions in (3.3.56) we also obtain that

$$\frac{\partial}{\partial \nu} w_l = g_l - w_{l,t}|_{\Gamma_1} \to \frac{\partial}{\partial \nu} w + w_t|_{\Gamma_1} - w^*_t|_{\Gamma_1} \text{ in } L_2(\Sigma_1).$$

Therefore, passing with the limit on (3.3.56), we obtain that the limit function w^* satisfies

$$\begin{aligned} w^*_{tt} &= \Delta w^* + f \quad \text{in } \Omega \times (0,\infty), \\ \frac{\partial}{\partial \nu} w^* + w^*_t &= \frac{\partial}{\partial \nu} w + w_t \quad \text{on } \Gamma_1 \times (0,\infty), \\ w^* &= 0 \quad \text{on } \Gamma_0 \times (0,\infty), \\ w^*(0) = w_0, &\quad w^*_t(0) = w_1 \quad \text{in } \Omega. \end{aligned} \tag{3.3.61}$$

Thus, the function w^* satisfies the same equation as w (see (3.3.37)). By the uniqueness of the initial boundary value problem for the linear wave equation, we conclude that $w^* = w$, which completes the proof of the theorem. □

We are in a position to state the main regularization result pertaining to the nonlinear problem (3.3.17).

Corollary 3.3.4. *Let w be any finite energy solution of nonlinear problem* (3.3.17) *such that*

$$\begin{aligned} w \in C([0,T], H^1(\Omega)) \cap C^1([0,T], L_2(\Omega)), \\ \frac{\partial}{\partial \nu} w, w_t, \in L_2(\Sigma_1), \\ f_0(w) \in L_1(0,T; L_2(\Omega)). \end{aligned} \tag{3.3.62}$$

Then, there exists a sequence of functions

$$w_l \in C([0,T]; H^2(\Omega)) \cap C^1([0,T]; H^1(\Omega)), \tag{3.3.63}$$
$$f_l \in C(0,T; H^1_{\Gamma_0}(\Omega)) \tag{3.3.64}$$

such that

$$w_{l,t,t} = \Delta w_l + f_l, \tag{3.3.65}$$
$$f_l \to f(w) \ \ in \ \ L_1(0,T; L_2(\Omega)), \tag{3.3.66}$$
$$w_l \to w \ \ in \ \ C([0,T]; H^1(\Omega)), \tag{3.3.67}$$
$$w_{l,t} \to w_t \ \ in \ \ C([0,T]; L_2(\Omega)), \tag{3.3.68}$$
$$w_{l,t}|_{\Gamma_1} \to w_t|_{\Gamma_1} \ \ in \ \ L_2(\Sigma_1), \tag{3.3.69}$$
$$\frac{\partial}{\partial \nu} w_l|_{\Gamma_1} \to \frac{\partial}{\partial \nu} w|_{\Gamma_1} \ \ in \ \ L_2(\Sigma_1). \tag{3.3.70}$$

Proof of Corollary 3.3.4. It suffices to apply the result of Theorem 3.3.3 with $f \equiv -f_0(w)$. □

Corollary 3.3.5. *Assume Assumption* 3.3.1 *and let w be any finite energy solution of a nonlinear problem* (3.3.17) *such that*

$$w \in C([0,T], H^1(\Omega)) \cap C^1([0,T], L_2(\Omega)),$$
$$\frac{\partial}{\partial \nu} w, w_t, \in L_2(\Sigma_1). \tag{3.3.71}$$

Then, the conclusion of Corollary 3.3.4 *holds true.*

Proof of Corollary 3.3.5. It suffices to note that under Assumption 3.3.1, Sobolev embeddings give

$$f_0(w) \in C([0,T]; L_2(\Omega)) \text{ for } w \in C([0,T]; H^1(\Omega)). \qquad \Box$$

Finally, we remark that the result of Corollary 3.3.4 applies to any solution postulated by Theorem 3.3.1. Thus, assuming Assumptions 3.3.1 and 3.3.2 will guarantee the existence of solutions to the nonlinear problem (3.3.17) such that the regularization scheme of Corollary 3.3.4 applies to it.

3.3.3 Preliminary PDE inequalities

We begin with two PDE inequalities that are critical for the proof of the main theorem. The first is the energy dissipation inequality, and the second is the so-called observability inequality, which reconstructs the information from the boundary into the interior.

Dissipativity inequality. We introduce the energy function for the system defined by

$$E(t) = E_w(t) + 2\int_\Omega F_0(w(t))d\Omega + 2\int_{\Gamma_1} F_1(w(t))d\Gamma_1, \tag{3.3.72}$$

where F_0 (resp., F_1) denote antiderivatives of f_0 (resp., f_1).

Lemma 3.3.2. *Let w be a finite energy solution of* (3.3.17) *and such that*

$$w \in C_{loc}(0,\infty;H^1(\Omega)), \quad w_t \in C_{loc}(0,\infty;L_2(\Omega)).$$

Moreover, assume the following boundary regularity takes place:

$$w_t \in L_2(0,\infty;L_2(\Gamma)), \quad \frac{\partial}{\partial\nu}w \in L_2(0,\infty;L_2(\Gamma)).$$

Then the following dissipativity relation holds:

$$E(t) + 2\int_0^t \int_{\Gamma_1} g(w_t)w_t d\Gamma_1 dt = E(0). \tag{3.3.73}$$

Proof of Lemma 3.3.2. We note that the statement of the lemma follows by the usual formal energy method, where we multiply the equation by the velocity and integrate by parts. However, this formal procedure needs a justification, since the weak solutions do not display enough regularity as required by PDE calculations. In such cases, one applies formal integration by parts to smooth solutions, which typically converge weakly to the original weak solution. However, the passage with the weak limit yields only the inequality

$$E(t) \le E(0),$$

which is insufficient for our purposes. To cope with this issue, we use critically the regularized problem introduced in the previous section along with the convergence established in Theorem 3.3.3.

Step 1. From Assumptions 3.3.1 and 3.3.2 imposed on nonlinearities and Sobolev's embeddings, it follows that

$$f_0(w) \in C_{loc}(0,\infty;L_2(\Omega)), f_1(w|_{\Gamma_1}) \in C_{loc}(0,\infty;L_2(\Gamma_1)). \tag{3.3.74}$$

Therefore, the solution of the original nonlinear problem can be viewed as a solution to the linear problem (3.3.37) with $f(t,x) \equiv f_0(w(t,x))$. The regularity of the $f_1(w)$ term posted in (3.3.74) together with the a priori boundary regularity postulated by Lemma 3.3.2 imply that the regularity requirements in (3.3.39) are met. Thus we are in a position to apply the result of regularization Theorem 3.3.3, which provides a smooth approximation to the solution of (3.3.37).

Step 2. We apply the standard energy (multipliers) argument to the approximated problem defined in Theorem 3.3.3 (i.e., we multiply both sides of the equation by $w_{l,t}$ and we apply the divergence theorem). This gives

$$E_{w_l}(t) - 2\int_0^t \int_{\Gamma_1} \frac{\partial}{\partial\nu}(w_l)w_{l,t} d\Gamma_1 dt - 2\int_0^t \int_\Omega f_l w_{l,t} d\Omega dt = E_{w_l}(0). \tag{3.3.75}$$

Using the convergence properties proved in Theorem 3.3.3 and passing with the limit on the equation gives

$$E_w(t) - 2\int_0^t \int_{\Gamma_1} \frac{\partial}{\partial\nu}(w)w_t d\Gamma_1 dt - 2\int_0^t \int_\Omega f_0(w)w_t d\Omega dt = E_w(0). \tag{3.3.76}$$

Going back to (3.3.17), reading off boundary conditions yields

$$E_w(t) + 2\int_0^t \int_{\Gamma_1} [g(w_t) + f_1(w|_\Gamma)] w_t d\Gamma_1 dt - 2\int_0^t \int_\Omega f_0(w) w_t d\Omega dt = E_w(0). \tag{3.3.77}$$

The final step is to write

$$\begin{aligned} 2f_0(w)w_t &= d/dt F_0(w), \\ 2f_1(w|_\Gamma)w_t|_{\Gamma_1} &= d/dt F_1(w|_{\Gamma_1}), \end{aligned} \tag{3.3.78}$$

which, together with (3.3.77) and after accounting for the definition of $E(t)$, gives the result stated by the Lemma 3.3.2. □

Observability inequality. The main goal of this subsection is to prove the following observability inequality valid for finite energy solutions of the original nonlinear problem (3.3.17).

Lemma 3.3.3. *Let w be a finite energy solution to* (3.3.17) *with the boundary regularity posted in* (3.3.39). *Then for any value of $T > T_0$ we have*

$$\int_0^T E(t)\,dt \le C(E(0)) \left[\int_{\Sigma_{1,T}} (g^2(w_t) + w_t^2 + w^2) d\Sigma_1 + \int_{Q_T} w^2 dQ_T \right]. \tag{3.3.79}$$

Proof of Lemma 3.3.3. *Step* 1. As before, we consider the approximated problem of Theorem 3.3.3 to which we will be applying the multipliers technique to obtain suitable observability estimates. A first step is an application of the standard, by now, multipliers technique that relies on multiplication of the original equation by $(x - x_0) \cdot \nabla w$. This gives

$$\begin{aligned} &\int_0^T [|w_l|_{1,\Omega}^2 + |w_{l,t}|_{0,\Omega}^2]\,dt \\ &\le C_T \left[\int_{\Sigma_1} [|w_{l,t}|^2 + |\nabla w_l|^2] d\Sigma_1 + \int_0^T [|f_l|_{0,\Omega}^2 + |w_l|_{0,\Omega}^2]\,dt \right]. \end{aligned} \tag{3.3.80}$$

A troublesome term is the tangential gradient of w on the boundary. This term is not bounded by the energy and needs to be "absorbed." We note that in the case when star-shaped conditions are imposed on the boundary, this term comes with the right sign and thus can be deleted. This is not the case in the absence of this geometric assumption. To deal with this issue, a crucial role is played by the following trace regularity result, proved in [131] by microlocal analysis methods, which is fundamental for the development:

$$\begin{aligned} &\int_\epsilon^{T-\epsilon} \int_{\Gamma_1} \left| \frac{\partial}{\partial \tau} w_l \right|^2 dx dt \\ &\le C_{T,\epsilon} \left[\int_{\Sigma_1} \left[|w_{l,t}|^2 + \left| \frac{\partial}{\partial \nu} w_l \right|^2 \right] d\Sigma_1 + \int_0^T [|f_l|_{0,\Omega}^2 + |w_l|_{0,\Omega}^2]\,dt \right], \end{aligned} \tag{3.3.81}$$

where $\epsilon > 0$ can be taken sufficiently small. By combining the inequality in (3.3.80) with the one in (3.3.81) and applying the energy argument (details are given in [120]) to dispense with ϵ, we obtain

$$\int_0^T [|w_l|_{1,\Omega}^2 + |w_{l,t}|_{0,\Omega}^2]\,dt$$

$$\leq C_T \left[\int_{\Sigma_1} \left[|w_{l,t}|^2 + |\frac{\partial}{\partial \nu} w_l|^2 \right] + \int_0^T [|f_l|_{0,\Omega}^2 + |w_l|_{0,\Omega}^2]\,dt \right]. \tag{3.3.82}$$

By using the convergence properties specified in Theorem 3.3.3 and passing through the limit we obtain

$$\int_0^T [|w|_{1,\Omega}^2 + |w_t|_{0,\Omega}^2]\,dt$$

$$\leq C_T \left[\int_{\Sigma_{1,T}} \left[|w_t|^2 + |\frac{\partial}{\partial \nu} w|^2 \right] + \int_0^T [|f_0(w)|_{0,\Omega}^2 + |w|_{0,\Omega}^2]\,dt \right]. \tag{3.3.83}$$

Reading off the boundary conditions in (3.3.17) we arrive at

$$\int_0^T [|w|_{1,\Omega}^2 + |w_t|_{0,\Omega}^2]\,dt \leq C_T \left[\int_{\Sigma_{1,T}} [g^2(w_t) + w_t^2 + f_1(w)^2] d\Sigma_{1,T} \right.$$

$$\left. + \int_0^T [|f_0(w)|_{0,\Omega}^2 + |w|_{0,\Omega}^2]\,dt \right]. \tag{3.3.84}$$

Step 2. To complete the proof we need to relate the energy norms to the Sobolev norms. This is very simple, however, and the result follows from Sobolev embeddings. Indeed, it is obvious that

$$E_w(t) \leq E(t).$$

By the properties of the nonlinear functions f_0, f_1 we infer that

$$\int_\Omega F_0(w) d\Omega + \int_{\Gamma_1} F_1(w) d\Gamma_1$$

$$\leq C \left[\int_\Omega [w^2 + |w|^{k_0+1}] d\Omega + \int_{\Gamma_1} [w^2 + |w|^{k_1+1}] d\Gamma_1 \right]. \tag{3.3.85}$$

By Sobolev embeddings,

$$H^1(\Omega) \subset L_{2n/n-2}(\Omega), \quad H^{1/2}(\Gamma) \subset L_{2n-2/n-2}(\Gamma), \qquad n > 2, \tag{3.3.86}$$

$$H^1(\Omega) \subset L_p(\Omega), \quad H^{1/2}(\Gamma) \subset L_p(\Gamma), \qquad n = 2, \tag{3.3.87}$$

and the dissipativity inequality in Lemma 3.3.2 together with the inequality (3.3.85), we infer

$$\int_\Omega F_0(w(t)) d\Omega + \int_{\Gamma_1} F_1(w(t)) d\Gamma_1 \leq C(E(0)) |w(t)|_{1,\Omega}^2. \tag{3.3.88}$$

Hence

$$E(t) \leq C(E(0))|w(t)|^2_{1,\Omega} + |w_t(t)|^2_{0,\Omega}. \tag{3.3.89}$$

Collecting (3.3.89) and (3.3.84) yields the following preliminary inequality:

$$\int_0^T E(t)\,dt \leq C(E(0)) \times \left[\int_{\Sigma_{1,T}} (g^2(w_t) + w_t^2 + f_1(w)^2) d\Sigma_1 + \int_{Q_T} (f_0^2(w) + w^2) dQ_T\right]. \tag{3.3.90}$$

Step 3. Our final step is to estimate the nonlinear functions on the right-hand side of inequality (3.3.90). This is done in Proposition 3.3.1.

Proposition 3.3.1. *Assume Assumptions* 3.3.1–3.3.4. *Let w be a finite energy solution to* (3.3.17) *with boundary regularity posted in* (3.3.39). *Then for any value of $T > T_0$ and $\epsilon > 0$ there exists a constant $C(\epsilon, E(0))$ such that*

$$\int_{\Sigma 1,T} f_1^2(w) \leq \epsilon |E(0)|^{\frac{k_1-1}{1-\epsilon k_1}} \int_0^T E(t)\,dt + C(\epsilon, E(0)) \int_{\Sigma 1,T} w^2 d\Sigma_{1,T}, \tag{3.3.91}$$

$$\int_{Q_T} f_0^2(w) \leq \epsilon |E(0)|^{\frac{k_0-1}{1-\epsilon k_1}} \int_0^T E(t)\,dt + C(\epsilon, E(0)) \int_{Q_T} w^2 dQ_T. \tag{3.3.92}$$

Proof of Proposition 3.3.1. It is enough to prove the proposition for the case $k_0, k_1 > 1$. The case $k_i = 1$ is trivial and in this case we obtain all the constants appearing in Proposition 3.3.1 independent of $E(0)$. Recalling Assumption 3.3.1 and applying interpolation inequalities on L_p spaces [191], i.e.,

$$|w|_{L_p} \leq |w|^{1-q}_{L_2} |w|^q_{L_r}, \qquad 1/p = 1 - q/2 + q/r, \tag{3.3.93}$$

with $p = 2k_1, r = 2k_1 + s,\ 0 < s < 1/2$, we obtain

$$\int_{\Gamma_1} f_1^2(w) d\Gamma_1 \leq C \int_{\Gamma_1} [w^2 + w^{2k_1}] d\Gamma_1 \leq C \left[\int_{\Gamma_1} [w^2 + |w|^{(1-q)2k_1}_{L_2(\Gamma_1)} |w|^{2qk_1}_{L_{2k_1+s}(\Gamma_1)}\right], \tag{3.3.94}$$

where $0 < q < 1$ is now given by

$$q \equiv 1 + \frac{s}{k_1(2 - 2k_1 - s)}.$$

Using Young's inequality,

$$ab \leq \frac{a^p \epsilon^{-p}}{p} + \frac{b^{\bar{p}} \epsilon^{\bar{p}}}{\bar{p}}, \qquad 1/p + 1/\bar{p} = 1,$$

with $p = 1/k_1(1-q), \bar{p} = 1/1 - k_1(1-q)$, we obtain

$$\int_{\Gamma_1} f_1^2(w) d\Gamma_1 \leq C \left[\int_{\Gamma_1} w^2 d\Gamma_1 + \frac{1}{\epsilon^{1/k_1(1-q)}} |w|^2_{L_2(\Gamma_1)} + \epsilon^{1/1-k_1(1-q)} |w|^{\frac{2qk_1}{1-k_1(1-q)}}_{L_{2k_1+s}(\Gamma_1)}\right]. \tag{3.3.95}$$

For $s < \frac{2n-2}{n-2} - 2k_1$ (which is positive by virtue of Assumption 3.3.1), Sobolev's embeddings combined with the trace theorem give

$$|w|_{L_{2k_1+s}(\Gamma_1)} \le C|w|_{H^{1/2}(\Gamma_1)} \le C[|\nabla w|^2_{0,\Omega} + |w|^2_{0,\Gamma_1}]. \tag{3.3.96}$$

Combining the above inequality with (3.3.95) yields

$$\int_{\Gamma_1} f_1^2(w)d\Gamma_1$$

$$\le C\left[(1+\epsilon^{-1/(1-q)k_1})|w|^2_{0,\Gamma_1} + \epsilon^{1/1-k_1(1-q)}\left[|\nabla w|_{0,\Omega}^{\frac{2qk_1}{1-k_1(1-q)}} + |w|_{L_2(\Gamma_1)}^{\frac{2qk_1}{1-k_1(1-q)}}\right]\right]. \tag{3.3.97}$$

We show that under the assumptions imposed on the problem, we have

$$|w(t)|_{1,\Omega} \le CE(t) \le CE(0). \tag{3.3.98}$$

Indeed, if $\Gamma_0 \neq \emptyset$, the assertion follows from the Poincaré inequality. Otherwise, according to Assumption 3.3.3, we have

$$\text{either } F_0(s) \ge cs^2 \text{ or } F_1(s) \ge cs^2,$$

hence either

$$E(t) \ge |\nabla w(t)|^2_{0,\Omega} + c|w(t)|^2_{0,\Omega} \ge c|w(t)|^2_{1,\Omega}$$

or else

$$E(t) \ge |\nabla w(t)|^2_{0,\Omega} + c|w(t)|^2_{0,\Gamma_1} \ge c|w(t)|^2_{1,\Omega},$$

which together with the dissipativity inequality gives the desired inequality in (3.3.98). Integrating the inequality (3.3.97) over $(0,T)$, applying

$$|w|_{0,\Gamma_1}^{\frac{2k_1q}{1-k_1(1-q)}} \le CE(0)^{\frac{k_1-1}{1-k_1(1-q)}}|w|^2_{0,\Gamma_1}$$

and rescaling $\epsilon = c\epsilon^{\frac{1}{1-k_1(1-q)}}$ gives

$$\int_{\Gamma_1} f_1^2(w)d\Gamma_1 \le \epsilon E(0)^{\frac{k_1-1}{1-k_1(1-q)}}\int_0^T E(t)\,dt + C(\epsilon)\int_0^T |w|^2_{0,\Gamma_1}dt \tag{3.3.99}$$

as desired for the first inequality in Proposition 3.3.1. The proof of the second part is the same and hence is omitted. □

Combining the results of inequality (3.3.90) in Step 2 with Proposition 3.3.1 gives the desired result in Lemma 3.3.3. □

3.3.4 Absorption of the lower-order terms

The main goal of this section is to absorb the lower-order term

$$\int_{\Sigma_{1,T}} w^2 d\Sigma_1 + \int_{Q_T} w^2 dQ_T$$

appearing in inequality (3.3.79) in Lemma 3.3.3. This is done by applying a nonlinear version of a compactness-uniqueness argument. The main result of this section is formulated in Lemma 3.3.4.

Lemma 3.3.4. *Let w be a finite energy solution to* (3.3.17) *with the boundary regularity posted in* (3.3.39). *Assume Assumptions* 3.3.1–3.3.4. *Then for any value of $T > T_0$, where T_0 is sufficiently large, we have*

$$\int_{\Sigma_{1,T}} w^2 d\Sigma_1 + \int_{Q_T} w^2 dQ_T \le C(E(0)) \int_{\Sigma_{1,T}} (g^2(w_t) + w_t^2) d\Sigma_1. \tag{3.3.100}$$

Proof of Lemma 3.3.4. We argue by contradiction. Let $w_l(t)$ be a sequence of solutions to (3.3.17) such that

$$\lim_{l\to\infty} \frac{\int_{\Sigma_{1,T}} w_l^2 d\Sigma_1 + \int_{Q_T} w_l^2 dQ_T}{\int_{\Sigma_{1,T}} (g^2(w_t) + w_t^2) d\Sigma_1} = \infty, \tag{3.3.101}$$

while the energy of all initial data $w_l(0), w_{l,t}(0)$ remains bounded by $E_l(0) \le M$. Therefore, from dissipativity Lemma 3.3.2 we infer $E_l(t) \le M$ for all positive t and, on a subsequence denoted by the same symbol, we obtain

$$\begin{aligned} w_l \to w \text{ weakly in } H^1(Q_T) \text{ and weakly* in } L_\infty(0,T;H^1(\Omega)),\\ w_l \to w \text{ strongly in } L_2(\Sigma_T). \end{aligned} \tag{3.3.102}$$

We consider two cases.

Case 1: $w \ne 0$. From the Aubin–Simon compactness result and from Assumptions 3.3.1–3.3.3 we infer that

$$\begin{aligned} f_0(w_l) \to f_0(w) \quad \text{in } \mathcal{D}'(Q),\\ f_1(w_l) \to f_1(w) \quad \text{strongly in } L_\infty(0,T;L_2(\Gamma)). \end{aligned} \tag{3.3.103}$$

From (3.3.101)

$$w_{l,t}, g(w_{l,t}) \to 0 \text{ in } L_2(\Sigma).$$

Passing with the limit on the original equation, we obtain

$$\begin{aligned} w_{tt} - \Delta w + f_0(w) = 0,\\ \frac{\partial}{\partial\nu} w = f_1(w), \quad w_t = 0 \text{ on } \Sigma_1,\\ w = 0 \text{ on } \Sigma_0, \end{aligned} \tag{3.3.104}$$

and for $v \equiv w_t$

$$\begin{aligned} v_{tt} - \Delta v + f_0{}'(w) v = 0,\\ \frac{\partial}{\partial\nu} v = v = 0 \text{ on } \Sigma_1,\\ v = 0 \text{ on } \Sigma_0. \end{aligned} \tag{3.3.105}$$

Now we consider the following three possibilities:

- If f_0 is linear, then the standard Holmgren uniqueness theorem yields $v \equiv 0$.
- If $\Gamma_0 = \emptyset$, we apply the uniqueness result of [86]. Indeed, since $w \in L_\infty(0,T;H^1(\Omega))$ implies $w \in L_{2n/n-2(Q)}$, and according to Assumption 3.3.2 $f_0'(w) \in L_n(Q)$, for $T > 2\text{diam}\Omega$ the result of Isakov [86] implies that $v = w_t = 0$.

- If $\Gamma_0 \neq \emptyset$, we apply a unique continuation property in [144], which also implies that $v = 0$.

Therefore, in all the cases considered, we have obtained that $w_t = 0$.

Remark 3.3.9. *We note that the unique continuation results referred to above play a critical role for the ultimate stability result. On the other hand, these results are highly nontrivial, and much of the progress in this direction was achieved only recently. Indeed, the issue is that we need a unique continuation property applicable to a wave equation with time- and space-dependent coefficients. While the classical Holmgren's unique continuation applies to a very general class of operators with analytic coefficients, very few results are available when it comes to variable coefficients with limited smoothness (as needed in the applications above). In addition to the references quoted above, we point out some other papers relevant to this topic. For the case when $\Gamma_0 = \emptyset$, one could use the unique continuation result given in [182]. However, this result is much less general than the one in [86], which allows variable coefficients of all first-order terms in the equation. Similarly, for the case $\Gamma_0 \neq \emptyset$, a unique continuation result in [91] is, in principle, applicable. The only drawback is that the result in [91] requires that the solutions be C^2. Instead, we work within the framework of H^1 solutions. This reduction of the regularity (from C^2 to H^1) was precisely achieved in [144].*

Thus our problem is reduced to an elliptic problem,

$$\Delta w = f_0(w),$$

$$\frac{\partial}{\partial \nu} w + f_1(w) = 0 \text{ on } \Gamma_1, \qquad w = 0 \text{ on } \Gamma_0. \tag{3.3.106}$$

Multiplying (3.3.106) by w and integrating by parts yields

$$\int_\Omega [|\nabla w|^2 + w f_0(w)] d\Omega + \int_{\Gamma_1} w f_1(w) d\Gamma = 0. \tag{3.3.107}$$

Hence, by the virtue of the sign condition imposed in the nonlinear functions, we conclude that

$$\nabla w = 0.$$

If $\Gamma_0 \neq \emptyset$, then the Poincaré inequality yields that $w \equiv 0$. If $\Gamma_0 = \emptyset$, we use Assumption 3.3.4 together with (3.3.107) to reach the same conclusion that $w = 0$. This is a contradiction with the assumption that $w \neq 0$.

Case 2: $w = 0$. Denote

$$c_l \equiv |w_l|_{0,\Sigma_1} + |w_l|_{0,Q},$$

$$\hat{w}_l \equiv {c_l}^{-1} w_l.$$

Clearly,

$$|\hat{w}_l|_{0,\Sigma_1} + |\hat{w}_l|_{0,Q} = 1. \tag{3.3.108}$$

Since $c_l \to 0$, by (3.3.101) we obtain that

$$w_{l,t} \to 0 \text{ in } L_2(\Sigma_1). \tag{3.3.109}$$

On the other hand, from the observability estimate in Lemma 3.3.3 and dissipativity Lemma 3.3.2 we infer

$$TE(T) \leq \int_0^T E(t)\,dt \leq C(E(0)) \left[\int_{\Sigma_{1,T}} (g^2(w_t) + w_t^2 + w^2) d\Sigma_1 + \int_{Q_T} w^2 dQ_T \right]. \tag{3.3.110}$$

Hence

$$E(t) \leq E(O) \leq C(E(0)) \left[\int_{\Sigma_{1,T}} (g^2(w_t) + w_t^2 + w^2) d\Sigma_1 + \int_{Q_T} w^2 dQ_T \right]. \tag{3.3.111}$$

Dividing both sides of (3.3.111) by c_l^2 and recalling (3.3.101) yields

$$|\nabla \hat{w}_l(t)|_{0,\Omega}^2 + |\hat{w_{l,t}}(t)|_{0,\Omega} \leq C_T(E(0)), \qquad 0 \leq t \leq T. \tag{3.3.112}$$

Hence, in a subsequence (denoted by the same symbol) we have

$$\begin{aligned} \hat{w}_l &\to \hat{w} \text{ weakly in } H^1(Q), \\ \hat{w}_l &\to \hat{w} \text{ strongly in } L_2(\Sigma) \cap L_2(Q). \end{aligned} \tag{3.3.113}$$

Moreover, $\hat{w}_l$ satisfies the equation

$$\begin{aligned} \hat{w_{l,tt}} &= \Delta \hat{w}_l - c_l^{-1} f_0(w_l), \\ \hat{w}_l &= 0 \text{ on } \Sigma_0, \\ \frac{\partial}{\partial \nu} \hat{w}_l + c_l^{-1} [g(w_{l,t}) &+ f_1 w_l] = 0 \text{ on } \Sigma_1. \end{aligned} \tag{3.3.114}$$

To pass with the limit on (3.3.114) we need to determine the limits for nonlinear terms. This is done in Proposition 3.3.2.

Proposition 3.3.2. *Under Assumptions* 3.3.1–3.3.4 *we obtain*

$$\begin{aligned} c_l^{-1} g(w_{l,t}) &\to 0 \text{ in } L_2(\Sigma_1), \\ c_l^{-1} f_0(w_l) &\to f_0'(0)\hat{w} \text{ in } L_2(Q), \\ c_l^{-1} f_1(w_l) &\to f_1'(0)\hat{w} \text{ in } L_2(\Sigma_1). \end{aligned} \tag{3.3.115}$$

Proof of Proposition 3.3.2. The first statement in the proposition is a direct consequence of (3.3.101). For the second part, we estimate

$$\begin{aligned} \Delta_l \equiv |f_0'(0)\hat{w}_l - c_l^{-1} f_0(w_l)|_{0,Q}^2 \leq & \int_{|w_l| \leq \epsilon} \hat{w}_l^2 |f_0'(0) - w_l^{-1} f_0(w_l)|^2 dQ \\ & + 2|f_0'(0)|^2 \int_{|w_l| \geq \epsilon} \hat{w}_l^2 dQ + 2 \int_{|w_l| \geq \epsilon} c_l^{-2} f_0^2(w_l) dQ. \end{aligned} \tag{3.3.116}$$

Recalling Assumption 3.3.2 we obtain

$$\Delta_l \leq |\hat{w}_l|_{0,Q}^2 \rho_\epsilon^2 + C \int_{|w_l| \geq \epsilon} c_l^{-2} [w_l^2 + w_l^{2k_0}] dQ, \tag{3.3.117}$$

where

$$\rho_\epsilon \equiv \sup_{|x|<\epsilon} |f_0'(0) - \frac{f_0(x)}{x}|,$$

and by Assumption 3.3.2 we have that

$$\rho_\epsilon \to 0 \text{ as } \epsilon \to 0.$$

Without loss of generality we assume that $k_0 > 1$ (otherwise, the proof is even simpler). Then, by (3.3.117), we obtain

$$\Delta_l \leq |\hat{w}_l|^2_{0,Q}\rho_\epsilon^2 + C\int_{|w_l|\geq\epsilon} c_l{}^{-2} w_l^{2k_0}[1+\epsilon^{2-2k_0}]dQ$$
$$\leq |\hat{w}_l|^2_{0,Q}\rho_\epsilon^2 + C_\epsilon c_l{}^{-2}|w_l|^{2k_0}_{L_{2k_0}(Q)} = \hat{w}_l|^2_{0,Q}\rho_\epsilon^2 + C_\epsilon c_l{}^{2k_0-2}|\hat{w}_l|^{2k_0}_{L_{2k_0}(Q)}. \quad (3.3.118)$$

Since $\hat{w}_l$ is bounded in $L_\infty(H^1(\Omega))$, by Sobolev's embeddings it is also bounded in $L_{2k_0}(Q)$. Then, as $l \to \infty$, we obtain

$$\limsup_{l\to\infty} \Delta_l \leq \sup_{l\to\infty} |\hat{w}_l|_{L_2(Q)}\rho_\epsilon^2, \quad (3.3.119)$$

and as $\epsilon \to 0$,

$$\lim_{l\to\infty} \Delta_l = 0,$$

which proves the second inequality in Proposition 3.3.2. The proof of the third inequality in (3.3.115) is the same and hence is omitted. □

Applying the result of Proposition 3.3.2 to (3.3.114) and passing with the limit gives

$$\begin{aligned} \hat{w_{tt}} &= \Delta\hat{w} - f_0'(0)(\hat{w}), \\ \hat{w} &= 0 \text{ on } \Sigma_0, \\ \frac{\partial}{\partial\nu}\hat{w} + f_1'(0)\hat{w} &= 0 \text{ on } \Sigma_1, \\ \hat{w}_t &= 0 \text{ on } \Sigma_1. \end{aligned} \quad (3.3.120)$$

Thus, $v = \hat{w}_t \in C[0,T; L_2(\Omega)]$ satisfies

$$\begin{aligned} \hat{v_{tt}} &= \Delta\hat{v} - f_0'(0)v, \\ v &= 0 \text{ on } \Sigma, \\ \frac{\partial}{\partial\nu}v &= 0 \text{ on } \Sigma_1. \end{aligned} \quad (3.3.121)$$

By the standard Holmgren uniqueness theorem, we conclude that for $T > T_0 > 2$ diameter of Ω,

$$v = \hat{w}_t \equiv 0. \quad (3.3.122)$$

Returning to (3.3.120) and taking advantage of (3.3.122) we obtain

$$\begin{aligned} \Delta\hat{w} - f_0'(0)(\hat{w}) &= 0, \\ \hat{w} &= 0 \text{ on } \Sigma_0, \\ \frac{\partial}{\partial\nu}\hat{w} + f_1'(0)\hat{w} &= 0 \text{ on } \Sigma_1. \end{aligned} \quad (3.3.123)$$

As in Case 1, we multiply (3.3.123) by $\hat{w}$, and we integrate by parts and obtain the conclusion $\hat{w} = 0$. This contradicts (3.3.108) and hence completes the proof of Lemma 3.3.4. □

3.3.5 Completion of the proof of the main theorem

By combining the results of Lemma 3.3.4 with the result of observability Lemma 3.3.3 we obtain the following.

Lemma 3.3.5. *Assume Assumptions* 3.3.1–3.3.4. *Let w be a finite energy solution to* (3.3.17) *with the boundary regularity posted in* (3.3.39). *Then for any value of $T > T_0$ with T_0 sufficiently large, we have*

$$\int_0^T E(t)\,dt \le C(E(0)) \left[\int_{\Sigma_{1,T}} (g^2(w_t) + w_t^2) d\Sigma_1 \right]. \tag{3.3.124}$$

Our final estimate is the following.

Lemma 3.3.6. *Under the assumptions of Lemma* 3.3.5 *and with a function p introduced in section* 3.1, *we have*

$$p(E(T)) + E(T) \le E(0). \tag{3.3.125}$$

Proof of Lemma 3.3.6. Denote

$$\Sigma_A \equiv \{u \in L_2(\Sigma); |u| \ge N \text{ a.e.}\},$$

$$\Sigma_B \equiv \Sigma_1 - \Sigma_A.$$

From Assumption 3.3.1,

$$\int_{\Sigma_A} [w_t^2 + g(w_t)^2] d\Sigma_A \le (M_1 + M_2^{-1}) \int_{\Sigma_A} w_t g(w_t) d\Sigma_A. \tag{3.3.126}$$

On the other hand, from the definition of the function h and the second part of Assumption 3.3.1 we infer

$$\int_{\Sigma_B} [w_t^2 + g(w_t)^2] d\Sigma_B \le \int_{\Sigma_B} h(w_t g(w_t)) d\Sigma_B. \tag{3.3.127}$$

By Jensen's inequality,

$$\begin{aligned} \int_{\Sigma_B} h(w_t g(w_t)) d\Sigma_B &\le \textit{meas } \Sigma_1 h\Big(\textit{meas } \Sigma_1{}^{-1} \int_{\Sigma_1} w_t g(w_t) d\Sigma_1 \Big) \\ &= \textit{meas } \Sigma_1 \tilde{h}\Big(\int_{\Sigma_1} w_t g(w_t) d\Sigma_1 \Big). \end{aligned} \tag{3.3.128}$$

Combining inequalities (3.3.126)–(3.3.128) with the result of Lemma 3.3.5 yields the estimate

$$E(T) \le C_T(E(0)) \Big((M_1 + M_2^{-1}) \int_{\Sigma_1} w_t g(w_t) d\Sigma_1 + \textit{meas } \Sigma_1 \tilde{h} \Big(\int_{\Sigma_1} h(w_t g(w_t)) d\Sigma_1 \Big) \Big). \tag{3.3.129}$$

Setting

$$K \equiv \frac{1}{C_T(E(0)) meas \Sigma_1}, \quad c \equiv \frac{M_1 + M_2{}^{-1}}{meas \Sigma_1}, \tag{3.3.130}$$

we obtain

$$p(E(T)) \leq \int_{\Sigma_1} w_t g(w_t) d\Sigma_1 = E(0) - E(T), \tag{3.3.131}$$

which gives the result of Lemma 3.3.6. □

To conclude the proof of the theorem, it suffices to combine the results of Lemma 3.3.6 applied repeatedly on intervals $[mT, (m+1)T]$ with that of Theorem 3.2.1 and the dissipativity inequality in Lemma 3.3.2. □

3.4 Nonlinear Plate Equations

Since the literature related to boundary controllability and stabilization of plates is very extensive, with many results available (see [99, 97] and the references therein), it is beyond the scope of these lectures to even mention all the existing results. For this reason we limit ourselves to nonlinear models and results, where the emphasis is put on theory that does not impose unnecessary restrictions on the geometry of the domain and makes a point to provide stability estimates with a minimal number of controls used as the stabilizers. This, of course, requires an introduction of microlocal analysis into the proofs, and the final estimates are no longer explicit (although are optimal). On the other hand, applications to structural acoustic and other interactive models call for results with a minimal number of restrictions, and this is a main motivation for our choice of models.

In what follows, we provide a brief account of stability results pertinent to several nonlinear plate problems with a nonlinear boundary feedback. For the proofs of these results, the reader is referred to the papers quoted in the text.

3.4.1 Modified von Karman equations

We begin with the formulation of our first nonlinear plate model [99], which is often called the modified von Karman equation.

Let Ω be a bounded domain in R^2 with sufficiently smooth boundary $\Gamma = \overline{\Gamma_0} \cup \overline{\Gamma_1}$. In $\Omega \times (0, \infty)$ we consider the following equation:

$$w_{tt} - \gamma \Delta w_{tt} + D\Delta^2 w + b(x) w_t + f(w) = \alpha[\mathcal{F}(w), w], \tag{3.4.132}$$

where $[\cdot, \cdot]$ denotes the von Karman bracket and $\mathcal{F}(w)$ is the Airy stress function introduced in section 2.4. For the convenience of the reader we recall these definitions:

$$[u, v] \equiv u_{xx} v_{yy} + u_{yy} v_{xx} - 2u_{xy} v_{xy},$$

$$\Delta^2 \mathcal{F} = -\frac{Eh}{2}[w, w] \text{ in } \Omega, \tag{3.4.133}$$

$$\mathcal{F} = \frac{\partial}{\partial \nu} \mathcal{F} = 0 \text{ on } \Gamma. \tag{3.4.134}$$

Function $b(\cdot) \in L_\infty(\Omega)$ is nonnegative and the constants $\alpha \geq 0, \gamma \geq 0$.

The nonlinear function $f(s)$ is assumed C^1 and satisfies the following dissipativity condition:

$$f(s)s \geq 0. \tag{3.4.135}$$

We note that the model (3.4.132), when equipped with appropriate homogeneous boundary conditions (clamped, hinged, or free) is either conservative ($b \equiv 0$) or is strongly stable ($b \geq 0$) or uniformly stable ($\gamma = 0, b(x) \geq b > 0$). The case of interest to us is when there is no uniform stability, hence the support of the function b is either very small or empty. In this case, uniform stability is achieved by introducing an appropriate boundary feedback. Since we are interested in finite energy solutions, i.e., $w \in H^2(\Omega)$, we consider boundary feedbacks acting via hinged or free boundary conditions. Feedbacks inserted into the clamped boundary conditions lead to less regular solutions (with the velocity in negative Sobolev's spaces) and as such are less relevant in the context of problems considered in these notes.

Hinged boundary conditions. With (3.4.132) we associate hinged boundary conditions

$$w = 0 \text{ on } \Gamma \times (0, \infty), \quad \Delta w = 0 \text{ on } \Gamma_0 \times (0, \infty),$$
$$\Delta w = -g\left(\frac{\partial}{\partial \nu} w_t\right) \text{ on } \Gamma_1 \times (0, \infty). \tag{3.4.136}$$

The following assumptions will be made on function g and the portion of the boundary Γ_0 that is not subject to dissipation.

Assumption 3.4.1. *Let $\boldsymbol{h}(x)$ be a vector field given by $\boldsymbol{h}(x) \equiv x - x_0$, $x_0 \in R^2$. We assume that*

$$\boldsymbol{h}\nu \leq 0 \text{ on } \Gamma_0. \tag{3.4.137}$$

Assumption 3.4.2. *Function $g(s)$ is assumed continuous on R, monotone, and zero at the origin and it satisfies the following growth condition: there exist positive constants $0 < m \leq M$ such that for $|s| \geq R$ with a constant R sufficiently large, we have*

$$m|s|^2 \leq (g(s), s)_{R^2} \leq M|s|^2. \tag{3.4.138}$$

To describe the topology for the decay rates, we introduce the following linear energy function:

$$E_\gamma(t) \equiv |w(t)|^2_{H^2(\Omega)} + |w_t(t)|^2_{L_2(\Omega)} + \gamma|\nabla w_t(t)|^2_{L_2(\Omega)}. \tag{3.4.139}$$

Existence and uniqueness of finite energy solutions to this problem follow from the results in section 2.4. Indeed, the case $\gamma > 0$ follows from Theorem 2.4.1 and case $\gamma = 0$ follows from Theorem 2.4.2. (See also Remark 2.4.3.) We recall that for the case $\gamma = 0$, the uniqueness of finite energy solutions is a highly nontrivial issue where sharp regularity of the Airy stress function plays a major role [62].

Our main goal is to establish uniform decay rates for the model under consideration.

Theorem 3.4.1 (see [87, 88]). *Assume that $b(x) > 0$ a.e. in Ω. Then, under Assumption 3.4.1, Assumption 3.4.2, and condition (3.4.135) all weak solutions corresponding to (3.4.132) with boundary conditions (3.4.136) obey the estimate*

$$E_\gamma(t) \leq S(t), \tag{3.4.140}$$

where the function $S(t) \to 0,\ t \to \infty$, *and it satisfies the nonlinear ODE given by* (3.3.29) *with function* g *given by* (3.3.22)–(3.3.26). *Moreover,* $S(t)$ *does not depend explicitly on* γ; *it depends, however, on* $E_\gamma(0)$.

Remark 3.4.1. *Given the information on the behavior of function* $g(s)$ *at the origin, the explicit decay rates can be commuted according to the algorithm given in Corollary* 3.3.2 *and Corollary* 3.3.3.

Remark 3.4.2. *A linear version of this result* ($f \equiv 0, \alpha = 0, b \equiv 0, g(s) = a^2 s$) *was established first in* ([82]; *see also* [76] *for Euler–Bernoulli models) and subsequently extended to nonlinear feedbacks* g *in* [88]. *In the case of linear equation* ($\alpha = 0$) *one can take* $b \equiv 0$. *The presence of light damping, represented by* $b(x)$, *is due to the lack of appropriate uniqueness of continuation result for PDEs with nonlocal operators (as represented by the Airy stress function).*

The following features of the result stated in Theorem 3.4.1 should be noted:

1. The decay rates in (3.4.140) do not depend on the parameter $\gamma \geq 0$. This is an important robustness result showing that the effectiveness of the feedback in (3.4.136) remains in force when the thickness of the plate goes to zero.
2. The stabilization occurs via moments only, and the feedback is applied to a portion of the boundary Γ. This is unlike most of the results published in the literature where, even in the linear case, two boundary feedbacks are required.
3. In the limit case, when $\gamma = 0$, the stabilizing feedback is of the form $\Delta w = -g(\frac{\partial}{\partial \nu} w_t)$. We note that this is a very different feedback with respect to a classical and nonlocal feedback $\Delta w = \frac{\partial}{\partial \nu} \Delta^{-1} w_t$ which stabilizes the linear model with $\gamma = 0$ with respect to the topology of $H_0^1(\Omega) \times H^{-1}(\Omega)$ [76, 104]. The advantage of the present feedback is that it is local and, more important, it has the desired robustness properties with respect to γ.
4. The nonlinear function $g(s)$ does not need to satisfy any growth condition at the origin.

The proof of Theorem 3.4.1 is carried out by combining the multipliers technique with methods of microlocal analysis [87, 88]. In fact, the latter are responsible for achieving the goals in points 1–4 above. In the case $\alpha > 0$, sharp regularity of the Airy stress function [62] plays an important role.

For the reader's orientation, we give here the observability inequality that is a corner stone of the proof of Theorem 3.4.1:

$$\int_0^T E_\gamma(t)\,dt + E_\gamma(T)$$
$$\leq C_T(E_\gamma(0)) \int_0^T \left[\left| g\left(\frac{\partial}{\partial \nu} w_t\right) \right|_{0,\Gamma_1}^2 + |w_t|_{0,\Gamma_1}^2 + \int_\Omega b(x) w_t^2 dx \right] dt, \qquad (3.4.141)$$

where $T > 0$ is sufficiently large.

Free boundary conditions. With (3.4.132) we associate the following free boundary conditions with a stabilizing feedback acting via moments and shears:

$$
\begin{aligned}
w = \frac{\partial}{\partial \nu} w &= 0 \text{ on } \Gamma_0 \times (0,\infty),\\
\Delta w + (1-\mu)B_1 w &= -g_1\left(\frac{\partial}{\partial \nu} w_t\right) \text{ on } \Gamma_1 \times (0,\infty),\\
\frac{\partial}{\partial \nu}\Delta w + (1-\mu)B_2 w - \gamma \frac{\partial}{\partial \nu} w_{tt} &= g_0(w_t) - \frac{\partial}{\partial \tau} g_2\left(\frac{\partial}{\partial \tau} w_t\right) \text{ on } \Gamma_1 \times (0,\infty),
\end{aligned} \tag{3.4.142}
$$

where boundary operators B_1, B_2 [99] are defined in section 2.4. Γ_0 is subject the same condition as in (3.4.137). Here we assume additionally that $\overline{\Gamma_0} \cap \overline{\Gamma_1} = \emptyset$. This assumption is necessitated by regularity considerations. Indeed, in the absence of the above condition, the elliptic problems may develop singularities. This makes it difficult to justify the PDE calculus. To cope with the singularities, an approach developed by Grisvard may be used in some cases, where an elementary (energy/multiplier) type of estimate suffice. However, in more complex microlocal setups, it is unlikely that microlocal analysis estimates can be justified by this method.

Functions g_i, $i = 0,1,2$, are continuous, monotone increasing, and zero at the origin and for $i = 1,2$ satisfy the growth condition (3.4.138). Instead, function g_0 satisfies

$$
g_0(s)s > 0, \qquad s \neq 0, \qquad ms^2 \leq g_0(s)s \leq M|s|^p, \qquad |s| \geq 1, \tag{3.4.143}
$$

where $p < \infty$ is arbitrary, if Ω is star shaped, and $p \leq 5$, in general.

The energy function is given by the formula

$$
E_\gamma(t) \equiv |w(t)|^2_{H^2(\Omega)} + |w_t(t)|^2_{L_2(\Omega)} + \gamma|\nabla w_t(t)|^2_{L_2(\Omega)} + a(w,w), \tag{3.4.144}
$$

where the bilinear form $a(w,w)$, topologically equivalent to $H^2(\Omega)$ norm, is given by

$$
\begin{aligned}
a(w,z) \equiv D\int_\Omega [&w_{x,x}z_{x,x} + w_{y,y}z_{y,y} + \mu w_{x,x}z_{y,y} + \mu w_{y,y}z_{x,x}\\
&+ 2(1-\mu)w_{x,y}z_{x,y}]d\Omega + l\int_{\Gamma_1} wz d\Gamma_1.
\end{aligned}
$$

Existence and uniqueness of finite energy solutions to this problem follow from the results in section 2.4. Indeed, the case $\gamma > 0$ follows from Theorem 2.4.1 and the case $\gamma = 0$ follows from Theorem 2.4.2.

In the case when $\gamma = 0$ with $f = 0, b = 0$ and functions g_i are linear, exponential decay rates for the energy function were derived in [99] under the star-shaped conditions imposed on Ω. In what follows we concentrate on nonlinear feedbacks and on the domains that do not need to be star shaped.

To formulate our result we introduce the functions

$$
\mathcal{H}_0(s) \equiv \mathcal{G}1(s) + \mathcal{G}2(s),
$$

where the functions $\mathcal{G}1, \mathcal{G}2$ are concave, strictly increasing functions, zero at the origin and such that the following inequalities are satisfied for $|s| \leq 1$:

$$
\begin{aligned}
\mathcal{G}1(sg_1(s)) &\geq s^2 + g_1^2(s), \qquad s \in R,\\
\mathcal{G}2(sg_2(s)) &\geq s^2 + g_2^2(s), \qquad s \in R.
\end{aligned}
$$

Due to the assumed monotonicity of the nonlinear functions g_i, one can easily construct functions $\mathcal{G}1, \mathcal{G}2$ with the properties listed above (see [120]).

Theorem 3.4.2 (see [78, 75]). *Assume that $b(x) > 0$ a.e. in Ω and either $\Gamma_0 \neq \emptyset$ or $l > 0$. Then, under Assumption 3.4.1, Assumption 3.4.2 imposed on g_1, g_2, and conditions (3.4.143), (3.4.135) imposed on g_0 and f, respectively, all weak solutions corresponding to (3.4.132) with boundary conditions (3.4.142) obey the estimate*

$$E_\gamma(t) \leq S(t), \tag{3.4.145}$$

where a real variable function $S(t)$ converges to zero as $t \to \infty$ and it obeys the ODE

$$S_t(t) + q(S(t)) = 0, \quad S(0) = E(0). \tag{3.4.146}$$

The (nonlinear) monotone increasing function $q(s)$ is determined entirely from the behavior at the origin of the nonlinear functions g_i, $i = 1, 2$, and it is given by the following algorithm:

$$q \equiv I - (I + p)^{-1}, \tag{3.4.147}$$

$$p \equiv (KI + \mathcal{H})^{-1}, \tag{3.4.148}$$

$$\mathcal{H} \equiv \mathcal{H}_0(\cdot/meas\Sigma_1), \tag{3.4.149}$$

where the constant K is proportional to $\frac{1}{meas\Sigma_1}(m^{-1} + M)$ with $\Sigma_1 \equiv \Gamma_1 \times (0, T)$. $S(t)$ does not depend explicitly on γ; it depends, however, on $E_\gamma(0)$.

The result stated above was shown in [78, 75] with $f = 0$. However, an extension to the case when $f \neq 0$ follows by adapting the arguments in [88].

In the case when $\gamma = 0$, it is expected that only shears should suffice for stabilization. In fact, such a result was already established for linear plates with linear feedback defined on star-shaped domains [99, 97]. The issue of whether the same holds for arbitrary (smooth) domains was an open problem until recently. Indeed, to dispense with one control without excessive geometric hypotheses, one needs to employ methods of microlocal analysis. This result was proved, in [109], by using microlocal analysis together with the trace decoupling technique introduced in [189].

Theorem 3.4.3 (see [109]). *Assume that either $\Gamma_0 \neq \emptyset$ or $l > 0$. Consider (3.4.132) with $\gamma = 0$, $f = 0, \alpha = 0$ and boundary conditions 3.4.142 with $g_1 = g_2 \equiv 0$. We assume that g_0 complies with the requirements in (3.4.143). Then, all weak solutions corresponding to this system obey the estimate*

$$E_{\gamma=0}(t) \leq S(t), \tag{3.4.150}$$

where the function $S(t) \to 0$, $t \to \infty$, and it satisfies the nonlinear ODE given by (3.3.29).

Remark 3.4.3. *Theorem 3.4.3 generalizes stability results obtained for this plate model earlier in the literature [99, 97] in the following two directions: (i) it does not require any geometric conditions imposed on the boundary Γ_1, and (ii) it dispenses with the growth conditions imposed customarily [97] on the nonlinear function g_0 at the origin.*

The techniques of [83, 84] can be combined with those presented in [109] to yield the same conclusions with $f \neq 0, \alpha > 0$ and, in the case $\gamma = 0$, with the stabilizers acting via shears only (i.e., $g_1, g_2 = 0$).

The following features of the results described by Theorem 3.4.2 and Theorem 3.4.3 should be noted:

1. The decay rates established in Theorem 3.4.2 for the energy function $E_\gamma(t)$ do not depend on the parameter $\gamma \geq 0$. This is an important robustness result showing that the effectiveness of the feedback in (3.4.142) remains effective when the thickness of the plate goes to zero.
2. For the case $\gamma = 0$, the stabilization occurs via shears only, where the feedback is applied to a portion of the boundary Γ.
3. There are no geometric conditions imposed on the controlled portion of the boundary Γ_1, and there are no restrictions imposed on the growth of the nonlinearity $g(s)$ at the origin.

3.4.2 Full von Karman system and dynamic system of elasticity

In this section we consider a full von Karman system (introduced in section 2.7), which is no longer represented by a scalar equation. It consists of an elastodynamic system strongly coupled with the Kirchhoff plate.

Here [99] the variables w and $u = (u_1, u_2)$ represent, respectively, the vertical and the in-plane displacement of a thin plate occupying a two-dimensional domain Ω with sufficiently smooth boundary $\Gamma = \overline{\Gamma_0} \cup \overline{\Gamma_1}$. We assume that $\overline{\Gamma_0} \cap \overline{\Gamma_1} = \emptyset$. On Γ_0, we assume that geometric condition (3.4.137) holds true.

The governing equations, defined on $\Omega \times (0, \infty)$, are given by

$$\begin{aligned} u_{tt} - \operatorname{div}[\mathcal{C}[\epsilon(u) + f(\nabla w)]] &= 0, \\ w_{tt} - \gamma\Delta w_{tt} + \Delta^2 w - \operatorname{div}[\mathcal{C}[\epsilon(u) + f(\nabla w)]\nabla w] &= 0 \end{aligned} \tag{3.4.151}$$

with Dirichlet boundary conditions on the portion of the boundary Γ_0

$$u = w = \nabla w = 0 \quad \text{on } \Gamma_0 \times (0, \infty).$$

The boundary conditions on the boundary Γ_1 are free and given by

$$\begin{aligned} \mathcal{C}[\epsilon(u) + f(\nabla w)]\nu + ku &= -g_0(u_t), \\ \Delta w + (1-\mu)B_1 w &= -g_1\left(\frac{\partial}{\partial \nu} w_t\right), \\ \frac{\partial}{\partial \nu}\Delta w + (1-\mu)B_2 w - \gamma\frac{\partial}{\partial \nu}w_{tt} - \mathcal{C}[\epsilon(u) + f(\nabla w)]\nu\nabla w &= -\frac{\partial}{\partial \tau}\left(g_2\left(\frac{\partial}{\partial \tau}w_t\right)\right). \end{aligned} \tag{3.4.152}$$

Here, the vector ν denotes the outward normal direction to the boundary. Similarly, we denote by $\frac{\partial}{\partial \tau}$ the tangential derivative on Γ. The constant $\gamma > 0$ is proportional to the square of the thickness of the plate and hence assumed small. The term $\gamma\Delta w_{tt}$ represents moments of inertia. D represents the flexural rigidity; the constant

$0 < \mu < 1/2$ is Poisson's modulus. The fourth-order tensor $\mathcal{C}$ is defined by

$$\mathcal{C}\epsilon \equiv \frac{E}{(1-2\mu)(1+\mu)}[\mu \; trace \; \epsilon \; I + (1-2\mu)\epsilon],$$

where the strain tensor is given by $\epsilon(u) \equiv 1/2(\nabla u + \nabla^T u)$. It can be easily verified that the tensor $\mathcal{C}$ is symmetric and strictly positive. The function f is given by

$$f(s) \equiv (1/2)s \otimes s, \qquad s \in R^2.$$

The nonlinear, monotone increasing functions g_i satisfy the same condition as in (3.4.138). Here, for simplicity of formulation and without loss of generality we may assume that $g_0(u) = [g_{01}(u_1), g_{02}(u_2)]$ and each component g_{0i} satisfies condition (3.4.138).

A natural energy function associated with this model is given by

$$E_\gamma(t) = E_k(t) + E_p(t); \quad E_{k,\gamma}(t) = \int_\Omega [|u_t|^2 + w_t^2 + \gamma|\nabla w_t|^2] d\Omega, \tag{3.4.153}$$

$$E_p(t) = \int_\Omega [\mathcal{C}N(u,w)N(u,w)] \, d\Omega + a(w,w) + k\int_{\Gamma_1} |u|^2 d\Gamma_1, \tag{3.4.154}$$

where the stress resultant $N(u,w)$ is given by

$$N(u,w) \equiv \epsilon(u) + f(\nabla w)$$

and

$$a(w,z) \equiv \int_\Omega [w_{x,x}z_{x,x} + w_{y,y}z_{y,y} + \mu w_{x,x}z_{y,y} + \mu w_{y,y}z_{x,x} + 2(1-\mu)w_{x,y}z_{x,y}]d\Omega + l\int_{\Gamma_1} wz d\Gamma_1.$$

It is well known that $E_p(t)$ is topologically equivalent to $H^2_{\Gamma_0}(\Omega) \times [H^1_{\Gamma_0}(\Omega)]^2 \times [L_2(\Omega)]^2$ topology. Here the spaces $H^1_{\Gamma_0}(\Omega)$ and $H^2_{\Gamma_0}(\Omega)$ denote the usual Sobolev spaces that incorporate zero boundary conditions on Γ_0.

Existence and uniqueness (the latter if g_i are linear) of finite energy solutions follow from the results in section 2.8. We recall that the uniqueness of finite energy solution is highly nontrivial and still is an open question in the case of nonlinear boundary dissipation.

Theorem 3.4.4 (see [107, 95]). *With reference to the system* (3.4.151), (3.4.152) *with* $\gamma > 0$, *the following apply.*

1. Finite energy solutions.

   (i) *There exists a global solution of finite energy. This is to say that for any initial data*

   $$u_0, u_1 \in [H^1_{\Gamma_0}(\Omega)]^2 \times [L_2(\Omega)]^2 \; w_0, w_1 \in H^2_{\Gamma_0}(\Omega) \times H^1_{\Gamma_0}(\Omega),$$

   there exists a weak solution such that

   $$(u,w) \in C([0,T]; [H^1(\Omega)]^2 \times H^2(\Omega)),$$
   $$(u_t, w_t) \in C([0,T]; [L_2(\Omega)]^2 \times H^1(\Omega)).$$

(ii) *If, in addition,*

$$(g_i(s), s)_{R^2} \geq m|s|^2, \qquad m > 0, \ |s| \geq 1, \tag{3.4.155}$$

then the additional boundary regularity is valid: $u_t, \nabla u, \nabla w_t \in L_2((0,T) \times \Gamma_1)$.

(iii) *Assuming, additionally, that the vector function* g_i *are linear, the solution in part* (i) *is unique and depends continuously on the initial data.*

2. Regular solutions. *Assume, in addition, that* $g_i \in C^1(R^2)$. *For any initial data subject to the regularity in part* 1 *and, in addition,*

$$u_0 \in [H^2(\Omega)]^2, \ u_1 \in [H^1_{\Gamma_0}(\Omega)]^2, \qquad w_0 \in H^3(\Omega), \ w_1 \in H^2(\Omega),$$

and subject to compatibility on the boundary, there exists a unique, global solution

$$(u, w) \in C([0,T]; [H^2(\Omega)]^2 \times H^3(\Omega)),$$
$$(u_t, w_t) \in C([0,T]; [H^1(\Omega)]^2 \times H^2(\Omega)),$$

where $T > 0$ *is arbitrary.*

Moreover, regular solutions depend continuously on the initial data in the topology of regular solutions (as defined above).

Remark 3.4.4. *The result of Theorem* 3.4.4 *remains, of course, valid with the homogeneous boundary conditions (i.e.,* $g_i = 0$*) and with other boundary conditions such as hinged or clamped* [106].

Remark 3.4.5. *Regularity of solutions to* (3.4.151) *expressed in fractional Sobolev's spaces is given in* [106].

Stabilization of dynamic full von Karman system, in the one-dimensional case, was treated in [103]. In the two-dimensional case, the first analysis of the dynamic full von Karman system (3.4.151) and linear boundary conditions was presented in [177]. However, the model considered in this reference dealt with the different boundary conditions imposed on u. In fact, these boundary conditions, instead of representing natural boundary conditions given by tractions, had built in the additional tangential derivatives. The presence of these tangential derivatives is a mathematical convenience facilitating the estimates. An additional disadvantage of this particular model is that to reconcile the analysis with the existing regularity of the solutions, the authors were forced to assume that the parameter representing these tangential derivatives is very small. Thus, the corresponding feedback is not very effective.

To overcome these difficulties, new methods were brought in. In fact, as before, microlocal analysis plays a critical role. In particular, sharp estimates for the traces of elastodynamic solutions established in [77] are essential. In addition to being able to treat the natural boundary conditions with nonlinear boundary feedback, we dispense altogether with geometric restrictions on the controlled portion of the boundary.

Below, we provide a brief description of the main stabilization result proved in [107].

To formulate our result we introduce the functions

$$\mathcal{H}_0(s) \equiv \mathcal{G}(s) + \mathcal{G}1(s) + \mathcal{G}2(s),$$

where the functions $\mathcal{G}, \mathcal{G}1, \mathcal{G}2$ are concave, strictly increasing functions, zero at the origin, and such that the following inequalities are satisfied for $|s| \leq 1$:

$$\mathcal{G}(\vec{s}g(\vec{s})) \geq |\vec{s}|^2 + |g_0(\vec{s})|^2, \qquad \vec{s} \in R^2,$$

$$\mathcal{G}1(sg_1(s)) \geq s^2 + g_1^2(s), \qquad s \in R,$$
$$\mathcal{G}2(sg_2(s)) \geq s^2 + g_2^2(s), \qquad s \in R.$$

Due to the assumed monotonicity of the nonlinear functions g_i, one can easily construct functions $\mathcal{G}, \mathcal{G}1, \mathcal{G}2$ with the properties listed above (see [120]).

Theorem 3.4.5 (see [107]). *Let u, w be solutions to the original system* (3.4.151), *and let us assume that either $l > 0, k > 0$ or else $\Gamma_0 \neq \emptyset$. Moreover, assume validity of Assumption* 3.4.1 *and Assumption* 3.4.2 *satisfied for each function g_{0i}, g_1, g_2. Then, there exists a constant $T_0 > 0$ such that the following estimate holds:*

$$E(t) \leq C(E(0))S(t/T_0 - 1)), \qquad t \geq T_0, \tag{3.4.156}$$

where a real variable function $S(t)$ converges to zero as $t \to \infty$ and it obeys the ODE

$$S_t(t) + q(S(t)) = 0, \quad S(0) = E(0). \tag{3.4.157}$$

The (nonlinear) monotone increasing function $q(s)$ is determined entirely from the behavior at the origin of the nonlinear functions g_i, and it is given by the algorithm

$$q \equiv I - (I + p)^{-1}, \tag{3.4.158}$$
$$p \equiv (KI + \mathcal{H})^{-1}, \tag{3.4.159}$$
$$\mathcal{H} \equiv \mathcal{H}_0(\cdot/meas\Sigma_1), \tag{3.4.160}$$

where the constant K is proportional to $\frac{1}{meas\Sigma_1}(m^{-1} + M)$ with $\Sigma_1 \equiv \Gamma_1 \times (0, T)$.

Remark 3.4.6. *In the formulation of Theorem* 3.4.5, *it is critical that $\gamma > 0$. Whether the result can be extended to the case $\gamma = 0$ is unclear. A major difficulty is the gap of the $1 + \epsilon$ derivative in differentiability of the nonlinear term. This, of course, has serious implications on the regularity of the solutions and hence on the justification of the corresponding estimates.*

Remark 3.4.7. *A critical role in the proofs is played by trace regularity results available in* [77] *(resp.,* [134]*) for elastodynamic systems (resp., Kirchhoff plates) and new unique continuation results given in* [86].

Remark 3.4.8. *The same result stated in Theorem* 3.4.5 *can be obtained for the model with the dissipation via hinged boundary conditions. That is, the boundary conditions in* (3.4.152) *for the plate can be replaced by*

$$w = 0, \; \Delta w = -g\left(\frac{\partial}{\partial \nu} w_t\right) \quad on \; \Gamma_1.$$

Needless to say that Theorem 3.4.5, when specialized to a linear model of elastic waves, provides a nonlinear boundary stabilization result applicable to such models. This result was first obtained in [78, 77]. For the reader's convenience we provide a full statement.

Consider the elastodynamic system

$$u_{tt} - \operatorname{div}[\mathcal{C}[\epsilon(u)]] = 0 \tag{3.4.161}$$

with Dirichlet boundary conditions on the portion of the boundary Γ_0

$$u = 0 \quad \text{on } \Gamma_0 \times (0, \infty).$$

The boundary conditions on the boundary Γ_1 are prescribed by boundary forces/ tractions and given by

$$\mathcal{C}[\epsilon(u)]\nu + ku = -g_0(u_t). \tag{3.4.162}$$

Nonlinear, monotone increasing vector function g_0 satisfies the same condition as in (3.4.138).

A natural energy associated with this system is given by

$$E_\gamma(t) = \int_\Omega [|u_t|^2 + [\mathcal{C}\epsilon(u(t))\epsilon(u(t))]]d\Omega + k\int_{\Gamma_1} |u|^2 d\Gamma_1. \tag{3.4.163}$$

Theorem 3.4.6. *Let u be a weak (finite energy) solution of* (3.4.151) *with boundary conditions* (3.4.162) *and assume that either $k > 0$ or else $\Gamma_0 \neq \emptyset$. Moreover, assume validity of Assumption* 3.4.1 *and Assumption* 3.4.2 *satisfied by each component g_{0i}.*

Then there exists a constant $T_0 > 0$ such that the following estimate holds:

$$E(t) \le C(E(0))S(t/T_0 - 1)), \qquad t \ge T_0, \tag{3.4.164}$$

where a real variable function $S(t)$ converges to zero as $t \to \infty$ and it obeys the ODE

$$S_t(t) + q(S(t)) = 0, \quad S(0) = E(0), \tag{3.4.165}$$

$$q \equiv I - (I + p)^{-1}, \tag{3.4.166}$$

$$p \equiv (KI + \mathcal{H})^{-1}, \tag{3.4.167}$$

$$\mathcal{H} \equiv \mathcal{H}_0(\cdot/meas\Sigma_1), \tag{3.4.168}$$

where the constant K is proportional to $\frac{1}{meas\Sigma_1}(m^{-1} + M)$ with $\Sigma_1 \equiv \Gamma_1 \times (0, T)$ and

$$\mathcal{H}_0(sg_0(s)) \ge s^2 + g_0(s)^2, \qquad |s| \le 1.$$

Remark 3.4.9. *Theorem* 3.4.6 *extends results known in the literature (see* [101, 1, 158] *and references therein) on several accounts. The most important fact is that Theorem* 3.4.6 *deals with natural boundary conditions prescribed by boundary tractions, rather than artificial boundary conditions containing the additional tangential derivatives that were considered in the literature. These additional tangential derivatives are not physically motivated and lead to boundary feedbacks that are not bounded on the energy space* [158]. *Paper* [1], *instead, considers the correct boundary conditions; however, the domain Ω is restricted to a spherical domain.*

The first version of Theorem 3.4.6 with linear dissipation was proved in [75]. Extensions of this result to nonlinear boundary feedbacks are given in [79]. Further extensions accounting for anisotropic elastic operators can be found in [80].

3.4.3 Nonlinear plates with thermoelasticity

We consider a model describing nonlinear oscillations of a plate subjected to thermal effects and accounting for in-plane accelerations. The resulting system is referred to as thermoelastic full von Karman system. References related to general von Karman systems are [103, 99, 90, 51, 165, 57]. Thermoelastic von Karman systems with Dirichlet boundary conditions were considered in [26]. The derivation of this model [26, 25] parallels the arguments presented in [102] for linear thermoelastic equations after incorporating the nonlinear strain-displacement relations. The model considered in this paper is based on [27], after taking into account the free (rather than the Dirichlet) boundary conditions.

Here, the variables w and $u = (u_1, u_2)$ represent, respectively, the vertical and in-plane displacement of a thin plate occupying a two-dimensional domain Ω with sufficiently smooth boundary $\Gamma = \overline{\Gamma_0} \cup \overline{\Gamma_1}$. We shall assume that $\overline{\Gamma_0} \cap \overline{\Gamma_1} = \emptyset$. The variables θ and ϕ describe the thermal strain resultants affecting the vertical displacement and in-plane displacements, respectively. More specifically, θ is the average (across the thickness of the plate) of thermal moment while ϕ is the average of thermal stress (see formulas (6.8) and (6.9) in [102]). The governing equations to be considered are given by

$$\begin{aligned} u_{tt} - \operatorname{div}[\mathcal{C}[\epsilon(u) + f(\nabla w)]] + \nabla\phi = 0 \text{ in } \Omega \times (0, \infty), \\ w_{tt} + \Delta^2 w - \operatorname{div}[\mathcal{C}[\epsilon(u) + f(\nabla w)]\nabla w + \phi\nabla w] + \Delta\theta = 0 \text{ in } \Omega \times (0, \infty) \end{aligned} \tag{3.4.169}$$

with Dirichlet boundary conditions on the portion of the boundary Γ_0

$$u = w = \nabla w = 0 \text{ on } \Gamma_0 \times (0, \infty).$$

The boundary conditions on the boundary Γ_1 are free and given by

$$\begin{aligned} \mathcal{C}[\epsilon(u) + f(\nabla w)]\nu + ku - \phi\nu = -g(u_t), \\ \Delta w + (1-\mu)B_1 w + \theta = 0, \\ \frac{\partial}{\partial\nu}\Delta w + (1-\mu)B_2 w - \mathcal{C}[\epsilon(u) + f(\nabla w)]\nu\nabla w + \frac{\partial}{\partial\nu}\theta - \phi\frac{\partial}{\partial\nu}w = 0. \end{aligned} \tag{3.4.170}$$

Here, the vector ν denotes the outward normal direction to the boundary. Similarly, we denote by $\frac{\partial}{\partial\tau}$ the tangential derivative on Γ. The temperature is described by the system of equations

$$\begin{aligned} \phi_t = \Delta\phi - \operatorname{div}u_t + \nabla w\nabla w_t \text{ in } \Omega \times (0, \infty), \\ \theta_t = \Delta\theta + \Delta w_t \text{ in } \Omega \times (0, \infty) \end{aligned} \tag{3.4.171}$$

with the boundary conditions imposed on ϕ, θ

$$\frac{\partial}{\partial\nu}\theta + \lambda_1\theta = 0, \quad \frac{\partial}{\partial\nu}\phi + \lambda_2\phi = 0 \text{ on } \Gamma \times (0, \infty),$$

where the constants $\lambda_i > 0$, $i = 1, 2$.

With (3.4.169) and (3.4.171) we associate the initial conditions

$$\begin{aligned} u(0) = u_0, \quad u_t(0) = u_1, \quad w(0) = w_0, \\ w_t(0) = w_1, \quad \theta(0) = \theta_0, \quad \phi(0) = \phi_0 \text{ in } \Omega. \end{aligned} \tag{3.4.172}$$

D represents the flexural rigidity, and the constant $0 < \mu < 1/2$ is Poisson's modulus. The fourth-order tensor $\mathcal{C}$ is defined by

$$\mathcal{C}\epsilon \equiv \frac{E}{(1-2\mu)(1+\mu)}[\mu \; trace \; \epsilon \; I + (1-2\mu)\epsilon],$$

where the strain tensor is given by $\epsilon(u) \equiv 1/2(\nabla u + \nabla^T u)$. It can be easily verified that the tensor $\mathcal{C}$ is symmetric and strictly positive. The function f is given by

$$f(s) \equiv (1/2) s \otimes s, \qquad s \in R^2,$$

and, we recall, the boundary operators B_i are defined by

$$B_1 \equiv 2\nu_1\nu_2 D^2_{x,y} - \nu_1^2 D^2_{y,y} - \nu_2^2 D^2_{x,x},$$

$$B_2 \equiv \frac{\partial}{\partial \tau}[(\nu_1^2 - \nu_2^2)D^2_{x,y} + \nu_1\nu_2(D^2_{y,y} - D^2_{x,x})] + lI.$$

The dissipation in the system is represented by a nonlinear vector function g that is assumed continuous, monotone increasing, zero at the origin, and of linear growth at the infinity. To state our main result, we define the energy functional associated with the plate model (3.4.169), (3.4.171) and given by

$$E(t) = E_k(t) + E_p(t), \qquad E_k(t) = \int_\Omega [|u_t|^2 + w_t^2] d\Omega, \tag{3.4.173}$$

$$E_p(t) = \int_\Omega [\mathcal{C}N(u,w)N(u,w) + |\theta|^2 + |\phi|^2] d\Omega + a(w,w) + k\int_{\Gamma_1} |u|^2 d\Gamma_1, \tag{3.4.174}$$

where the stress resultant $N(u,w)$ is given by

$$N(u,w) \equiv \epsilon(u) + f(\nabla w)$$

and

$$a(w,z) \equiv D\int_\Omega [w_{x,x}z_{x,x} + w_{y,y}z_{y,y} + \; muw_{x,x}z_{y,y} + \mu w_{y,y}z_{x,x} + 2(1-\mu)w_{x,y}z_{x,y}] d\Omega + l\int_{\Gamma_1} wz d\Gamma_1.$$

It is well known that $E_p(t)$ is topologically equivalent to $H^2(\Omega) \times [H^1(\Omega)]^2 \times [L_2(\Omega)]^2$ topology. In particular, the following inequalities resulting from Korn's inequality and Sobolev's embeddings are used frequently:

$$|u|^2_{H^1(\Omega)} \leq C\left[|N(u,w)|^2_{L_2(\Omega)} + |\nabla w|^4_{L_4(\Omega)} + k\int_{\Gamma_1} |u|^2 d\Gamma_1\right],$$
$$\leq C\left[|N(u,w)|^2_{L_2(\Omega)} + |w|^4_{H^2(\Omega)} + k\int_{\Gamma_1} |u|^2 d\Gamma_1\right], \tag{3.4.175}$$

where $k > 0$ whenever $\Gamma_0 = \emptyset$.

Existence and uniqueness of weak (finite energy) solutions for this model follows from Theorem 2.7.1.

We are concerned with stability properties of the dynamics given by (3.4.169) and (3.4.171).

As is well known, the presence of thermal effects provides some dissipation for the energy. Thus, it should be expected that to achieve the uniform decays for the energy, one may not need the additional mechanical dissipation. In fact, this is the case for many thermoelastic plates, where uniform decay rates were proved without any mechanical dissipation; see [89, 180, 181, 5, 6, 9].

In the case of the full von Karman system, the situation is different. This is due to the presence of the elastodynamic component (equation for u), which is known to not have uniform decay rates for the energy even in the linear case [55, 145, 94].

Thus some sort of mechanical dissipation is necessary. Our main goal is to show that the boundary dissipation introduced on the velocity of the trace of u only suffices to obtain the uniform stability of the overall system.

It should be noted that the results stated below do not require any geometric constraints (star-shaped type) imposed on the dissipative portion of the boundary and do not require any feedback acting on the tangential derivatives of u. This is in contrast with the literature dealing with a problem of boundary stabilization of elastodynamic linear systems; see [1, 158] and references therein.

To achieve this we use sharp regularity results developed for the traces of elastodynamic solutions [78, 77], which are then extended to provide the needed sharp regularity of boundary traces for the thermoelastic von Karman system—a result of independent PDE interest.

The fact that rotational forces are not present in the model greatly affects the analysis. Indeed, this term, represented by $\gamma\Delta w_{tt}$, $\gamma > 0$, added to the equation for w provides a regularizing effect for the velocity of w, which is critical to the arguments in the previous literature (both for the well-posedness and for the stabilization). In fact, to our best knowledge, there is no result in this direction dealing with the case $\gamma = 0$. On the other hand, by adding thermal dissipation, there are new phenomena coming to play, and the regularizing effect on a subcomponent of the system comes from the analyticity of thermoelastic semigroups. This will have a critical bearing on the treatment of the nonlinear terms in the equation.

To formulate our result, we introduce the function $\mathcal{H}(s)$ which is assumed concave, strictly increasing functions, zero at the origin, and such that the following inequalities are satisfied for $|s| \leq 1$:

$$\mathcal{H}(sg(s)) \geq s^2 + |g(s)|^2.$$

Such a function can be easily constructed in view of the monotonicity assumption imposed on g [120]. We are ready to state the main results.

Theorem 3.4.7 (see [110]). *Let u, w, θ, ϕ be a weak solution to the original system given by* (3.4.169), (3.4.171). *Assume that the constants l, k are positive whenever $\Gamma_0 = \emptyset$. Moreover, assume validity of Assumption* 3.4.1 *and Assumption* 3.4.2 *satisfied by components of vector function g. Then, the estimate*

$$E(t) \leq C(E(0))s(t/T_0 - 1)), \qquad t \geq T_0, \tag{3.4.176}$$

holds, where the real variable function $s(t)$ converges to zero as $t \to \infty$, and it *satisfies the ODE*

$$s_t(t) + q(s(t)) = 0, \qquad s(0) = E(0). \tag{3.4.177}$$

The (nonlinear) monotone increasing function $q(s)$ is determined entirely from the behavior at the origin of the nonlinear functions g and it is given by the following algorithm:

$$q \equiv I - (I+p)^{-1}, \tag{3.4.178}$$

$$p \equiv (I + \mathcal{H}_0)^{-1}\left(\frac{\cdot}{K}\right), \tag{3.4.179}$$

$$\mathcal{H}_0 \equiv \mathcal{H}(\cdot/meas\Sigma_1), \tag{3.4.180}$$

where $\Sigma_1 \equiv \Gamma_1 \times (0,T)$ and the constant $K > 0$ may depend on $E(0)$.

Remark 3.4.10. *Note that the decay rates established in Theorem 3.4.7 can be computed explicitly by solving the nonlinear ODE system (3.4.177), once the behavior of $g(s)$ at the origin is specified. If the nonlinear function g is bounded from below by a linear function, then it can be shown that the decay rates predicted by Theorem 3.4.7 are exponential. This is to say that there exist positive constants C, ω, possibly depending on $E(0)$ and such that*

$$E(t) \leq Ce^{-\omega t} \quad for\ t > 0.$$

If, instead, this function has a polynomial growth (resp., exponential decay) at the origin, then the decay rates are algebraic (resp., logarithmic) (see [120]).

Remark 3.4.11. *It is evident that the result of Theorem 3.4.7, when specialized and restricted to the case when $w \equiv 0$, $\theta \equiv 0$, provides the same decay rates for the corresponding system of linear thermoelastodynamic system with a nonlinear boundary feedback; see [79, 77], where uniform decay rates were obtained for (3.4.169) with $\phi \equiv Q \equiv w \equiv 0$. However, in this case, the final result is less interesting, since the same boundary dissipation provides uniform decay rates for the mechanical part of the system. Thus, there is no advantage in using thermal effects as the means for stabilization. The situation is, of course, very different when the second equation for w is added. This addition not only makes the overall system highly nonlinear, but also the boundary dissipation available for the model does not produce any decay rates for the mechanical part of the system (i.e., in the absence of thermal effects). Thus, in this latter case, the effect of thermoelasticity is critical for uniform stability.*

Remark 3.4.12. *The same result as formulated in Theorem 3.4.7 can be obtained with other (hinged or clamped) boundary conditions imposed on the plate equation.*

The proof of Theorem 3.4.7 is based on the following three main ingredients: (i) application of the multipliers method introduced in [5] (where the uniform decay was proved for two-dimensional linear thermoelastic plates); (ii) methods introduced in [107], based on microlocal analysis, where the nonlinear full von Karman system was treated with the mechanical boundary dissipation; (iii) analytic estimates valid for the subcomponent of the systems which results from the analyticity of semigroups arising in linear thermoelastic plates with free boundary conditions and established recently in [136, 137, 138]. (The analyticity of linear thermoelastic plates with hinged or clamped boundary conditions has been known for some time [160, 138].) For the reader's orientation we state below a result pertaining to the said analyticity.

With reference to the linear problem

$$\begin{aligned} w_{tt} + \Delta^2 w + \Delta\theta = 0 \ \text{ in } \ \Omega \times (0,\infty), \\ \theta_t = \Delta\theta + \Delta w_t \ \text{in} \ \Omega \times (0,\infty) \end{aligned} \tag{3.4.181}$$

with the boundary conditions

$$\begin{aligned} w = \nabla w = 0 \ \text{ on } \Gamma_0 \times (0,\infty), \\ \Delta w + (1-\mu)B_1 w + \theta = 0, \\ \frac{\partial}{\partial\nu}\Delta w + (1-\mu)B_2 w + \frac{\partial}{\partial\nu}\theta - \phi\frac{\partial}{\partial\nu}w = 0 \text{ on } \Gamma_1 \times (0,\infty), \\ \frac{\partial}{\partial\nu}\theta + \lambda\theta = 0 \text{ on } \Gamma \times (0,\infty), \end{aligned} \tag{3.4.182}$$

we have the following lemma.

Lemma 3.4.1 (see [136, 137]). *The semigroup associated with dynamics described in* (3.4.181) *and* (3.4.182) *is analytic on* $H^2(\Omega) \times L_2(\Omega) \times L_2(\Omega)$.

Remark 3.4.13.

1. *We note that to guarantee the decay rates for the energy function, the presence of the damping term* u_t *on the boundary is necessary. Indeed, this follows from the fact that thermal dissipation in a two-dimensional linear system of elasticity may provide only strong stability and not the exponential stability; see* [145, 94]. *For this reason, it is necessary to introduce dissipation acting at least on the solenoidal part of the horizontal displacement.*
2. *The presence of thermal resultant* ϕ *is not essential for validity of decay rates stated in Theorem* 3.4.7. *In fact, the decay rates are still valid if the term* ϕ *is neglected in the model. The analysis in this case is even simpler.*
3. *The results of Theorem* 3.4.7 *can be easily extended to cover the cases of other boundary conditions, including hinged or clamped boundary conditions corresponding to the vertical displacements* w *and the Dirichlet or Neumann boundary conditions corresponding to thermal variables. For these cases, the analysis is much simpler, as other (than free) sets of boundary conditions satisfy Lopatinski conditions—a fact that greatly facilitates the arguments.*
4. *The fact that our model does not account for the moments of inertia (i.e.,* $\gamma = 0$ *in the notation of* [99]*) is critical to the analysis. Indeed, the extension of the results presented here to the case* $\gamma > 0$ *meets with substantial technical difficulties, which are due to the strong nonlinear coupling between the wave and the plate equations. (The corresponding results valid for each subcomponent separately are available in the literature* [6, 107]*.)*

   The only other result (that we are aware of) pertaining to uniform stability of full von Karman thermoelastic systems in the case $\gamma > 0$ *is* [26], *where, however, the dissipation acts on the solenoidal components of the velocity vector* u *in the interior (rather than on the boundary).*

   Another possibility (when $\gamma > 0$*) is to add mechanical dissipation on the velocity traces of the vertical displacements* w*. In that case, the uniform stability results are straightforward consequences of the techniques presented in*

[107]. *However, this type of result is of limited interest since the beneficial role of thermoelasticity does not come into play.*

3.5 Comments and Open Problems

The following comments and open questions can be formulated in the context of the material presented in this chapter.

- We recall from section 3.3 that in the case of boundary stabilization of the wave equation, the nonlinear function g is assumed to have a linear bound at infinity (although there are no restrictions on the growth at the origin). A natural question to ask is whether this linear growth at infinity is really necessary. It was shown in Chapter 2 that this assumption is not needed for the well-posedness of finite energy solutions. However, it is necessary for the purpose of obtaining uniform decay rates; see [161, 202].
- In the case of stabilization of the wave equation, via Neumann boundary conditions, with the uncontrolled portion of the boundary satisfying zero Dirichlet data, it is necessary to assume that the two portions of the boundary are disjoint. This assumption is necessitated by regularity considerations and a possibility of development of singularities in the corresponding elliptic problems. To cope with these singularities, in the case of the pure wave equation, with a linear boundary feedback and star-shaped domains, one may use special inequalities developed by Grisvard [69]. Whether this technique can be applied to a full nonlinear problem defied on domains that are not necessarily star shaped is, at present, an open problem.
- Another natural question that arises in the context of the stabilization problem for the wave equation presented in section 3.3 (particularly relevant in the context of structural acoustic problems) is the following: Can we stabilize with hard walls, i.e., with the Neumann uncontrolled boundary? A quick look at the classical, by now, proof of this result immediately reveals the difficulties. Classical, radial vector fields do not work at all unless an uncontrolled part of the boundary is flat. In fact, what is needed is a more sophisticated version of multipliers techniques that calls for nonradial vectors fields. At present, positive results are obtained for convex (or concave) uncontrolled boundaries [144]. It is unclear, however, whether these geometric restrictions can be removed. The above problem is open even in the case of a simple linear wave equation with a linear feedback.
- We can ask, of course, the same questions as above in the context of plate equations treated in section 3.4. This will not be repeated. However, plate dynamics are more complex, and stabilization problems lead to several new and often open questions.

  One of the fundamental issues is the ability to control or stabilize with one stabilizer only. While this issue is almost trivial in the one-dimensional case (e.g., beams), it is the major problem in higher dimensions. Here one faces two main obstacles. The first is the ability to obtain the so-called hard stability estimates, which may be polluted by the lower-order terms. The second obstacle is the lack of an appropriate unique continuation theorem, which would allow one to eliminate these lower-order terms. The second issue is a research topic on its own, so we simply refer to [86], where many new results developed in the last decade are presented. Regarding the first issue, a new trace decoupling technique introduced in [189] is fundamental. While the problem of controlling with one stabilizer acting via hinged or clamped boundary conditions (see Theorem 3.4.1) is somewhat simpler, with several results available [88], the real test is the case of free boundary conditions which do not satisfy the Lopatinski condition. In fact, for the free boundary conditions, the decoupling technique of [189] (perfectly compatible with microlocal analysis methods) was used critically to show that the Euler–Bernoulli plate, defined on a non-star-shaped domain, can be stabilized with one control only acting via shears [109] (see Theorem 3.4.2). Also, the same decoupling technique shows (with additional microlocal argument) that it is not possible to stabilize, in general, a two-dimensional plate by acting via bending moments only (in the free boundary conditions). This last result is in strong contrast with the one-dimensional problems, where such stabilization can be shown by very elementary methods; see [157] and the references therein. Thus, the one-dimensional dynamics are not good testing beds for this kind of problem.

For the Kirchhoff plate ($\gamma > 0$), defined on a two- (or higher-) dimensional domain, it seems very unlikely that one could stabilize with one control only acting in the free boundary conditions. Again, such stabilization is possible (elementary) for the one-dimensional models and for other types of boundary conditions such as hinged or clamped; see [157] and the references therein.

- We recall that for the model of the modified von Karman equation without any geometric restrictions imposed on a controlled portion of the boundary we require (see Theorem 3.4.1) a presence of a light interior damping. This requirement was necessitated by the lack of an appropriate unique continuation theorem which would allow one to dispense with such damping. Indeed, while there is a rich theory of unique continuation results pertaining to various PDE dynamics [86], these results proved by Carleman's estimates require intrinsically that the operators involved are local (differential operators). As we know, this is not the case with modified von Karman equations, which are defined by nonlocal operators. Thus, the issue of unique continuation property from the boundary for equations with nonlocal operators is an open and difficult problem.
- While in the case of modified von Karman equations, full stabilization theory is available (see Theorems 3.4.1 and 3.4.2) regardless of the value of the parameter $\gamma \geq 0$ (i.e., in the presence of the rotational inertia), this is not the case for the full von Karman system. The case $\gamma = 0$ leads to much less regular solutions (the gap in differentiability is $1 + \epsilon$), and this is the cause of very major difficulties. What saves the situation in the case of modified (scalar) von Karman equations with $\gamma = 0$ is the sharp regularity of the Airy stress function [120]. There is no analogue of similar results valid for the full von Karman system, since there is no Airy's stress function. Thus, in the case of the full von Karman system, the stabilization problem for the case $\gamma = 0$ is an open and difficult problem.
- Models considered in this lecture involve PDEs with constant coefficients in the principal part. For models with variable coefficients in the principal part, we refer to [22, 189, 143].

Chapter 4

Uniform Stability of Structural Acoustic Models

4.1 Orientation

In this chapter we focus our attention on uniform stability properties of structural acoustic models, which is an example of coupled PDE structures with an interface. While this topic is of independent mathematical interest, it has particular importance in the context of the material presented in this volume. Indeed, our ultimate goal is to construct a coherent theory of optimal control problems associated with interactive models. Of particular significance is the so-called infinite horizon optimal control problem, which is intimately related to properties of uniform stabilizability or exact controllability of the system. In the case of structural acoustic models, these properties are not inherent and they need to build in the model. In typical control applications, relevant to structural acoustic models, the control function acts as point control. On the other hand, it is known that the structural acoustic model cannot be stabilized by means of actuators defined through control operators B which model point controls (typically delta functions). Moreover, it is known that the free system (i.e., $u = 0$) is strongly stable [7] but not uniformly stable (see [155]). There are two reasons for the lack of uniform stability, one geometric and the other topological. By geometric, we refer to the fact that according to the propagation of rays of geometric optics, in order to obtain uniform stability, it is necessary to have at least half the boundary of Ω subjected to the dissipative action. The more critical and more difficult issue to deal with is the topological one. This has to do not so much with the placement of the actuators but rather with the type of controls applied. In fact, it is well known that the action of point control (of major interest in applications) is too weak to provide the decay rates in the right topology (finite energy states) for the hyperbolic component of the system. The same way, controllability properties with point control are very weak and they may hold at most in a very weak approximate sense (see [193, 194]).

For these reasons, it is necessary to introduce passive control in the form of some damping. In what follows we study models equipped with the absorbing boundary conditions active on the walls of the acoustic cavity. These models arise naturally in practice, where some sort of damping on the boundary is always present or easily achievable; see [58] for concrete models with porous walls represented by boundary damping.

Regarding the wall (plate equation), depending on the structural model describing the wall, we consider several forms of the damping. We focus on three major types: internal, boundary, and thermal. While in physical models several types of damping mechanisms may be present simultaneously, to emphasize the role played by each type of damping we prefer to consider them separately.

Internal damping, treated in section 4.2, corresponds to either viscous damping or structural damping, including the Kelvin–Voight type of model. The structural damping is a very strong damping with additional regularizing effects. (The uncoupled plate equation is governed by an analytic semigroup.)

Boundary damping, discussed in section 4.3, is a much weaker form of the damping. However, boundary damping is very attractive from the physical and mathematical points of view.

Thermal damping, considered in section 4.4, involves an additional coupling of the plate equation with the heat equation. Depending on whether rotational moments are included in the model, this damping may produce an analytic or a hyperbolic effect on the dynamics.

4.2 Internal Damping on the Wall

In this section, we consider the model involving internal damping on the wall (i.e., the parameter $\alpha > 0$ in the case of the model introduced in section 2.6 and described by (2.6.40), (2.6.39). In this case, the component of the structure, if uncoupled, is already uniformly stable. However, it is well known that this stability is not transferred onto the entire model. In fact, the whole system is only strongly stable [7].

More specifically, we will be dealing with the following model. Let $\Omega \subset R^n, n \geq 2$, be either a smooth or a convex bounded domain with a boundary Γ consisting of two parts Γ_0 and Γ_1 such that $\overline{\Gamma_0} \cap \overline{\Gamma_1} = Q$; see Figure 1.

Acoustic medium with damping.

$$\begin{aligned} z_{tt} &= c^2 \Delta z \quad \text{in } \Omega \times (0,\infty), \\ \frac{\partial}{\partial \nu} z + g_2(z_t) + dz &= 0 \quad \text{on } \Gamma_1 \times (0,\infty), \\ \frac{\partial}{\partial \nu} z + g_1(z_t) &= v_t \quad \text{on } \Gamma_0 \times (0,\infty), \\ z(0) = z_0 \in H^1(\Omega), &\quad z_t(0) = z_1 \in L_2(\Omega). \end{aligned} \tag{4.2.1}$$

Structural wall.

$$\begin{aligned} \mathcal{M}\, \mathbf{v}_{tt} + \mathcal{A}\mathbf{v} + \alpha \mathcal{A}^{\theta} g_3(\mathcal{A}^{\theta}\mathbf{v}_t) + \mathcal{C}\rho z_t|_{\Gamma_0} &= 0, \\ \mathbf{v}(0) = \mathbf{v}_0 \in D(\mathcal{A}^{1/2}), \quad \mathbf{v}_t(0) = \mathbf{v}_1 &\in D(\mathcal{M}^{\ 1/2}), \end{aligned} \tag{4.2.2}$$

where, we recall (see section 2.6), $\mathcal{M}$, $\mathcal{A}$ are self-adjoint, densely defined, and positive operators $\mathcal{A} : [L_2(\Gamma_0)]^s \to [L_2(\Gamma_0)]^s$ and $\mathcal{M} : [L_2(\Gamma_0)]^s \to [L_2(\Gamma_0)]^s$. The

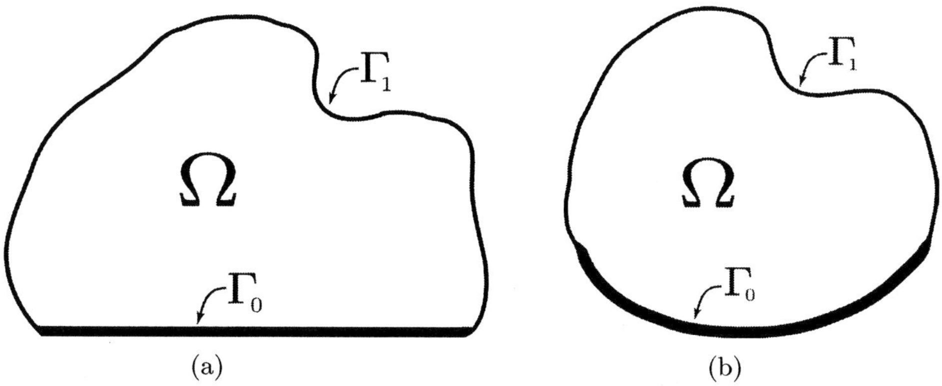

Figure 1: Cross sections of Ω: (a) $s = 1$, flat Γ_0. (b) $s = 3$, curved Γ_0.

operator $\mathcal{C}$ is simply defined as a projection on the last coordinate of vector $\vec{\mathbf{v}}$. Except for shells, $s = 3$, $\mathcal{C} = I$. The boundary damping, introduced in the acoustic medium, is described by the nonlinear functions $g_i, i = 1-2$, such that $g_i : R \to R$ are continuous, monotone increasing, and zero at the origin. The same assumptions are imposed on the function g_3 describing internal damping on the wall. The parameter $0 \le \theta \le 1/2$ describes the amount of the internal damping on the wall. Indeed, with $\mathcal{M} \equiv I$, $\theta = 1/2$ corresponds to a structural, Kelvin–Voight type of damping. If, instead, $\theta = 1/4$, this is still a structural damping, but in a weaker form often referred to as a square root damping [185, 47]. For $\theta = 0$, we deal with a viscous damping. In all cases, this damping provides uniform stability for the uncoupled wall model (with $\mathcal{M} = I$). However, for $\theta \ge 1/4$ the (uncoupled) plate model is also analytic.

Regarding the well-posedness of solutions corresponding to the system described by (4.2.1), (4.2.2), we are in a position to apply Theorems 2.6.1 and 2.6.2. These yield the following.

Theorem 4.2.1. *For all initial data* $z_0, z_1, \mathbf{v}_0, \mathbf{v}_1 \in H^1(\Omega) \times L_2(\Omega) \times D(\mathcal{A}^{1/2}) \times D(\mathcal{M}^{1/2})$ *there exists a unique, weak, and global solution to* (4.2.1), (4.2.2) *such that*

$$(z, z_t, \mathbf{v}, \mathbf{v}_t) \in C([0,T]; H^1(\Omega) \times L_2(\Omega) \times D(\mathcal{A}^{1/2}) \times D(\mathcal{M}^{1/2})),$$

where T *can be taken to be arbitrary.*

Our main goal is to establish uniform decay rates for this model. To achieve this, we need to impose some growth conditions on the nonlinear functions g_i.

Assumption 4.2.1.

$$sg_i(s) > 0 \ for \ s \neq 0, \qquad ms^2 \le g_i(s)s \le Ms^2$$

for some positive constants m, M *and large values of* $|s|$, $i = 1-3$.

We also introduce the notation

$$E_{z,\mathbf{v}}(t) \equiv |\nabla z|^2_{0,\Omega} + d|z|^2_{0,\Gamma_1} + |z_t|^2_{0,\Omega} + |\mathcal{A}^{1/2}\mathbf{v}|^2_{0,\Gamma_0} + |\mathcal{M}^{1/2}\mathbf{v}_t|^2_{0,\Gamma_0}.$$

Elementary energy estimate gives

$$E_{z,\mathbf{v}}(t)\,dt + 2\int_0^t \left[\int_{\Gamma_0} [\alpha g(\mathcal{A}^\theta \mathbf{v}_t)\mathcal{A}^\theta \mathbf{v}_t + g_1(z_t)z_t]dx + \int_{\Gamma_1} g_2(z_t)z_2 dx\right] ds = E_{z,\mathbf{v}}(0),$$

which implies that the energy is dissipative, whenever $\alpha \geq 0$.

To state our stability results, we recall some notation from section 3.3.

$h_i(s)$ denotes a real-valued function, defined for $s > 0$, concave, strictly increasing, $h_i(0) = 0$ and such that

$$h_i(sg(s)) \geq s^2 + g_i^2(s) \quad \text{for } |s| \leq 1. \tag{4.2.3}$$

Let $h = h_1 + h_2 + h_3$ $\tilde{h} \equiv h(\frac{x}{meas\partial\Sigma_0})$, $x \geq 0$, where $\partial\Sigma_0 \equiv (0,T) \times \partial\Gamma_0$ and T is a given constant. Since $\tilde{h}$ is monotone increasing, we always have that $I + \tilde{h}$ is invertible. Define

$$p(x) \equiv (I + \tilde{h})^{-1}(Kx), \qquad x > 0, \tag{4.2.4}$$

where K is a positive constant to be given later. Then p is a positive, continuous, strictly increasing function with $p(0) = 0$. Let

$$q(x) \equiv x - (I + p)^{-1}(x), \qquad x > 0. \tag{4.2.5}$$

Since $p(x)$ is positive, increasing, so is $q(x)$.

The main stability result is formulated below.

Theorem 4.2.2 (see [2, 8]). *Consider the system consisting of* (4.2.1) *and* (4.2.2) *with* $\mathcal{M} = I$, *under Assumption* 4.2.1 *with* $\alpha > 0, d > 0$. *Moreover, assume* $0 \leq \theta \leq 1/2$ *and* $\mathcal{D}(\mathcal{A}^\theta) \subseteq \mathcal{D}(\mathcal{M}^{1/2})$. *Then,*

(i) *If* $g(s)s \geq m_0 s^2$ *for some positive constant* m_0 *and* $|s| \leq 1$, *there exist positive constants* $C > 0, \omega > 0$ *such that*

$$E_{z,\mathbf{v}}(t) \leq Ce^{-\omega t}E_{z,\mathbf{v}}(0).$$

(ii) *In the case of unspecified growth of* g *at the origin, the decay rates for the energy function are uniform but not necessarily exponential. The decay rates can be explicitly evaluated from the solution of the nonlinear ODE*

$$S_t + q(S(t)) = 0, \qquad t > 0, \; s(0) = E(0),$$

where the nonlinear function $q(s)$, *given by* (4.2.5), *is monotone increasing and depends on the growth of the nonlinear dissipation* g *at the origin. The following estimate holds:*

$$E_{z,\mathbf{v}}(t) \leq S(t), \qquad t > T_0,$$

and $S(t) \to 0$, *when* $t \to \infty$.

Remark 4.2.1. *As we see from the result stated above, conditions that guarantee the uniform decay rates are more demanding than those required for the well-posedness. Additional growth conditions, imposed on the dissipation* g *at infinity, are needed.*

Remark 4.2.2. *One may take $d = 0$ in the model (4.2.1). In this case, the results of the theorem should be formulated on an appropriate quotient space (see [41, 40] for related work). However, this point is not essential to the mathematical analysis.*

Remark 4.2.3. *Note that the decay estimates in Theorem 4.2.2 do not require any geometric conditions imposed on the domain, nor do they require any growth conditions at the origin or regularity requirements imposed on the nonlinear function g. This is in contrast with most of the literature pertaining to boundary stability ([97, 99] and references therein).*

The proof of Theorem 4.2.2 follows from [8] (see also a previous linear version of this result in [2]). In the special case, when the domain $\Omega \in R^2$ is rectangular, the boundary damping is linear, and the wall's dynamic is represented by Euler–Bernoulli equations with Kelvin–Voight damping, a version of the result of Theorem 4.2.2 was proved in [58].

It is possible to extend the result of Theorem 4.2.2 to the case when the wall is described by a nonlinear model of the type considered in section 2.7. In fact, uniform stability for nonlinear shells subject to structural damping was demonstrated in [117]. Combining the technique of [117] with the one presented in this section will prove uniform stability for the corresponding structural acoustic problem with nonlinear equation for the wall.

One could also consider cases when there is no damping on Γ_1, provided that Γ_1 satisfies the following geometric constraints.

Assumption 4.2.2.

$$(x - x_0)\nu \leq 0 \text{ on } \Gamma_1,$$

where ν is an outward normal to the boundary and $x_0 \in R^n$.

In such case, the corresponding boundary conditions on Γ_1 would take either the Dirichlet form,

$$z = 0 \text{ on } \Gamma_1 \times (0, \infty), \tag{4.2.6}$$

or the Neumann form,

$$\frac{\partial}{\partial \nu} z = 0 \text{ on } \Gamma_1 \times (0, \infty). \tag{4.2.7}$$

In the first (Dirichlet) case, a potential drawback of such a setup, however, is that solutions to appropriate elliptic problems may develop singularities, unless Γ_0 and Γ_1 are separated or appropriate cone conditions are satisfied on Ω [69, 68]. While this additional separation assumption is still acceptable in the case of shell models describing the wall, it is highly undesirable when considering classical plate theory (when the operator $\mathcal{A}$ corresponds to a biharmonic operator). Physical considerations force, in such a case, Γ_0 to be flat. It is possible, in principle, to avoid this difficulty by working with singular solutions and, in particular, to use Grisvard's estimates. However, the corresponding analysis will become undoubtedly more complicated. See also the last comment in section 3.5.

Instead, in the second, Neumann, case one does not need to make any assumption forcing the closures of Γ_0 and Γ_1 not to intersect. However, the proof of the corresponding decay rates is more involved (unless Γ_1 is flat). The most recent advances in the area of Carleman's estimates, based on work of Tataru, provide estimates that can handle this new situation—at least in the case when Γ_1 is "convex" [112].

Remark 4.2.4. *An interesting problem is that of stabilization of the structural model but without the damping on the interface* Γ_0. *Thus, one would take*

$$\frac{\partial}{\partial\nu}z + dz + g(z_t) = 0 \text{ on } \Gamma_1 \times (0,\infty), \tag{4.2.8}$$

$$\frac{\partial}{\partial\nu}z = v_t \text{ on } \Gamma_0 \times (0,\infty). \tag{4.2.9}$$

This configuration is much more challenging from a mathematical point of view, since the presence of the damping on the interface provides extra regularizing effect, which is very helpful in carrying the estimates. In the absence of this damping, the analysis is much more involved. It requires rather subtle regularity theory of the traces of the solutions to the wave equation. This program was carried out successfully for structural models with thermoelasticity [114] *as well as acoustic interactions modeled by structurally damped shells* [39]. *It is believed that the same technique will be successful in proving similar decay rates for the structural acoustic problem referred to above*

4.3 Boundary Damping on the Wall

The main goal of this section is to dispense with the assumption on the existence of internal (including structural) damping on the wall, i.e., $\alpha > 0$. Since in this case the structure (wall) may be unstable, it is necessary to introduce some other form of damping acting on the wall. From the physical and also the mathematical point of view, the most attractive form of the damping is boundary damping, and this is the focus of this section. In what follows we concentrate on the most challenging case of the Euler–Bernoulli plate equation with only one boundary stabilizer acting either as a moment or a force. This model serves as a prototype for other structural acoustic interactions. Indeed, a similar analysis can be applied to other models of plates, including Kirchhoff plates [41, 42, 40].

4.3.1 Model

In this section we provide results on uniform stabilizability of a three-dimensional structural acoustic model describing pressure in an acoustic chamber with flexible walls.

The original model of the structure is conservative and thus not stable. Although in practical situations some damping may be present in the model, we wish to account for the worst scenario and the limit case, where there is no damping at all. There are various ways to achieve uniform stabilizability in this case. We are predominantly interested in physically attractive boundary stabilization, where the stabilization is achieved by introducing some form of the dissipation on the boundary. Our main goal is to show that either viscous or boundary damping added to the wave equation and boundary dissipation applied via shear forces only at the edge of the plate suffices to provide uniform decay rates of the natural energy function associated with the model. Moreover, we are able to dispense altogether with geometric restrictions (of the star-shaped type) which are imposed routinely in the context of boundary stabilization of waves and plates (see [102, 99, 97] and references therein). A strong motivation for stability results without any geometric restrictions comes also from the optimal design theory and shape optimization problems where any such restrictions become an additional constraint on the problem. The necessary PDE estimates are obtained by combining the multipliers method with methods of microlocal analysis. The model to be considered is described below.

Let $\Omega \in R^3$ be an open bounded domain with boundary Γ which is either sufficiently smooth or convex. This latter assumption guarantees that solutions to elliptic problems with smooth data have at least $H^2(\Omega)$ regularity. We assume that Γ consists of two portions Γ_1, Γ_0, each being simply connected. The acoustic medium is described by the wave equation in the variable z, while the quantity ρz_t describes the acoustic pressure, ρ being the density of the fluid. The structural vibrations of the elastic wall (structure) are described by the variable w representing the vertical displacement of the plate. The variable w will then satisfy a suitable plate equation defined on the manifold Γ_0. In the present setup we take Γ_0 to be flat and refer to [108, 117] for the case where Γ_0 is curved and modeled by a shell equation.

The PDE model in the variables z and w is as follows:

$$\begin{aligned} z_{tt} &= c^2\Delta z - d_1 z_t \quad \text{in } \Omega \times (0,\infty), \\ \frac{\partial}{\partial \nu} z + l_0 z + d_2 z_t &= 0 \quad \text{on } \Gamma_1 \times (0,\infty), \\ \frac{\partial}{\partial \nu} z + d_2 z_t &= w_t \quad \text{on } \Gamma_0 \times (0,\infty), \end{aligned} \tag{4.3.10}$$

$$w_{tt} + \Delta^2 w + \rho z_t|_{\Gamma_0} + b(x) w_t = 0 \quad \text{in } \Gamma_0 \times (0,\infty). \tag{4.3.11}$$

The boundary conditions prescribed on $\partial\Gamma_0$ are either hinged or free with a nonlinear boundary feedback:

$$w = 0, \quad \Delta w = -g\left(\frac{\partial}{\partial \nu} w_t\right) \quad \text{on } \partial\Gamma_0 \times (0,\infty) \qquad [\text{hinged}] \tag{4.3.12}$$

$$\Delta w + B_1 w = 0, \quad \frac{\partial}{\partial \nu}\Delta w + B_2 w = g(w_t) \text{ on } \partial\Gamma_0 \times (0,\infty) \qquad [\text{free}]. \tag{4.3.13}$$

The boundary operators B_1, B_2 are given by (see [99, 102])

$$B_1 w = -(1-\mu)[2n_1 n_2 w_{x,y} - n_1^2 w_{y,y} - n_2^2 w_{x,y}], \tag{4.3.14}$$

$$B_2 w = (1-\mu)\frac{\partial}{\partial \tau}[(n_1^2 - n_2^2) w_{x,y} + n_1 n_2 (w_{y,y} - w_{x,x})] + l w, \tag{4.3.15}$$

where $0 < \mu < 1$ is Poisson's modulus and the constants l, l_0 are positive. n_1, n_2 are the coordinates of the unit vector normal to the boundary $\partial\Gamma_0$, and $\frac{\partial}{\partial \tau}$ denotes the derivative in tangential direction.

The constant c^2 denotes as usual the speed of sound in the fluid. The nonnegative constants d_i represent damping coefficients (viscous damping and boundary damping). We assume that $d_1 + d_2 > 0$ and $b(x) \geq 0$ a.e. A continuous, monotone increasing function g such that $g(0) = 0$ represents a nonlinear friction acting on the edge of the plate (wall).

The z-problem models the acoustic model while the w-problem models the displacement of the elastic wall or structure. With the model (4.3.10), (4.3.11) we associate the following initial conditions subject to appropriate compatibility conditions on the boundary for the hinged case.

$$\begin{aligned} z(t=0) &= z_0 \in H^1(\Omega), \quad z_t(t=0) = z_1 \in L_2(\Omega), \\ w(t=0) &= w_0 \in H^2(\Gamma_0), \quad w_t(t=0) = w_1 \in L_2(\Gamma_0). \end{aligned} \tag{4.3.16}$$

The coupling between the structure and the acoustic medium is represented by the trace operator which acts on the trace $z_t|_{\Gamma_0}$ of z_t. The quantity $\rho z_t|_{\Gamma_0}$ represents the back pressure against the moving wall Γ_0.

We note that this model is a special case of a more general abstract model (2.6.42) introduced in section 2.6. Indeed, in our case we have $\mathbf{v} = v = w$ and $\mathcal{M} = I, \mathcal{C} = I, g_2 = g, \mathcal{A} = \Delta^2$ subject to either hinged or free homogeneous boundary conditions. The Green's map G is given by

$Gg \equiv v$ iff $\Delta^2 w = 0$, and $v = 0, \Delta v = g$ on $\partial\Gamma_0$ for the hinged boundary condition

and $\Delta v + B_1 v = 0$, $\frac{\partial}{\partial \nu}\Delta v + B_2 v = g$ on $\partial\Gamma_0$ for the free boundary condition.

4.3.2 Formulation of the results

A natural energy function associated with the model (4.3.10), (4.3.11) is the function $E(t) \equiv E_z(t) + E_w(t)$, where

$$\begin{aligned} E_z(t) &= \int_\Omega [z_t^2(t) + |\nabla z(t)|^2] d\Omega + l_0 \int_{\Gamma_1} z^2(t) d\Gamma_1, \\ E_w(t) &= \int_{\Gamma_0} w_t^2(t) d\Gamma_0 + a(w,w) + l \int_{\partial\Gamma_0} w^2(t) d\partial\Gamma_0. \end{aligned} \tag{4.3.17}$$

Here $a(w,z) = \int_{\Gamma_0} \Delta w \Delta z dx$ for the hinged boundary condition case, while for the free boundary condition [99],

$$a(w,z) \equiv \int_{\Gamma_0} [w_{x,x} z_{x,x} + w_{y,y} z_{y,y} + \mu(w_{x,x} z_{y,y} + w_{y,y} z_{x,x}) + 2(1-\mu) w_{x,y} z_{x,y}] dx.$$

It is well known that for $l, l_0 > 0$, $a(w,w) + l \int_{\partial\Gamma_0} w^2 d\partial\Gamma_0$ and $\int_\Omega |\nabla z^2| d\Omega + l_0 \int_{\Gamma_1} z^2 d\Gamma_1$ are equivalent to $H^2(\Gamma_0)$ and $H^1(\Omega)$ topology. Thus, the energy function is equivalent to the topology of

$$\begin{aligned} Y &\equiv H^1(\Omega) \times L_2(\Omega) \times (H^2(\Gamma_0) \cap H_0^1(\Omega)0 \times L_2(\Gamma_0) \text{ [hinged boundary condition]}, \\ Y &\equiv H^1(\Omega) \times L_2(\Omega) \times H^2(\Gamma_0) \times L_2(\Gamma_0) \text{ [free boundary condition]}. \end{aligned} \tag{4.3.18}$$

Regarding the well-posedness of the solutions, we have the following result, which states that initial data of finite energy produce unique solutions of finite energy. More precisely, we have the following proposition.

Proposition 4.3.1. *Let z_0, z_1, w_0, w_1 be an element of Y. Then, there exists a unique solution to* (4.3.10), (4.3.11) *with the boundary conditions* (4.3.12) *(resp.,* (4.3.13)*) and initial conditions* (4.3.16) *that belongs to $C([0,\infty); Y)$.*

Proposition 4.3.1 follows from Theorem 2.6.2.

If additional assumptions are imposed on the regularity of the dissipation g, then one can prove that more regular initial data will produce more regular solutions. However, at this point, we do not make any additional assumptions regarding the smoothness of g (besides mere continuity). For stability results, the following assumption regarding g is made.

Assumption 4.3.1. *There exist positive constants $0 < m < M < \infty$ such that*

$$ms^2 \leq g(s)s \leq M|s|^{p+1} \textit{ for } |s| > 1, \tag{4.3.19}$$

where $1 \leq p \leq 5$ in general and $1 \leq p \leq \infty$ in the case Γ_0 is star shaped, i.e., the following geometric condition holds: $(x - x_0) \cdot \nu > 0$, $x \in \partial\Gamma_0$, $x_0 \in R^2 > 0$. In the case of free boundary conditions (4.3.13) *we also assume that either $b(x) > 0$ a.e. on Γ_0 or else Ω is star shaped.*

Let $h(s)$ be a real-valued function, defined for $s > 0$, concave, strictly increasing, $h(0) = 0$ and such that

$$h(sg(s)) \geq s^2 + g^2(s) \text{ for } |s| \leq 1. \tag{4.3.20}$$

Let $\tilde{h} \equiv h(\frac{x}{meas\partial\Sigma_0})$, $x \geq 0$, where $\partial\Sigma_0 \equiv (0,T) \times \partial\Gamma_0$ and T is a given constant. Since $\tilde{h}$ is monotone increasing, we always have that $I + \tilde{h}$ is invertible. Define

$$p(x) \equiv (I + \tilde{h})^{-1}(Kx), \qquad x > 0, \tag{4.3.21}$$

where K is a positive constant to be given later. Then p is a positive, continuous, strictly increasing function with $p(0) = 0$. Let

$$q(x) \equiv x - (I + p)^{-1}(x), \qquad x > 0. \tag{4.3.22}$$

Since $p(x)$ is positive, increasing, so is $q(x)$. Our main result, established in [109], deals with uniform decay rates obtained for this model.

Theorem 4.3.1. *Consider the system consisting of* (4.3.10), (4.3.11) *with boundary conditions given by* (4.3.12) *(resp.,* (4.3.13)*) and $d_1 + d_2 > 0$. Assumption* 4.3.1 *is in force, where for the hinged boundary condition we assume additionally $p = 1$. Then, with the space Y, defined in* (4.3.18), *we obtain*

(i) *If $g(s)s \geq m_0 s^2$ for some positive constant m_0 and $|s| \leq 1$, there exist positive constants $C > 0, \omega > 0$, such that*

$$|(z(t), z_t(t), w(t), w_t(t))|_Y \leq Ce^{-\omega t}|(z(0), z_t(0), w(0), w_t(0))|_Y.$$

In the superlinear case, $p > 1$, the decay rates ω may depend on the initial energy $E(0)$, i.e., $\omega = \omega(E(0))$.

(ii) *If the growth of g is unspecified at the origin, the obtained decay rates for the solutions are uniform (not necessarily exponential). The decay rates can be explicitly evaluated from the solution of the nonlinear ODE*

$$S_t + q(S(t)) = 0, \qquad t > 0,\ S(0) = E(0), \tag{4.3.23}$$

where the nonlinear function $q(s)$, defined by (4.3.22) with a constant K depending on $m, M, E(0)$, is monotone increasing and depends on the growth of the nonlinear dissipation g at the origin (through the function h given in (4.3.20)). The obtained decay rates are governed by the inequality

$$E(t) \leq S\left(\frac{t}{T_0} - 1\right), \qquad t > T_0, \tag{4.3.24}$$

for some positive constant $T_0 > 0$. We always have that $S(t) \to 0$ as $t \to \infty$.

The main steps of the proof of Theorem 4.3.1, based on [109], are given in sections 4.3.2–4.3.6.

Remark 4.3.1. *If $g(s)$ is of a polynomial growth at the origin, i.e., $g(s)s \geq a|s|^{r+1}$, $|s| \leq 1$ for some positive constant a and $r > 1$, then $E(t) \leq C(E(0))\frac{1}{t^{\frac{2}{r-1}}}$. If $r < 1$ we have $E(t) \leq C(E(0))\frac{1}{t^{\frac{2r}{1-r}}}$. If, instead, the nonlinear function g decays slower than an exponential, the corresponding decay rates are logarithmic (see section 2.3).*

In the case of the Kirchhoff plate modeling the wall, uniform decay rates for the structural acoustic model were obtained in [41, 40]. Indeed, it is shown in [41] that the energy of structural acoustic interaction with the undamped Kirchhoff plate and hinged boundary conditions where the stabilizer acts via moment only decays exponentially to zero. Moreover, the decay rates do not depend on the presence of rotational forces in the model, i.e., the value of the parameter $\gamma \geq 0$. A similar result can be proved for Kirchhoff plate with free boundary conditions where moments and shears act as the stabilizers.

Remark 4.3.2.

1. *In the absence of the damping in the acoustic chamber, the damping present in the plate model is sufficient to provide (via coupling) a strong stability only (i.e., decay rates depending on the initial conditions) [155, 7, 164]. In fact, this phenomenon was recognized earlier in the context of hybrid systems [156].*
2. *Techniques presented below are flexible enough to accommodate different sets of boundary conditions imposed on Γ. Indeed, one could also consider the absorbing boundary conditions imposed on Γ_0 only and $\frac{\partial}{\partial \nu} z = 0$ on Γ_1. In this case (if also $d_1 = 0$), convexity of Γ_1 is required in addition to standard geometric condition $(x - x_0) \cdot \nu \leq 0$ on Γ_1.*

   Another possibility is to consider the plate equation defined on two (or more) flat segments of the boundary Γ with the absorbing boundary conditions for the wave equation on these parts of the boundary while the zero displacement (i.e., $z = 0$) is imposed on the remaining parts of the boundary. This kind of model arises in the context of modeling porous walls (see [58]). In such *a scenario one would need to impose the geometric condition* $(x - x_0) \cdot \nu \leq 0$ *on a portion of the boundary that is not subject to the dissipation. However, in these two cases, special attention should be paid to regularity issues, including potential singularities in elliptic solutions; see the last comment in section* 3.5. *Since we are mostly interested in problems with minimal geometric restrictions (an important aspect in shape optimization theory), our main emphasis is on the model* (4.3.10), (4.3.11).

3. *The presence of positive constants* l, l_0 *guarantees that the constant functions are not in the spectrum of the corresponding elliptic operator. The same goal can be achieved in several different ways, for instance, by adding a constant to the Laplacian or by considering the problem defined on an appropriate quotient space. This point is not essential for the analysis of the problem and different scenarios and models can be easily accommodated.*
4. *In the case of a two-dimensional chamber, the analysis is much simpler. Indeed, in this case, the plate equation is replaced by the one-dimensional beam equation. It is known that for the one-dimensional beam models, stabilization can be achieved by using either the shear forces or the moments only (see* [46, 157]*). Accordingly, one can obtain analogous results for the corresponding structural acoustic problem. The analysis is much simplified, since there is no need for microlocal estimates corresponding to the plate equation.*
5. *One could also consider the dissipation acting only on a portion of the boundary corresponding to the plate. This, however, will require geometric conditions satisfied on the uncontrolled part of the boundary* $\partial\Gamma_0$.
6. *The problem considered here can be easily generalized to include nonlinear (rather than assumed linear) dissipation in the acoustic chamber. Indeed, we could take* (i) *nonlinear absorbing boundary conditions modeled by a monotone increasing function* $g_1(z_t|_{\Gamma_0})$ *subject to linear growth conditions at infinity as in Assumption* 4.2.1, *and* (ii) *a nonlinear viscous damping modeled by a monotone increasing function* $g_2(z_t)$ *subject to a polynomial growth condition at infinity. This can be accomplished by incorporating the analysis of section* 3.3 *with the one presented in this section.*

It should be interesting to note that as a bypass product of the proof of Theorem 4.3.1 we obtain the following result pertaining to the uniform stabilizability of Euler–Bernoulli plates with free boundary conditions.

Theorem 4.3.2. *Consider* (4.3.11) *with* $\rho = 0$ *and with the boundary conditions given either by* (4.3.12) *or* (4.3.13) *where the nonlinear function is subject to Assumption* 4.3.1 *of Theorem* 4.3.1. *Then, the conclusion of Theorem* 4.3.1 *remains valid with* $E(t)$ *replaced by* $E_w(t)$. *In particular, we have*

$$E_w(t) \leq S\left(\frac{t}{T_0} - 1\right), \qquad t > T_0, \text{ and } S(t) \text{ is the solution of } (4.3.23).$$

Remark 4.3.3. *For the case of hinged boundary conditions, Theorem* 4.3.2 *was proved in* [104, 75] *with a linear feedback and in* [88] *with a nonlinear feedback. Free boundary conditions are treated in* [109].

The result in Theorem 4.3.2 generalizes earlier results of the literature [99, 97] in the following two directions: (i) geometric conditions imposed on the boundary are not required, and (ii) there is no need for growth conditions imposed customarily [97] on the nonlinear function g at the origin.

Regarding the geometric constraints (star-shaped-type conditions), which are typically required in the literature ([99, 97] and references therein), these are related to a lack of sufficient a priori regularity of the traces of the solutions on the boundary. To compensate for this lack of regularity, one imposes the geometric constraints that give a correct sign, in the requisite inequalities controlling boundary traces. To dispense altogether with the geometric conditions, one needs to establish much sharper regularity results for the traces involved. This approach, which involves microlocal analysis techniques, has been successfully carried out for several problems involving stabilization of waves and plates (see [131, 134, 120] and also [22], where techniques of propagation were employed for the case of wave equation). The novelty and technical difficulty of the problem considered in Theorem 4.3.2 is that the stabilization occurs only via one boundary condition with no geometric conditions imposed on the boundary. Handling of these will require microlocal analysis.

Comments about the problem. The results described in section 4.2 deal with internal—including structural—damping added to the wall model. In the case of structural damping present in the model, the component of the (uncoupled) system corresponding to the plate equation represents an analytic semigroup [184, 183, 48]. This provides, in addition to strong stability properties for the plate equation, a lot of regularity properties that facilitate the analysis of stability for the entire structure. The situation is different when the plate model does not account for structural damping. The coupling between the structure and the acoustic medium is more sensitive to mathematical analysis and becomes a source of technical difficulties. The terms modeling the interaction on the interface are prescribed by the appropriate trace operators which are not bounded by the terms determined by the energy function. This creates difficulties at the level of obtaining the observability estimates, which are well recognized in the literature. To cope with this problem, we use sharp trace regularity results applied to the second-order hyperbolic equations together with the infinite speed of propagation displayed by Petrovsky-type systems (Euler–Bernoulli equations). It is interesting to note that different arguments and a different mechanism are responsible for the estimates in the case of viscous damping applied to the wave equation (case $d_1 > 0$) as opposed to the case of boundary damping (case $d_2 > 0$).

The presence of boundary damping applied to the plate equation (necessary to provide stability properties for the plate) requires careful handling. This can best be seen by looking at the problem of boundary stabilization (with either shear forces or moments) of the plate alone. In the absence of geometric conditions imposed on the geometry of the plate, the usual multipliers method will not suffice to provide the desired estimates. This is because the multipliers estimates require the bounds on the second-order traces of the solutions to the plate equation, which, in turn, are not bounded by the energy terms. To cope with this issue, we use methods of microlocal analysis where we develop appropriate trace theory for the solutions to Euler–Bernoulli equations with free boundary conditions. In this step, a crucial role is played by trace decoupling methods introduced by Tataru [189]. As a consequence, our final result, when specialized to the plate equation (see Theorem 4.3.2), supplies uniform stability estimates for the boundary feedback acting either via moments or shears only and in the absence of geometric conditions. This provides, as a bypass product, a new contribution to the theory of stability of Euler–Bernoulli plate equations with free boundary conditions which extends the results obtained earlier in [99, 97] in two directions: (i) there is no need for star-shaped geometric conditions imposed on the plate model, and (ii) there is no need to assume growth conditions on the nonlinear feedback $g(w_t)$ at the origin. Of course, the trade-off is that we lose control of the constants involved in the estimates, which leads to the nonexplicit decay estimates (in contrast with [99, 97], where the decay rates are explicit).

We finally note that in the case of the one-dimensional beam equation, uniform stability can be achieved by applying dissipation either in the shear forces or in the boundary moments. (See [46] and also the recent contribution [157], where the one-dimensional beam equations with nonconstant coefficients have been treated.) However, in the two-dimensional case, as considered here, microlocal analysis of the problem reveals that by using moments only, one is bound to lose the regularity in the tangential sector (nonexistent for the one-dimensional problems) which, in turn, prevents one from obtaining the uniform stability estimates.

The remainder of this section is devoted to the proof of the main theorem in the case of free boundary conditions. We select this case as being more difficult (than hinged boundary conditions) and mathematically more interesting. This is because the Lopatinski condition is not satisfied for free boundary conditions.

The outline of the proof is as follows. In the next section we provide several PDE estimates, obtained by multipliers methods, which constitute a preliminary step of the analysis. In the subsequent subsection we develop trace theory for the solutions to Euler–Bernoulli equations with free boundary conditions. This, when combined with sharp regularity results obtained for the wave equation, leads to the observability and stability estimates valid for the entire structure, which is the main technical ingredient of the proof.

4.3.3 Preliminary multipliers estimates

In this subsection we derive several preliminary PDE estimates that are a starting point of the analysis. We begin with a fundamental energy identity. We use the following notation:

$$|w|_{s,D} \equiv |w|_{H^s(D)}, \qquad \langle u, v\rangle_D \equiv \int_D uv dD,$$

where $H^s(D)$ denotes the usual Sobolev's spaces of order s; see [151]. For $s < 0$ we define $H^s(D)$ by duality with respect to the $L_2(D)$ inner product. In what follows we take in (4.3.11) (without loss of generality) the value of $\rho = 1$.

Lemma 4.3.1. *Let z, w be a solution of* (4.3.10), (4.3.11) *with the boundary conditions* (4.3.13). *Then, for $0 \le s \le t$, we have*

$$E(t) + 2\int_s^t [d_1|z_t(\tau)|^2_{0,\Omega} + d_2|z_t(\tau)|^2_{0,\Gamma} + |b^{1/2}w_t|^2_{0,\Gamma_0} + \langle g(w_t(\tau)), w_t(\tau)\rangle_{\partial\Gamma_0}]d\tau = E(s). \tag{4.3.25}$$

Proof of Lemma 4.3.1 is standard and it follows from the usual energy method where we multiply (4.3.10) by z_t and (4.3.11) by w_t and integrate by parts using

Green's formula. The only delicate point is a justification of the calculus for non-smooth solutions. But this can be done by the same approximation-regularization argument as the one used in subsection 3.3.2.

In our next step we apply the multipliers method to equations describing the wave and the plate dynamics. This method is by now standard and has been applied in the literature in the context of stabilization of waves and plates (see [99, 97, 102] and references therein).

Lemma 4.3.2. *Let z be a weak solution to (4.3.10) and let $T > 0$ be an arbitrary constant.*

Then, there exists a constant $C > 0$ such that

$$\int_0^T E_z(t)\,dt \leq C[E_z(0) + E_z(T)]$$

$$+ C\int_0^T \left[d_1|z_t(t)|_{0,\Omega}^2 + (d_2+1)|z_t(t)|_{0,\Gamma}^2 + |w_t|_{0,\Gamma_0}^2 + \left|\frac{\partial}{\partial \tau} z(t)\right|_{0,\Gamma}^2 \right] dt, \qquad (4.3.26)$$

where τ denotes the tangential direction to the boundary Γ.

The proof of Lemma 4.3.2 follows by applying the multipliers $(x - x_0)\cdot \nabla z$ and z to an appropriate smooth approximation of (4.3.10) (see subsection 3.3.2). The passage with the limit reconstructs the estimates for the original weak (finite energy) solution of (4.3.10).

Lemma 4.3.2 will be used for the case of the absorbing boundary conditions (i.e., $d_2 > 0$). When only the viscous damping is present in the model (4.3.10), $(d_2 = 0)$, we need a different estimate, which is provided in Lemma 4.3.9 stated later.

Our next result deals with preliminary estimates for the plate equation.

Lemma 4.3.3. *Let w be a weak solution to (4.3.11) and let $T > 0$ be an arbitrary constant.*

Then, there exists a constant $C > 0$ such that for any $\epsilon_0 > 0$ we have,

1.

$$\int_0^T E_w(t)\,dt \leq \epsilon_0[E_w(0) + E_w(T)] + C|z_t|_{\Gamma_0}|^2_{[H^\delta(0,T;H^{1-2\delta}(\Gamma_0))]'}$$

$$+ C\int_0^T \left[\left|\left(\frac{\partial}{\partial\nu}\right)^2 w(t)\right|^2_{0,\partial\Gamma_0} + \left|\left(\frac{\partial}{\partial\tau}\right)^2 w(t)\right|^2_{0,\partial\Gamma_0} + \left|\frac{\partial}{\partial\tau}\frac{\partial}{\partial\nu} w(t)\right|^2_{0,\partial\Gamma_0} \right] dt$$

$$+ C(E(0)^{1/2}+1)\int_0^T \int_{\partial\Gamma_0} g(w_t)w_t d\partial\Sigma_0$$

$$+ C\tilde{h}\left(\int_0^T \int_{\partial\Gamma_0} g(w_t)w_t d\partial\Sigma_0\right) + C_{\epsilon_0}|w|^2_{L_\infty(0,T;H^1(\Gamma_0))}. \qquad (4.3.27)$$

Here the constant $0 \leq \delta < 1$ and the duality in $[H^\delta(0,T;H^{1-2\delta}(\Gamma_0))]'$ is defined with respect to the $L_2(\Sigma_0)$ inner product. If the function g is linearly bounded at infinity, then the constant $C(E(0)^{1/2}+1)$ does not depend on $E(0)$.

2. *If Γ_0 is star shaped, then the right-hand side of the inequality in* (4.3.27) *does not have the second derivatives of w on $\partial\Gamma_0$.*

Proof. Application of the multipliers $h \cdot \nabla w$ and w to a smooth approximation of (4.3.11) and subsequent passage on the limit (see [83], where the regularization technique of subsection 3.3.2 is presented in the context of plate equations) gives

$$\int_0^T E_w(t)\,dt \le C|w_t|_{0,\Gamma_0}|w|_{1,\Gamma_0}|_0^T + C\int_0^T \langle z_t, h\nabla w + w\rangle_{\Gamma_0} dt$$
$$+ C\int_0^T \left[\langle g(w_t), h\cdot\nabla w\rangle_{\partial\Gamma_0} + \langle g(w_t), w\rangle_{\partial\Gamma_0} + |w_t|^2_{0,\partial\Gamma_0} \right.$$
$$\left. + \left|\left(\frac{\partial}{\partial\nu}\right)^2 w(t)\right|^2_{0,\partial\Gamma_0} + \left|\left(\frac{\partial}{\partial\tau}\right)^2 w(t)\right|^2_{0,\partial\Gamma_0} + \left|\frac{\partial}{\partial\tau}\frac{\partial}{\partial\nu} w\right|^2_{0,\partial\Gamma_0}\right], dt. \quad (4.3.28)$$

Remark 4.3.4. *If Γ_0 is star shaped, then the inequality in* (4.3.28) *is valid without the second derivatives of w on $\partial\Gamma_0$; see* [109].

Splitting the product in the first term on the right-hand side of (4.3.28), rescaling by ϵ_0, and applying the estimates

$$\int_0^T \langle z_t, h\nabla w + w\rangle_{\Gamma_0} dt \le C|z_t|_{[H^\delta(0,T;H^{1-2\delta}(\Gamma_0))]'}|w|_{H^\delta(0,T;H^{2-2\delta}(\Gamma_0))},$$
$$|w|^2_{H^\delta(0,T;H^{2-2\delta}(\Gamma_0))} \le C|w|^{2(1-\delta)}_{L_2(0,T;H^2(\Gamma_0))}|w|^{2\delta}_{H^1(0,T;L_2(\Gamma_0))}$$
$$\le C[|w|^2_{L_2(0,T;H^2(\Gamma_0))} + |w|^2_{H^1(0,T;L_2(\Gamma_0))}] \le C\int_0^T E_w(t)\,dt \quad (4.3.29)$$

(where we have used interpolation inequality and Young's inequality) gives

$$\int_0^T E_w(t)\,dt \le \epsilon_0[E_w(0) + E_w(T)] + C_\epsilon|z_t|_{\Gamma_0}|^2_{[H^\delta(0,T;H^{1-\delta}(\Gamma_0))]'}$$
$$+ \epsilon\int_0^T E_w(t)\,dt + C\int_0^T \left[\langle g(w_t), h\cdot\nabla w\rangle_{\partial\Gamma_0} + \langle g(w_t), w\rangle_{\partial\Gamma_0} + |w_t|^2_{0,\partial\Gamma_0} \right.$$
$$\left. + \left|\left(\frac{\partial}{\partial\nu}\right)^2 w(t)\right|^2_{0,\partial\Gamma_0} + \left|\left(\frac{\partial}{\partial\tau}\right)^2 w(t)\right|^2_{0,\partial\Gamma_0} + \left|\frac{\partial}{\partial\tau}\frac{\partial}{\partial\nu} w\right|^2_{0,\partial\Gamma_0}\right] dt$$
$$+ C_{\epsilon_0}|w|^2_{L_\infty(0,T;H^1(\Gamma_0))}. \quad (4.3.30)$$

Hölder's inequality applied to the first inner product term on the right-hand side of (4.3.30) with $q^{-1} + \bar{q}^{-1} = 1$ gives

$$\int_0^T |\langle g(w_t), h\cdot\nabla w\rangle_{\partial\Gamma_0}|dt \le C|g(w_t)|_{L^q(\partial\Sigma_0)}|\nabla w|_{L^{\bar{q}}(\partial\Sigma_0)}$$

by Sobolev's embeddings applied with any finite $\bar{q}$

$$\le C|g(w_t)|_{L^q(\partial\Sigma_0)}|w|_{C(0,T;H^2(\Gamma_0))} \le C(E(0))^{1/2}|g(w_t)|_{L^q(\partial\Sigma_0)}. \quad (4.3.31)$$

We split $\partial\Sigma_0$ into two parts $\partial\Sigma_A$ and $\partial\Sigma_B$ where

$$\partial\Sigma_0 = \partial\Sigma_A \cup \partial\Sigma_B \text{ and } \partial\Sigma_A \equiv \{(x,t) \in \partial\Sigma_0;\ |w(t,x)| \geq 1\}.$$

Specializing (4.3.31) to $\partial\Sigma_A$, applying (4.3.19), and selecting $p(q-1) \leq 1$ gives

$$\begin{aligned}
&\langle g(w_t), h\cdot\nabla w\rangle_{\partial\Sigma_A} \leq C(E(0))^{1/2}|g(w_t)|_{L^q(\partial\Sigma_A)}\\
&\leq C(E(0))^{1/2}\left(\int_{\partial\Sigma_A}|g(w_t)||g(w_t)|^{q-1}d\partial\Sigma_A\right)^{\frac{1}{q}}\\
&\leq C(E(0))^{1/2}\left(\int_{\partial\Sigma_A}|g(w_t)||w_t|^{p(q-1)}d\partial\Sigma_A\right)^{\frac{1}{q}}\\
&\leq C(E(0))^{1/2}\left(\int_{\partial\Sigma_A}g(w_t)w_t d\partial\Sigma_0\right)^{\frac{1}{q}} \leq C(E(0))^{1/2}\int_{\partial\Sigma_0}g(w_t)w_t d\partial\Sigma. \qquad (4.3.32)
\end{aligned}$$

Remark 4.3.5. *If the function g is of linear growth at infinity (i.e., $p=1$), then the estimate in* (4.3.32) *is simplified and it does not depend on the initial energy* $E(0)$.

To estimate the contribution on $\partial\Sigma_B$ we simply apply the Cauchy–Schwarz inequality and the trace theorem (see [151]).

$$|\langle g(w_t), h\cdot\nabla w\rangle_{\partial\Gamma_0}| \leq C_\epsilon|g(w_t)|^2_{0,\partial\Gamma_0} + \epsilon|w|^2_{1,\partial\Gamma_0} \leq C_\epsilon|g(w_t)|^2_{0,\partial\Gamma_0} + \epsilon|w|^2_{2,\Gamma_0}.$$

Hence

$$|\langle g(w_t), h\cdot\nabla w\rangle_{\partial\Sigma_B}| \leq C_\epsilon\int_{\partial\Sigma_B}|g(w_t)|^2 d\partial\Sigma_B + \epsilon\int_0^T E_w(t)\,dt. \qquad (4.3.33)$$

Combining (4.3.32) and (4.3.33), we arrive at

$$\begin{aligned}
|\langle g(w_t), h\cdot\nabla w\rangle_{\partial\Sigma_0}| \leq C_\epsilon\int_{\partial\Sigma_B}|g(w_t)|^2 d\partial\Sigma_B + \epsilon\int_0^T E_w(t)\,dt&\\
+ C(E(0))^{1/2}\int_{\partial\Sigma_0}g(w_t)w_t d\partial\Sigma_0.& \qquad (4.3.34)
\end{aligned}$$

Similarly,

$$\begin{aligned}
|\langle g(w_t), w\rangle_{\partial\Sigma_0}| \leq C_\epsilon\int_{\partial\Sigma_B}|g(w_t)|^2 d\partial\Sigma_B + \epsilon\int_0^T E_w(t)\,dt&\\
+ C(E(0))^{1/2}\int_{\partial\Sigma_0}g(w_t)w_t d\partial\Sigma_0.& \qquad (4.3.35)
\end{aligned}$$

Finally, applying (4.3.19) we also obtain

$$\int_{\partial\Sigma_0}|w_t|^2 d\partial\Sigma_0 \leq m^{-1}\int_{\partial\Sigma_A}g(w_t)w_t d\partial\Sigma_A + \int_{\partial\Sigma_B}|w_t|^2 d\partial\Sigma_B. \qquad (4.3.36)$$

From (4.3.34), (4.3.35), (4.3.36) and recalling properties of function h (see (4.3.20)) together with Jensen's inequality, we obtain

$$\begin{aligned}
&\int_0^T [\langle g(w_t), h \cdot \nabla w\rangle_{\partial\Gamma_0} + \langle g(w_t), w\rangle_{\partial\Gamma_0} + |w_t|^2_{0,\partial\Gamma_0}]dt \\
&\le \epsilon \int_0^T E_w(t)\,dt + C(E(0)^{1/2} + 1)\int_{\partial\Sigma_0} g(w_t)w_t d\partial\Sigma_0 \\
&\qquad + C_\epsilon \int_{\partial\Sigma_B} |w_t|^2 + |g(w_t)|^2 d\partial\Sigma_B \\
&\le \epsilon \int_0^T E_w(t)\,dt + C(E(0)^{1/2} + 1)\int_{\partial\Sigma_0} g(w_t)w_t d\partial\Sigma_0 \\
&\qquad + C_\epsilon \int_{\partial\Sigma_0} h(g(w_t)w_t)d\partial\Sigma_0) \\
&\le \epsilon \int_0^T E_w(t)\,dt + C(E(0)^{1/2} + 1)\int_{\partial\Sigma_0} g(w_t)w_t d\partial\Sigma_0 \\
&\qquad + C\tilde{h}\left(\int_{\partial\Sigma_0} g(w_t)w_t d\partial\Sigma_0\right).
\end{aligned} \tag{4.3.37}$$

Inserting (4.3.37) into (4.3.30) and selecting small ϵ gives the desired inequality in (4.3.27). The second part in Lemma 4.3.3 follows from Remark 4.3.4. □

Our next step is to eliminate the second-order traces of solution w appearing in (4.3.27). This part is more technical and requires microlocal analysis estimates. Of course, if Γ_0 is star shaped, we do not need these estimates (see Remark 4.3.4), and the results presented in the next sections are not relevant.

4.3.4 Microanalysis estimate for the traces of solutions of Euler–Bernoulli equations and wave equations

This subsection provides sharp regularity estimates for the second-order traces of the solutions to linear wave equations and Euler–Bernoulli equations with free boundary conditions prescribed on the boundary. These results, besides being critical for the proof of our main Theorem 4.3.1, are also of independent PDE interest. For this reason, we collect all relevant estimates. To make the exposition in this subsection independent on the rest of the section, we change notation from Γ_0 to Ω and from $\partial\Gamma_0$ to Γ. Thus, we consider the Euler–Bernoulli equation defined on a two-dimensional bounded domain Ω with smooth boundary Γ,

$$w_{tt} + \Delta^2 w = f \quad \text{in } \Omega \times (0,T) \equiv Q, \tag{4.3.38}$$

and with free boundary conditions prescribed on the boundary $\Gamma \times (0,T) \equiv \Sigma$:

$$\begin{aligned}
\Delta w + B_1 w &= 0, \\
\frac{\partial}{\partial\nu}\Delta w + B_2 w &= g.
\end{aligned} \tag{4.3.39}$$

The boundary operators are given by

$$B_1 w = -(1-\mu)[2n_1n_2 w_{x,y} - n_1^2 w_{y,y} - n_2^2 w_{x,y}], \tag{4.3.40}$$

$$B_2 w = (1-\mu)\frac{\partial}{\partial\tau}[(n_1^2 - n_2^2)w_{x,y} + n_1n_2(w_{y,y} - w_{x,x})], \tag{4.3.41}$$

where $0 < \mu < 1$ is Poisson's modulus. The main goal in this subsection is to provide estimates for the second-order traces on the boundary of the solutions to (4.3.38), (4.3.39) in terms of the velocity traces and the lower-order terms (i.e., terms below the energy level). This is accomplished in two steps. In the first step we relate the second spatial derivatives on the boundary to all velocity traces (normal and tangential components). In the second step, we eliminate the normal components of the velocity. It is precisely this factor that is critical to achieve stabilization via forces only. Both results referred to above are obtained by microlocal analysis methods. For the second step we use the decoupling method developed in [189]. We recall that estimates of a similar nature were already used in section 3.3 in the context of the wave equation. Since these estimates will be used again in this section, we recall them below.

Main trace estimates. Before we formulate our main result, we need to introduce some notation. By $H^{r,s}(Q)$ we denote, following Hormander (see [73, 74]), anisotropic Sobolev's spaces that, roughly speaking, have r derivatives in the normal direction to the boundary with the values in $H^s(\Sigma)$. More precisely, by introducing local coordinates in the neighborhood of the boundary Σ and identifying (locally) $Q = ((0,\infty) \times \Sigma)$, so that $(0,\infty)$ corresponds to the normal direction,

$$H^{r,s}(Q) \equiv \bigcap_{i=0}^{r} H^{r-i}(0,\infty; H^{s+i}(\Sigma)) \cap H^{r+s}(Q), \qquad r \geq 0, s \in R.$$

We also use anisotropic Sobolev spaces [86], denoted by $H^s_a(\Sigma)$, which are equivalent for $s \geq 0$ to $L_2(0,T;H^s(\Gamma)) \cap H^{\frac{1}{2}s}(0,T;L_2(\Gamma))$. For $s < 0$, we define $H^s_a(\Sigma)$ spaces by duality with respect to the $L_2(\Sigma)$ inner product.

The main results of this subsection are as follows.

Theorem 4.3.3. *Let w be a solution to (4.3.38) with the boundary conditions as in (4.3.39). Let $T > 0$ be arbitrary and let α be an arbitrary small constant such that $\alpha < \frac{T}{2}$. Then, we have*

$$\int_\alpha^{T-\alpha} \int_\Gamma \left[\left| \left(\frac{\partial}{\partial\nu}\right)^2 w \right|^2 + \left| \left(\frac{\partial}{\partial\tau}\right)^2 w \right|^2 + \left| \frac{\partial}{\partial\nu}\frac{\partial}{\partial\tau} w \right|^2 \right] d\Gamma dt$$
$$\leq C_{T,\alpha}[|w_t|^2_{0,\Sigma} + |g|^2_{H_a^{-1}(\Sigma)} + |f|^2_{H^{0,-3/2}(Q)} + lot(w)]. \qquad (4.3.42)$$

The lower-order terms ($lot(w)$) denote the terms below the energy level. This is to say

$$lot(w) \leq C \int_0^T [|w(t)|^2_{2-\epsilon,\Omega} + |w_t(t)|^2_{-\frac{1}{2}\epsilon,\Omega}]\, dt,$$

where $\epsilon > 0$ can be taken less than $1/2$.

With reference to the wave equation (4.3.10) we have the following result.

Theorem 4.3.4. *Let z be a solution to (4.3.10) with the boundary conditions as in (4.3.13). Let $T > 0$ be arbitrary and let α be an arbitrary small constant such that $\alpha < \frac{T}{2}$. Then, we have*

$$\int_{\alpha}^{T-\alpha}\int_{\Gamma}\left|\frac{\partial}{\partial\tau}z\right|^2 \leq C_{T,\alpha}\left[\int_0^T\left[\left|\frac{\partial}{\partial\nu}z\right|_{0,\Gamma}^2+|z_t|_{0,\Gamma}^2\right]dt+lot(z)\right]$$
$$\leq C_{T,\alpha}\left[\int_0^T[|w_t|_{0,\Gamma_0}^2+(1+d_2)|z_t|_{0,\Gamma}^2]\,dt+lot(z)\right], \qquad (4.3.43)$$

where

$$lot(z)\leq C\int_0^T[|z|_{1-\epsilon,\Omega}^2+|z_t|_{-\epsilon,\Omega}^2]\,dt, \qquad \epsilon>0.$$

The proof of Theorem 4.3.4 follows from Lemma 7.1 in [131] (which gives precisely the first estimate on the right-hand side of (4.3.43)), applied to the problem (4.3.10)).

Instead, the proof of Theorem 4.3.3 is technical and is given in [109]. It follows by combining the results of Lemmas 4.3.4 and 4.3.5, which are given below. The details, which are lengthy, can be found in [109].

A first step in the analysis is the estimate for the second-order traces of w by all velocity traces (in the normal and tangential directions).

Lemma 4.3.4. *Let w be a solution to (4.3.38) with the boundary conditions as in (4.3.39). Let $T>0$ be arbitrary and let α be an arbitrary small constant less than $\frac{T}{2}$. Then, the following inequality holds:*

$$\int_{\alpha}^{T-\alpha}\int_{\Gamma}\left[\left|\left(\frac{\partial}{\partial\nu}\right)^2 w\right|^2+\left|\left(\frac{\partial}{\partial\tau}\right)^2 w\right|^2+\left|\frac{\partial}{\partial\nu}\frac{\partial}{\partial\tau}w\right|^2\right]d\Gamma dt$$
$$\leq C_{T,\alpha}\left[\left|\frac{\partial}{\partial\nu}w_t\right|_{H_a^{-1}(\Sigma)}^2+|w_t|_{0,\Sigma}^2+|g|_{H_a^{-1}(\Sigma)}^2+|f|_{H^{0,-3/2}(Q)}^2+lot(w)\right], \qquad (4.3.44)$$

where

$$lot(w)\leq\int_0^T[|w(t)|_{2-\epsilon,\Omega}^2+|w_t|_{-\frac{1}{2}\epsilon,\Omega}^2]\,dt \quad \textit{and} \quad \epsilon<\frac{1}{2}.$$

Lemma 4.3.4 follows from Theorem 2.3 in [134].

The key ingredient in handling the issue of stabilizing with only the one boundary condition is the estimate that allows one to express $\frac{\partial}{\partial\nu}w_t$ in terms of free boundary conditions (i.e., g), traces of the velocity $w_t|_\Gamma$, and the lower-order terms. To accomplish this we use the decoupling technique introduced and developed in [189]. Here we note that if the boundary moments are also allowed as the means of stabilization, then one does not need to carry this step, and the estimate stated in Lemma 4.3.4 is sufficient.

Lemma 4.3.5. *Let w be a solution to (4.3.38) with the boundary conditions as in (4.3.39). Let $T>0$ be arbitrary and let α be an arbitrary small constant such that $\alpha<\frac{T}{2}$. Then, we have*

$$\left|\frac{\partial}{\partial\nu}w_t\right|_{H_a^{-1}(\Sigma_\alpha)}^2\leq C_{T,\alpha}[|w_t|_{0,\Sigma}^2+|g|_{H_a^{-1}(\Sigma)}+|f|_{H^{0,-3/2}(Q)}^2+lot(w)], \qquad (4.3.45)$$

where $\Sigma_\alpha \equiv (\alpha, T-\alpha)\times\Gamma$ *and, we recall,* $H_a^{-1}(\Sigma_\alpha)$ *is the dual to* $H_a^1(\Sigma_\alpha) = H^{1/2}((\alpha, T-\alpha); L_2(\Gamma)) \cap L_2((\alpha, T-\alpha); H^1(\Gamma))$*. The duality is taken with respect to the* $L_2(\Sigma_\alpha)$ *inner product.*

The proof of Lemma 4.3.5 given in [109] is technical and is based on [189].

4.3.5 Observability estimates for the structural acoustic problem

In this subsection we combine the trace estimates obtained for the plate with those obtained for the waves.

We combine the estimates obtained in Lemmas 4.3.2 and 4.3.3 with those obtained in Theorems 4.3.3 and 4.3.4. Indeed, collecting the results of Lemma 4.3.2, applied on $(\alpha, T-\alpha)$ and Theorem 4.3.4, yields the following.

Lemma 4.3.6. *Let* z *be a solution to* (2.6.39) *and let* $T>0$ *be an arbitrary constant. Let* $0<\alpha<\frac{T}{2}$*. Then, there exists a constant* $C>0$ *such that*

$$\int_\alpha^{T-\alpha} E_z(t)\,dt \le C[E_z(\alpha)+E_z(T-\alpha)]$$

$$+ C_{T,\alpha}\int_0^T [d_1|z_t(t)|_{0,\Omega}^2 + (d_2+1)|z_t(t)|_{0,\Gamma}^2 + |w_t|_{0,\Gamma_0}^2]\,dt + C_{T,\alpha} lot(z). \quad (4.3.46)$$

Remark 4.3.6. *Note that inequality* (4.3.46) *has still on the right-hand side the energy level coupling term* $|w_t|_{0,\Gamma_0}^2$*. This term cannot be absorbed by the compactness-uniqueness argument (it is not a lower-order term), and it will have to be eliminated from the inequality. It is at this point where the infinite speed of propagation of the Euler–Bernoulli model comes to play.*

Collecting the results of Lemma 4.3.3 and Theorem 4.3.3 applied in the context of (4.3.11) and noting that $|z_t|_{H^{0,-3/2}(\Sigma_0)} \le C|z|_{\Gamma_0}|_{0,\Sigma_0} + lot(z)$ yields the following.

Lemma 4.3.7. *Let* w *be a solution to* (4.3.11) *and let* $T>0$ *be an arbitrary constant. Let* $0<\alpha<\frac{T}{2}$*. Then, there exists a constant* $C>0$ *such that with any* $0\le\delta<1$ *we have*

$$\begin{aligned}\int_\alpha^{T-\alpha} E_w(t)\,dt &\le \epsilon_0[E_w(\alpha)+E_w(T-\alpha)]\\ &+ C_{T,\alpha}\rho[|z_t|_{\Gamma_0}|^2_{[H^\delta(0,T;H^{1-2\delta}(\Gamma_0))]'} + |z|_{\Gamma_0}|^2_{0,\Sigma_0}\\ &+ lot(z)] + C(E(0)^{1/2}+1)\int_0^T\int_{\partial\Gamma_0} g(w_t)w_t d\partial\Sigma_0\\ &+ C\tilde{h}\left(\int_0^T\int_{\partial\Gamma_0} g(w_t)w_t d\partial\Sigma_0\right)\\ &+ C_{T,\alpha}[|w_t|^2_{0,\partial\Sigma_0} + |g(w_t)|^2_{H_a^{-1}(\partial\Sigma_0)}\\ &+ lot(w)] + C_{\epsilon_0}|w|^2_{L_\infty(0,T;H^1(\Gamma_0))}. \end{aligned} \quad (4.3.47)$$

If the function g *is of linear growth at infinity, then the constant* $C(E(0)^{1/2}+1)$ *does not depend on* $E(0)$*. Moreover, if* Γ_0 *is star shaped, then the term* $|g(w_t)|^2_{H_a^{-1}(\partial\Sigma_0)} + lot(w)$ *does not appear in* (4.3.47).

Applying inequality (4.3.46) on a shorter time interval $(2\alpha, T-2\alpha)$, noting that

$$\int_{\alpha}^{T-\alpha} |w_t|_{0,\Gamma_0}^2 dt \le \int_{\alpha}^{T-\alpha} E_w(t)\, dt,$$

and using the estimate in (4.3.47) applied with $\delta = 0$ to eliminate the coupling term $|w_t|_{0,\Gamma_0}^2$ from (4.3.46) gives

$$\begin{aligned}\int_{2\alpha}^{T-2\alpha} E_z(t)\, dt &\le C[E_z(2\alpha) + E_z(T-2\alpha)] \\ &\quad + C_{T,\alpha}\epsilon_0[E_w(\alpha) + E_w(T-\alpha)] \\ &\quad + C_{T,\alpha}\int_0^T [d_1|z_t(t)|_{0,\Omega}^2 + (d_2+1)|z_t(t)|_{0,\Gamma}^2]\, dt + C_{T,\alpha} lot(z) \\ &\quad + C_{T,\alpha}\Bigg[\int_0^T |z_t|_{0,\Gamma_0}^2 dt + (E(0)^{1/2}+1)\int_0^T\int_{\partial\Gamma_0} g(w_t)w_t d\partial\Sigma_0 \\ &\quad + \tilde{h}\left(\int_0^T\int_{\partial\Gamma_0} g(w_t)w_t d\partial\Sigma_0\right) \\ &\quad + |w_t|_{0,\partial\Sigma_0}^2 + |g(w_t)|_{H_a^{-1}(\partial\Sigma_0)}^2 + lot(w)\Bigg] \\ &\quad + C_{\epsilon_0,T,\alpha}|w|_{L_\infty(0,T;H^1(\Gamma_0))}^2. \qquad (4.3.48)\end{aligned}$$

From Lemma 4.3.1 we easily obtain

$$\begin{aligned}\left(\int_0^{\alpha} + \int_{T-\alpha}^T\right) E_w(t)\, dt &\le 2\alpha(E(T) \\ + \int_0^T [d_1|z_t(t)|_{0,\Omega}^2 + d_2|z_t(t)|_{0,\Gamma}^2 &+ |b^{1/2}w_t|_{0,\Gamma_0}^2 + \langle g(w_t(t)), w_t(t)\rangle_{\partial\Gamma_0}]\, dt). \qquad (4.3.49)\end{aligned}$$

Similarly,

$$\begin{aligned}\left(\int_0^{2\alpha} + \int_{T-2\alpha}^T\right) E_z(t)\, dt &\le 4\alpha(E(T) \\ + \int_0^T [d_1|z_t(t)|_{0,\Omega}^2 + d_2|z_t(t)|_{0,\Gamma}^2 &+ |b^{1/2}w_t|_{0,\Gamma_0}^2 + \langle g(w_t(t)), w_t(t)\rangle_{\partial\Gamma_0}]\, dt). \qquad (4.3.50)\end{aligned}$$

Selecting in (4.3.48) $\epsilon_0 = C_{T\alpha}{}^{-1}$, inserting (4.3.50) (resp., (4.3.49)) into (4.3.48) (resp., (4.3.47) applied with $\delta = 0$), and taking advantage once more of the energy identity in Lemma 4.3.1 yields

$$\begin{aligned}\int_0^T E_z(t)\, dt &\le C(\alpha+1)E(T) \\ &\quad + C_{T,\alpha}\int_0^T [d_1|z_t(t)|_{0,\Omega}^2 + (d_2+1)|z_t(t)|_{0,\Gamma}^2 + |b^{1/2}w_t|_{0,\Gamma_0}^2 \\ &\quad + \langle g(w_t(t)), w_t(t)\rangle_{\partial\Gamma_0}]\, dt + C_{T,\alpha} lot(z) + C_{T,\alpha}\Bigg[\int_0^T |z_t|_{0,\Gamma_0}^2 dt\end{aligned}$$

$$+ (E(0)^{1/2}+1)\int_0^T\int_{\partial\Gamma_0} g(w_t)w_t d\partial\Sigma_0 + \tilde{h}\left(\int_0^T\int_{\partial\Gamma_0} g(w_t)w_t d\partial\Sigma_0\right)$$

$$\left. + |w_t|^2_{0,\partial\Sigma_0} + |g(w_t)|^2_{H_a^{-1}(\partial\Sigma_0)} + lot(w)\right] + C_{T,\alpha}|w|^2_{L_\infty(0,T;H^1(\Gamma_0))} \quad (4.3.51)$$

and

$$\int_0^T E_w(t)\,dt \le C(\alpha+1)E(T) + C_{T,\alpha}\int_0^T [(1+d_2)|z_t|^2_{0,\Gamma} + d_1|z_t|^2_{0,\Omega} + |b^{1/2}w_t|^2_{0,\Gamma_0}]\,dt$$

$$+ (C(E(0))^{1/2}+C\alpha)\int_0^T\int_{\partial\Gamma_0} g(w_t)w_t d\partial\Sigma_0 + C\tilde{h}\left(\int_0^T\int_{\partial\Gamma_0} g(w_t)w_t d\partial\Sigma_0\right)$$

$$+ C_{T,\alpha}[|w_t|^2_{0,\partial\Sigma_0} + |g(w_t)|^2_{H_a^{-1}(\partial\Sigma_0)} + lot(w) + |w|^2_{L_\infty(0,T;H^1(\Gamma_0))}]. \quad (4.3.52)$$

Adding the inequality in (4.3.52) to inequality (4.3.51) gives

$$\int_0^T E(t)\,dt \le \int_0^T [E_z(t)+E_w(t)]\,dt \le C(\alpha+1)[E(T)]$$

$$+ C_{T,\alpha}\int_0^T [d_1|z_t(t)|^2_{0,\Omega} + (d_2+1)|z_t(t)|^2_{0,\Gamma} + |b^{1/2}w_t|^2_{0,\Gamma_0}]\,dt + C_{T,\alpha}lot(z)$$

$$+ C_{T,\alpha}\left[(E(0)^{1/2}+\alpha)\int_0^T\int_{\partial\Gamma_0} g(w_t)w_t d\partial\Sigma_0 + C\tilde{h}\left(\int_0^T\int_{\partial\Gamma_0} g(w_t)w_t d\partial\Sigma_0\right)\right]$$

$$+ C_{T,\alpha}[|w_t|^2_{0,\partial\Sigma_0} + |g(w_t)|^2_{H_a^{-1}(\partial\Sigma_0)} + lot(w) + |w|^2_{L_\infty(0,T;H^1(\Gamma_0))}]. \quad (4.3.53)$$

Our next step is to estimate $|w_t|^2_{0,\partial\Sigma_0} + |g(w_t)|^2_{H_a^{-1}(\partial\Sigma_0)}$. We note that if Γ_0 is star shaped, there is no need to estimate $|g(w_t)|^2_{H_a^{-1}(\partial\Sigma_0)}$. It is precisely this term that forces us to assume, in the general case, that $p \le 5$ in Assumption 4.3.1.

Here we employ an argument similar to those used before. We notice first that by interpolation theory of Sobolev's spaces and Sobolev's embeddings,

$$H_a^1(\partial\Sigma) \subset H^{\theta/2}(0,T;H^{1-\theta}(\partial\Gamma_0)) \subset L_{\frac{2}{1-\theta}}(0,T;L_{\frac{1}{\theta-1/2}}(\partial\Gamma_0)), \quad (4.3.54)$$

where the last inclusion will be used only for $1/2 < \theta < 1$.

We show below that by splitting $\partial\Sigma_0$ into $\partial\Sigma_A$ and $\partial\Sigma_B$ (introduced before), we obtain the following inequalities valid with any function $\phi \in H_a^1(\partial\Sigma_0)$:

$$\left|\int_{\partial\Sigma_A} g(w_t)\phi d\partial\Sigma_A\right| \le C|\phi|_{H_a^1(\partial\Sigma_0)}\int_{\partial\Sigma_0} g(w_t)w_t d\partial\Sigma_0,$$

$$\left|\int_{\partial\Sigma_B} g(w_t)\phi d\partial\Sigma_B\right| \le |\phi|_{L_2(\partial\Sigma_B)}|g(w_t)|_{L_2(\partial\Sigma_B)}. \quad (4.3.55)$$

Indeed, the second inequality in (4.3.55) is obvious. Thus it suffices to establish the first inequality only. Let $\partial\Gamma_A(t) \equiv \{x \in \partial\Gamma_0; w_t(t,x) \ge 1\}$. By Hölder's inequality applied with $\frac{1}{r} + \frac{1}{\bar{r}} = 1$,

$$\int_{\partial\Sigma_A} g(w_t)\phi d\partial\Sigma_A = \int_0^T \int_{\partial\Gamma_A} g(w_t(t,x))\phi(t,x) d\partial\Gamma_A(t)\, dt$$
$$\leq \int_0^T \left(\int_{\partial\Gamma_A(t)} |g(w_t(t))|^r d\partial\Gamma_A(t) \right)^{1/r} |\phi(t)|_{L_{\bar{r}}(\partial\Gamma_0)} dt,$$

using the growth condition imposed on g by Assumption 4.3.1,

$$\leq C \int_0^T \left(\int_{\partial\Gamma_A(t)} |g(w_t(t))||w_t(t)|^{p(r-1)} d\partial\Gamma_A(t) \right)^{1/r} |\phi(t)|_{L_{\bar{r}}(\partial\Gamma_0)} dt, \tag{4.3.56}$$

selecting $r = \frac{p+1}{p} > 1$, $\bar{r} = p+1$, and using once more Hölder's inequality,

$$\leq C \left[\int_0^T \left(\int_{\partial\Gamma_A(t)} g(w_t(t)) w_t(t) d\partial\Gamma_A(t) \right)^{q/r} dt \right]^{\frac{1}{q}} \left[\int_0^T |\phi(t)|^{\bar{q}}_{L_{\bar{r}}(\partial\Gamma_0)} dt \right]^{\frac{1}{\bar{q}}},$$

selecting $\frac{1}{\theta - 1/2} = \bar{r} = p+1, \bar{q} = \frac{2}{1-\theta} = \frac{4(p+1)}{p-1}$, $q = \frac{4(p+1)}{3p+5}$, and recalling (4.3.54),

$$\leq C \left[\int_0^T \left(\int_{\partial\Gamma_A(t)} g(w_t(t)) w_t(t) d\partial\Gamma_A(t) \right)^{\frac{4p}{3p+5}} dt \right]^{\frac{3p+5}{4(p+1)}} |\phi|_{H^1_a(\partial\Sigma_0)}, \tag{4.3.57}$$

since for $p \leq 5$, $\frac{4p}{3p+5} \leq 1$, by Jensen's inequality

$$\leq C \left[\int_0^T \int_{\partial\Gamma_A(t)} g(w_t) w_t(t) d\partial\Gamma_A(t)\, dt \right]^{\frac{p}{p+1}} |\phi|_{H^1_a(\partial\Sigma_0)}$$
$$\leq C \int_0^T \int_{\partial\Gamma_A(t)} g(w_t) w_t(t) d\partial\Gamma_A(t)\, dt |\phi|_{H^1_a(\partial\Sigma_0)}, \tag{4.3.58}$$

which implies the desired inequality in the first part of (4.3.55).

From (4.3.55) it follows that

$$|g(w_t)|_{H_a^{-1}(\partial\Sigma_0)} \leq |g(w_t)|_{L_2(\partial\Sigma_B)} + C \int_{\partial\Sigma_0} g(w_t) w_t d\partial\Sigma_0. \tag{4.3.59}$$

But recalling the properties of function h,

$$|w_t|^2_{0,\partial\Sigma_B} + |g(w_t)|^2_{0,\partial\Sigma_B} \leq \tilde{h}\left(\int_0^T \int_{\partial\Gamma_0} g(w_t) w_t d\partial\Sigma_0 \right), \tag{4.3.60}$$

and combining (4.3.60), (4.3.55), (4.3.59), and (4.3.25), we obtain

$$|g(w_t)|^2_{H_a^{-1}(\partial\Sigma_0)} + |w_t|^2_{0,\partial\Sigma_0} \leq CE(0)^{1/2} \int_{\partial\Sigma_0} g(w_t) w_t d\partial\Sigma_0$$
$$+ \tilde{h}\left(\int_{\partial\Sigma_0} g(w_t) w_t d\partial\Sigma_0 \right). \tag{4.3.61}$$

This, together with the inequality in (4.3.53) and (4.3.25), leads to the following estimate:

$$TE(T) \leq \int_0^T E(t)\,dt \leq C(\alpha+1)E(T) + C_{T,\alpha}\int_0^T [d_1|z_t(t)|^2_{0,\Omega} + (d_2+1)|z_t(t)|^2_{0,\Gamma} + |b^{1/2}w_t|^2_{0,\Gamma_0}]\,dt + C_{T,\alpha}\left[((E(0))^{1/2}+1)\int_0^T\int_{\partial\Gamma_0} g(w_t)w_t d\partial\Sigma_0 + \tilde{h}\left(\int_0^T\int_{\partial\Gamma_0} g(w_t)w_t d\partial\Sigma_0\right)\right] + C_{T,\alpha}[lot(w) + |w|_{L_\infty(0,T;H^1(\Gamma_0))} + lot(z)]. \tag{4.3.62}$$

Selecting now $\alpha < T/2$ (but independent on T), T large enough, so that $T > 2C(\alpha+1)+1$, we obtain the main observability estimate for a structural acoustic model given by (4.3.10) and (4.3.11).

Lemma 4.3.8. *Let $T > T_0$ be sufficiently large. Then*

$$\int_0^T E(t)\,dt + E(T) + E(0) \leq C_T\int_0^T [d_1|z_t(t)|^2_{0,\Omega} + (d_2+1)|z_t(t)|^2_{0,\Gamma} + |b^{1/2}w_t|^2_{0,\Gamma_0}]\,dt + C_T\left[((E(0))^{1/2}+1)\int_0^T\int_{\partial\Gamma_0} g(w_t)w_t d\partial\Sigma_0 + \tilde{h}\left(\int_0^T\int_{\partial\Gamma_0} g(w_t)w_t d\partial\Sigma_0\right)\right] + C_T[lot(w,z)], \tag{4.3.63}$$

where we have denoted

$$lot(w,z) \equiv lot(z) + lot(w) + |w|^2_{L_\infty(0,T;H^1(\Gamma_0))}.$$

Lemma 4.3.8 provides the observability estimate that will be critical in proving the result of Theorem 4.3.1 for the case of boundary dissipation active on Γ, i.e., $d_2 > 0$. In the case of viscous dissipation alone (i.e., $d_2 = 0$ and $d_1 > 0$), we need a different estimate, which is given below.

Lemma 4.3.9 (see [109]). *Let $T > 0$ and $d_1 > 0, d_2 = 0$. Then*

$$\int_0^T E(t)\,dt + E(T) + E(0) \leq C_T\int_0^T d_1|z_t(t)|^2_{0,\Omega}dt + C_T\left[((E(0))^{1/2}+1)\int_0^T\int_{\partial\Gamma_0} g(w_t)w_t d\partial\Sigma_0 + \tilde{h}\left(\int_0^T\int_{\partial\Gamma_0} g(w_t)w_t d\partial\Sigma_0\right)\right] + C_T lot_1(w,z), \tag{4.3.64}$$

where we denote $lot_1(w,z) \equiv lot_1(z) + lot(w) + |w|^2_{L_\infty(0,T;H^1(\Gamma_0))}$, *with*

$$lot_1(z) \equiv |z_t|_{\Gamma_0}|^2_{[H^\delta(0,T;H^{1-2\delta}(\Gamma_0))]'} + \int_0^T |z|^2_{1-\epsilon,\Omega} dt + |z|^2_{L_\infty((0,T);L_2(\Omega))}.$$

4.3.6 Completion of the proof of Theorem 4.3.1

Since we are equipped with the observability estimate given by Lemmas 4.3.8 and 4.3.9, the remaining arguments are mostly routine, with the exception of some rather delicate regularity considerations in the case $d_2 = 0$. Indeed, by applying the compactness-uniqueness argument we can absorb the lower-order terms $lot(w,z)$. The compactness-uniqueness argument used for this problem is mostly routine except for a few technical points regarding the first terms in $lot_1(z)$. For this we refer to [109].

Lemma 4.3.8 (resp., Lemma 4.3.9) and the compactness-uniqueness argument give

$$lot(w,z) \le C_{T,E(0)} \left[\int_0^T |z_t(t)|_\Gamma|^2_{0,\Gamma} dt + \langle g(w_t), w_t \rangle_{\partial\Sigma_0} \right]. \tag{4.3.65}$$

Combining (4.3.65) with the inequality in Lemma 4.3.8 gives

$$\begin{aligned} &\int_0^T E(t)\,dt + E(T) + E(0) \\ &\le C_{T,E(0)} \int_0^T [d_1|z_t(t)|^2_{0,\Omega} + (d_2+1)|z_t(t)|^2_{0,\Gamma} + |b^{1/2}w_t|^2_{0,\Gamma_0}]\,dt \\ &\quad + (C_T(E(0))^{1/2} + C_T\alpha) \int_0^T \int_{\partial\Gamma_0} g(w_t)w_t d\partial\Sigma_0 \\ &\quad + C_T\tilde{h}\left(\int_0^T \int_{\partial\Gamma_0} g(w_t)w_t d\partial\Sigma_0 \right). \end{aligned} \tag{4.3.66}$$

Denoting

$$\mathcal{F} \equiv \int_0^T [d_1|z_t(t)|^2_{0,\Omega} + d_2|z_t(t)|^2_{0,\Gamma} + |b^{1/2}w_t|^2_{0,\Gamma_0}]\,dt + \int_0^T \int_{\partial\Gamma_0} g(w_t)w_t d\partial\Sigma_0$$

and taking account of the fact that $d_2 > 0$ (this is critical), the inequality in (4.3.66) reads

$$\int_0^T E(t)\,dt + E(T) + E(0) \le C_{T,E(0)}\mathcal{F} + C_{T,E(0)}\tilde{h}(\mathcal{F}). \tag{4.3.67}$$

To continue with the proof of Theorem 4.3.1 we apply the energy relation (4.3.25) to (4.3.67). This yields

$$TE(T) \le [C_{T,(E(0)^{1/2}} + C_T\tilde{h}](\mathcal{F}) = C_{T,(E(0)^{1/2}}[1+\tilde{h}](E(0) - E(T)) \tag{4.3.68}$$

and

$$KE(T) \le [1+\tilde{h}](E(0) - E(T)) \tag{4.3.69}$$

for some constant $K(E(0))$. Applying the inverse of $1+\tilde{h}$ to both sides of (4.3.69) gives

$$[1+\tilde{h}]^{-1}(KE(T)) \leq E(0) - E(T) \tag{4.3.70}$$

and

$$[1+\tilde{h}]^{-1}(KE(T)) + E(T) \leq E(0). \tag{4.3.71}$$

This gives

$$p(E(T)) + E(T) \leq E(0). \tag{4.3.72}$$

The final conclusion of the main Theorem 4.3.1 follows from Theorem 3.2.1. □

4.4 Thermal Damping

Structural acoustic models considered in the previous sections assumed an existence of the passive damping on the wall Γ_0 of a structure. This damping was in the form of either structural damping or viscous/boundary damping acting along the edges of the plate. On the other hand, it is well known and acknowledged in the engineering literature that structural damping is rather rare and its modeling is poorly understood. Moreover, dampers (in particular structural dampers) placed on the active wall may produce a local overdamping effect, undermining the effectiveness of active controllers. For these reasons, it is desirable to consider other models that do not account for structural damping placed on the active (vibrating) walls and, ideally, for no other forms of damping affecting the interface between the acoustic and structural media. Although the actual physical models may exhibit some residual damping, to be on the safe side we prefer not to account for its presence. The goal of this section is to address the problem by using a model that includes thermal effects in the wall and dispenses altogether with the need for any type of damping affecting the active wall or the interface (including structural damping). The interest in studying this problem, besides strong physical motivation, is that mathematical analysis of these models is much more subtle and it involves new trace estimates developed in the context of wave and plate theory.

4.4.1 Model

We consider a structural acoustic model consisting of an acoustic chamber with a combination of flexible (vibrating) and hard walls. The PDE model describing the structure is given by a system of equations consisting of a wave equation coupled with a thermal plate equation. At the interface of the two regions, the coupling between the wave and the plate occurs.

Let $\Omega \in R^3$ be an open bounded domain with boundary Γ which is assumed either sufficiently smooth or convex. The boundary $\Gamma = \overline{\Gamma_0} \cup \overline{\Gamma_1}$ is connected. Here Γ_0, Γ_1 are regions in R^3, relatively open with respect to Γ. (See Figure 1.)

The pressure in the chamber (acoustic medium) is defined on a spatial domain Ω, while the displacement of the flexible walls is defined on a flat segment Γ_0. Γ_1 represents a hard wall, which is subject to frictional forces.

In addition to classical notation used for Sobolev's spaces, we use $H^1_\gamma(\Gamma_0) = H^1(\Gamma_0), \gamma > 0$, $L_2(\Gamma_0)$, $\gamma = 0$, with inner product

$$(\omega_1, \omega_2)_{H^1_\gamma(\Gamma_0)} \equiv (\omega_1, \omega_2)_{L^2(\Gamma_0)} + \gamma(\nabla\omega_1, \nabla\omega_2)_{L^2(\Gamma_0)} \quad \text{for all } \omega_1, \omega_2 \in H^1(\Gamma_0). \tag{4.4.73}$$

The notation $H^1_{\gamma,0}(\Gamma_0)$ is used if in the definition above the $H^1(\Gamma_0)$ component is replaced by $H^1_0(\Gamma_0)$.

The PDE model considered consists of the wave equation in the variable z (where the quantity ρz_t is the acoustic pressure and ρ is the density of the fluid),

$$\begin{aligned} z_{tt} &= c^2\Delta z \quad \text{in} \quad \Omega \times (0,\infty), \\ \frac{\partial}{\partial\nu} z + dz &= -g(z_t) \quad \text{on} \quad \Gamma_1 \times (0,\infty), \\ \frac{\partial}{\partial\nu} z &= w_t \quad \text{on} \quad \Gamma_0 \times (0,\infty), \\ z(0) = z_0 \in H^1(\Omega), &\qquad z_t(0) = z_1 \in L^2(\Omega), \end{aligned} \tag{4.4.74}$$

and the elastic equation representing the displacement w of the wall subject to thermal effects (see, e.g., [99]),

$$\left.\begin{aligned} &w_{tt} - \gamma\Delta w_{tt} + \Delta^2 w = -\Delta\theta - \rho z_t \\ &\theta_t - \Delta\theta = \Delta w_t \end{aligned}\right\} \quad \text{on} \quad \Gamma_0 \times (0,\infty), \tag{4.4.75}$$

equipped either with clamped or with hinged or free boundary conditions on $\partial\Gamma_0 \times (0,\infty)$,

$$w = \frac{\partial}{\partial\nu} w = \theta = 0 \quad \text{[clamped boundary condition]}, \tag{4.4.76}$$

$$w = \Delta w = \theta = 0 \quad \text{[hinged boundary condition]}, \tag{4.4.77}$$

$$\left.\begin{aligned} &\Delta w + (1-\mu)B_1 w + \theta = 0 \\ &\tfrac{\partial}{\partial\nu}\Delta w + (1-\mu)B_2 w - \gamma\tfrac{\partial}{\partial\nu} w_{tt} + \tfrac{\partial}{\partial\nu}\theta = 0 \\ &\tfrac{\partial}{\partial\nu}\theta + \lambda\theta = 0 \end{aligned}\right\} \quad \text{[free boundary condition]}. \tag{4.4.78}$$

The vector ν (resp., $\tilde{\nu}$) denotes the unit normal vector to the boundary Γ (resp., $\partial\Gamma_0$) and τ (resp., $\tilde{\tau}$) denotes the unit tangential direction to Γ (resp., $\partial\Gamma_0$). The constant λ is assumed positive. Here θ is the temperature, c^2 is the speed of sound as usual, and ρ represents back pressure acting on the wall. The function g, representing a potential damping (friction), is assumed continuous, monotone increasing, and zero at the origin. The constant γ accounts for rotational forces and here is taken to be small and nonnegative. The boundary operators B_1, B_2 are given by

$$B_1 \equiv 2\nu_1\nu_2 D^2_{x,y} - \nu_1^2 D^2_{y,y} - \nu_2^2 D^2_{x,x},$$

$$B_2 \equiv \frac{\partial}{\partial\tau}[(\nu_1^2 - \nu_2^2)D^2_{x,y} + \nu_1\nu_2(D^2_{y,y} - D^2_{x,x})] + lw,$$

where $[\nu_1, \nu_2] = \tilde{\nu}$.

It should be noted that the presence of the parameter γ in (4.4.75) changes the character of dynamics. Indeed, the uncoupled thermoelastic plate is of a hyperbolic type, when $\gamma > 0$, and of an analytic type, when $\gamma = 0$ [137, 136].

Our goal is to show that the energy of the entire system, given by (4.4.74), (4.4.75), decays to zero at the uniform rate. For convenience, and without loss of generality, we choose $c = \rho = 1$. Also, in what follows, we assume that the constants l, d are positive. This assumption is not essential to mathematical analysis of the problem, but it guarantees that the constants functions are properly damped. In the absence of this restriction, one needs either to formulate the result on an appropriate quotient space or to account for other (lower-order) terms in the equation that prevent zero value to be a member of the spectrum of the corresponding linear elliptic operator.

An important feature of this model is that we do not assume any source of structural (Kelvin–Voight type) or other type of damping affecting the flexible (vibrating) wall and the interface Γ_0. In particular, we do not require any dissipation on the interface boundary Γ_0. A related model accounting for the damping applied to the interface Γ_0 and no damping (or limited damping) on Γ_1 was considered in [112, 113]. Our aim is to show that the thermal effects alone and boundary dissipation affecting only the hard walls of the acoustic medium stabilize the system. From the mathematical standpoint, the fact that we do not assume any damping affecting Γ_0 is critical and it contributes to new mathematical issues. It is not only that we have less damping available in the system, and hence stability is less likely to occur; more important, the lack of the damping on Γ_0 has important implications regarding the regularity of solutions. To appreciate this point, it is enough to notice that the presence of the damping on the wall Γ_0 provides a priori L_2 regularity on the trace of the pressure $z_t|_{\Gamma_0}$, which is the coupling term between the wall and the acoustic medium. If there is no damping on Γ_0, the term $z_t|_{\Gamma_0}$ is not even defined (recall $z_t \in L_2(\Omega)$). Thus, a fundamental task is to provide appropriate estimates (in appropriate negative norms) for this term, as well as for the tagential derivatives of z on Γ_0.

To accomplish this goal we use differential multipliers developed in the context of stability analysis for the wave equation [109] together with the operator multiplier method introduced in [5, 6]. Our analysis is different for the two cases $\gamma > 0$ and $\gamma = 0$. The newly developed sharp trace theory for both the wave and the plate equations plays a critical role in our arguments. In the case $\gamma = 0$, we exploit analyticity of the thermoelastic semigroup, which property is now known to hold for all boundary conditions, including the free boundary conditions [137, 136, 160].

4.4.2 Statement of main results

We begin with a preliminary result stating that the system is well posed. To this end we define the following spaces Y which differentiate between clamped and hinged boundary conditions and free boundary conditions:

$$Y \equiv H^1(\Omega) \times L^2(\Omega) \times H^2(\Omega) \times H^1_{\gamma,0}(\Gamma_0) \times L^2(\Gamma_0)$$
$$\text{[clamped or hinged boundary condition]},$$
$$Y \equiv H^1(\Omega) \times L^2(\Omega) \times H^2(\Omega) \times H^1_{\gamma}(\Gamma_0) \times L^2(\Gamma_0)$$
$$\text{[free boundary condition]}.$$

Theorem 4.4.1 (well-posedness). *Let Ω be a bounded open domain in R^3 with boundary Γ as previously described. For all initial data $y_0 = [z_0, z_1, w_0, w_1, \theta_0] \in Y$, the solution $y(t) = [z, z_t, w, w_t, \theta]$ of the model (4.4.74), (4.4.75) exists in $C([0,\infty); Y)$ and is unique.*

Proof. Since the problem under consideration is maximal dissipative, conclusions of Theorem 4.4.1 follow from the results in Chapter 2, in particular, Theorem 2.6.1 and Theorem 2.6.2. □

To formulate our main result on stability, we introduce some notation. We define the energy functional associated with the model and given by

$$E_\gamma(t) = E_{k,\gamma}(t)+E_p(t), \;\; E_{k,\gamma}(t) = \int_\Omega |z_t|^2 d\Omega + \int_{\Gamma_0} [|w_t|^2+\gamma|\nabla w_t|^2] d\Gamma_0, \tag{4.4.79}$$

$$E_p(t) = \int_\Omega |\nabla z|^2 d\Omega + d\int_{\Gamma_1} z^2 d\Gamma_1 + \int_{\Gamma_0} |\theta|^2 d\Gamma_0 + a(w,w), \tag{4.4.80}$$

where in the case of clamped or hinged boundary conditions we have $a(w,z) \equiv \int_{\Gamma_0} \Delta w \Delta z dx$, and for the free case,

$$a(w,z) \equiv \int_{\Gamma_0} [w_{x,x}z_{x,x} + w_{y,y}z_{y,y} + \mu w_{x,x}z_{y,y} + \mu w_{y,y}z_{x,x} + 2(1-\mu)w_{x,y}z_{x,y}]d\Gamma_0 + l\int_{\partial\Gamma_0} wz d\partial\Gamma_0.$$

It is well known that $a(w,w)$ is topologically equivalent to the $H^2(\Gamma_0)$ [99] norm. We also introduce the function $h(s)$, which is an assumed concave, strictly increasing function, zero at the origin, and such that the following inequalities are satisfied for $|s| \le 1$:

$$h(sg(s)) \ge s^2 + |g(s)|^2.$$

Such a function can be easily constructed in view of the monotonicity assumption imposed on g [120].

Our main result is the following.

Theorem 4.4.2 (uniform stability). *Let Ω be a bounded open domain in R^3 with boundary Γ as previously described. Assume that the nonlinear function g satisfies*

$$sg(s) > 0, \qquad s \neq 0, \qquad ms^2 \le g(s)s \le Ms^2, \qquad |s| \ge 1. \tag{4.4.81}$$

Then, with the constant $\gamma \ge 0$, every weak solution $y(t) = [z, z_t, w, w_t, \theta]^{\mathrm{T}}$ of (4.4.74), (4.4.75) with any set of boundary conditions (clamped, hinged, or free) imposed on the plate model decays uniformly to zero. This is to say, the estimate

$$E_\gamma(t) \le C_\gamma S(t/T_0 - 1)), \qquad t \ge T_0, \tag{4.4.82}$$

holds, where the real variable function $S_\gamma(t)$ (which may depend on γ) converges to zero as $t \to \infty$ and it satisfies the following ODE:

$$\frac{d}{dt}S_\gamma(t) + q_\gamma(S_\gamma(t)) = 0, \qquad S_\gamma(0) = E_\gamma(0). \tag{4.4.83}$$

The (nonlinear) monotone increasing function $q_\gamma(s)$ *is determined entirely from the behavior at the origin of the nonlinear functions* g *and it is given by the following algorithm:*

$$q_\gamma \equiv I - (I + p_\gamma)^{-1}, \tag{4.4.84}$$

$$p_\gamma \equiv (I + h_0)^{-1}\left(\frac{\cdot}{K}\right), \tag{4.4.85}$$

$$h_0(x) \equiv h(x/mes(0,T) \times \Gamma_1), \tag{4.4.86}$$

where $\Sigma_1 \equiv \Gamma_1 \times (0,T)$ *and the constant* $K > 0$ *may depend on* $\gamma \geq 0$.

The proof of Theorem 4.4.2 is given in [114] for the case of free boundary conditions and in [146] for the case of clamped or hinged boundary conditions.

Remark 4.4.1. *Note that the decay rates established in Theorem* 4.4.2 *can be computed explicitly by solving the nonlinear ODE system* (4.4.83), *once the behavior of* $g(s)$ *at the origin is specified. If the nonlinear function* g *is bounded from below by a linear function, then it can be shown that the decay rates predicted by Theorem* 4.4.2 *are exponential. This is to say that there exist positive constants* C, ω, *possibly depending on* $E(0), \gamma$ *and such that*

$$E_\gamma(t) \leq Ce^{-\omega t}E_{\gamma_0} \quad for\ t > 0.$$

If, instead, this function has a polynomial growth (resp., exponentially decaying) at the origin, then the decay rates are algebraic (resp., logarithmic). This can be verified by solving explicitly the ODE problem (4.4.83) *(see* [120]*).*

Remark 4.4.2. *Note that we do not assume any geometric conditions imposed on* Γ_1. *Moreover, we do not impose any conditions on the growth of the nonlinearity at the origin.*

The result formulated in Theorem 4.4.2 can be extended to the following situations:

1. The support of damping g may be strictly included in Γ_1 [114]. In this case, one needs to impose certain geometric conditions (convexity) on a subportion of the boundary Γ_1, where there is no dissipation.
2. Nonlinear effects (e.g., of the von Karman type) can also be included to the plate model. (See [115] for the case of free boundary conditions and [146] for the case of clamped or hinged boundary conditions.)
3. Different configurations for the placement of the damping g can be considered. In particular, damping may be placed on the interface only. This will require appropriate geometric conditions to be satisfied on a portion of the boundary not subject to dissipation [148, 146, 113, 112].
4. In the case of clamped or hinged boundary conditions with the damping acting on the interface, the analysis is simpler and leads to decay rates that are uniform with respect to the parameter γ [148, 112].

The remainder of this section is devoted to the proofs of stability result in Theorem 4.4.2. We carry on by following [114] with full details the proof for the free case, which is mathematically more demanding than other cases. However, first, to orient the reader, we point out the main features of the problem under study:

- The plate equation is subject to neither structural nor mechanical damping (which induces strong stability properties for the plate). As a result, the structural acoustic model considered displays either hyperbolic-hyperbolic characteristics (when $\gamma > 0$) or hyperbolic-parabolic (when $\gamma = 0$).
- There is no mechanical damping on the interface Γ_0.
- In the case of free boundary conditions, the PDE equation describing the plate model does not satisfy the Lopatinski condition. As a consequence, the natural regularity of boundary traces (critical to the analysis) is much weaker. Handling this requires rather subtle analysis, particularly at the level of accounting for the effects of thermoelasticity. The analysis of the same problem with either hinged or clamped boundary conditions is simpler [113, 146].
- The plate model under consideration changes the character of dynamics (from analytic to hyperbolic) depending on the value of γ. This forces a completely different treatment of the problem (mainly at the level of treating the traces of solutions) for these two different classes of dynamics.
- The coupling between the medium and the wall, represented by trace operator $z_t|_{\Gamma_0}$, induces a strong coupling (rather than a weak coupling), which produces in the estimates uncontrolled, above energy level, terms. In the absence of any damping affecting the wall Γ_0, mathematical analysis of these terms requires a very special treatment. In particular, all machinery of sharp trace regularity for the waves and plates needs to be brought in. These results are formulated in the next subsection.

4.4.3 Sharp trace regularity results

In this section we collect several trace regularity results, recently established in the literature, for the wave and plate equations. These results, instrumental for the proof of Theorem 4.4.2, are sharp and do not follow from the usual trace-type theorems.

We adopt the following notation:

$$|w|_{s,\Omega} \equiv |w|_{H^s(\Omega)}, \quad (u,v)_\Omega \equiv \int_\Omega uv d\Omega.$$

The same notation is used with Ω replaced by Γ, etc. For $s < 0$ the negative Sobolev's spaces are defined as duals (pivotals) to $H^{-s}(\Omega)$ with respect to $L_2(\Omega)$ topology.

The constant C will always denote a generic constant, different at different occurrences. The same refers to small constants ϵ, which at different places may denote different quantities.

We begin with the wave equation.

$$z_{tt} = \Delta z \quad \text{in} \quad \Omega \times (0,T). \tag{4.4.87}$$

Lemma 4.4.1. *Let z be a solution to (4.4.87) with the interior regularity*

$$z \in C(0,T; H^1(\Omega)) \cap C^1(0,T; L_2(\Omega))$$

and the following boundary regularity $\frac{\partial}{\partial\nu}z \in L_2((0,T)\times\Gamma)$. *Let* $T>0$ *be arbitrary and let* α *be an arbitrary small constant such that* $\alpha < \frac{T}{2}$. *Then, we have*

$$\int_\alpha^{T-\alpha}\left[\left|\frac{\partial}{\partial\tau}z\right|^2_{0,\Gamma_1}\right]dt \le C_{T,\alpha}\left[\int_0^T\left[\left|\frac{\partial}{\partial\nu}z\right|^2_{0,\Gamma} + |z_t|^2_{0,\Gamma_1}\right]dt + lot(z)\right] \tag{4.4.88}$$

where $lot(z) \le C_\delta[\sup_{t\in[0,T]}|z(t)|^2_{1-\delta,\Omega} + |z_t(t)|^2_{-\delta,\Omega}]$ *for any* $\delta > 0$,

$$\int_0^T [|z_t|^2_{-4/5,\Gamma} + |z_t|^2_{-1/2,\Gamma_0}]\,dt$$
$$\le C_T\left[|z(0)|^2_{1,\Omega} + |z_t(0)|^2_{0,\Omega} + \int_0^T\left|\frac{\partial}{\partial\nu}z\right|^2_{0,\Gamma} dt + |z_t|^2_{0,\Gamma_1}\right]. \tag{4.4.89}$$

We note that the above inequalities do not follow from standard trace theory and the assumed interior regularity. These are sharp independent trace regularity results. Inequality (4.4.88) follows from techniques in Lemma 7.1 in [131]. The sharp Neumann regularity results, stated in (4.4.89), are given in Theorems A and C in [130] (see also [127, 188]). The first estimate in Lemma 4.4.1 was already used in section 2.3 in the context of stabilization of the wave equation.

Our next estimates deal with the plate equation. We consider the Kirchhoff plate

$$\begin{aligned} w_{tt} - \gamma\Delta w_{tt} + \Delta^2 w &= f_t + k \quad \text{on } \Gamma_0\times(0,T),\\ \Delta w + (1-\mu)B_1 w &= g_1 \quad \text{on } \partial\Gamma_0\times(0,T),\\ \frac{\partial}{\partial\nu}\Delta w + (1-\mu)B_2 w - \gamma\frac{\partial}{\partial\nu}w_{tt} &= g_2 \quad \text{on } \partial\Gamma_0\times(0,T). \end{aligned} \tag{4.4.90}$$

Lemma 4.4.2. *With reference to* (4.4.90), $\gamma>0$, *and taking* $f=0$, *the following regularity holds:*

$$|w(t)|^2_{2,\Gamma_0} + |w_t(t)|^2_{1,\Gamma_0} \le C_{\gamma,T}\left[|w(0)|^2_{2,\Gamma_0} + |w_t(0)|^2_{1,\Gamma_0}\right.$$
$$\left. + \int_0^T [|g_1|^2_{1/2,\partial\Gamma_0} + |g_2|^2_{-1/2,\partial\Gamma_0} + |k|^2_{-1,\Gamma_0}]\,dt\right]. \tag{4.4.91}$$

The regularity due to k, g_2 follows from a standard semigroup argument. However, this is not the case with regularity due to g_1. Indeed, we note that standard regularity results for Kirchhoff plates require $g_1 \in H^{1/2}((0,T)\times\partial\Gamma_0)$, rather than 1/2 derivative in the space only. The proof of Lemma 4.4.2, based on microlocal analysis arguments, is given in [139].

Lemma 4.4.3. *With reference to* (4.4.90), $\gamma>0$, *and taking* $g_1 = g_2 = 0, k = 0$, *the following regularity holds. There exists a positive constant* $\rho > 0$ *such that*

$$\int_0^T |w|^2_{3/2+\rho,\partial\Gamma_0}dt \le C_{T,\gamma}\left[|w(0)|^2_{2,\Gamma_0} + |w_t(0)|^2_{1,\Gamma_0}\right.$$
$$\left. + \left[\int_0^T |f|_{\rho,\Gamma_0}dt\right]^2 + \sup_{t\in[0,T]}|f(t)|^2_{-1+\rho,\Gamma_0}\right]. \tag{4.4.92}$$

The value of ρ depends on the geometry Γ_0. However, for general smooth domains ρ can always be taken positive $0 < \rho < 1/5$.

We note that the standard regularity result gives the above inequality with $\rho = 0$. The proof of the lemma is given in [114].

Our next result deals with the case $\gamma = 0$. In this case, the linear thermoelastic system is governed by an analytic semigroup and we have higher interior regularity of the corresponding solutions [137].

Lemma 4.4.4. *Let $\gamma = 0$. We consider the original system given by* (4.4.74), (4.4.75),

$$\int_0^T [|w(t)|^2_{3,\Gamma_0} + |w_t(t)|^2_{1,\Gamma_0}]\,dt \le C_T[E_w(0) + |z(0)|^2_{1,\Omega} + |z_t(0)|^2_{0,\Omega}] + C_T \int_0^T [|w_t|^2_{-1,\Gamma_0} + |z|^2_{0,\Gamma_1} + |g(z_t)|^2_{0,\Gamma_1}]\,dt, \quad (4.4.93)$$

$$\int_0^T [|w(t)|^2_{3-\delta,\Gamma_0} + |w_t(t)|^2_{1-\delta,\Gamma_0}]\,dt \le \epsilon[E_w(0) + |z(0)|^2_{1,\Omega} + |z_t(0)|^2_{0,\Omega}] + C_{T,\epsilon,\delta} \int_0^T [|w_t|^2_{-1,\Gamma_0} + |w|^2_{1,\Gamma_0}]\,dt + C_T \int_0^T [|g(z_t)|^2_{0,\Gamma_1} + |z|^2_{0,\Gamma_1}]dt, \qquad \delta > 0. \quad (4.4.94)$$

4.4.4 Uniform stabilization: Proof of Theorem 4.4.2

Preliminaries. Our goal is to show the uniform stability of the coupled PDE system (4.4.74), (4.4.75). We begin with a preliminary energy identity that illustrates the fact that the system is dissipative.

Proposition 4.4.1. *With respect to the system of equations* (4.4.74), (4.4.75), *the following energy equality holds for all $0 \le s \le T$:*

$$E_\gamma(T) = E_\gamma(s) + 2\int_s^T (g(z_t)z_t)_{\Gamma_1} dt + 2\int_s^T |\nabla\theta|^2_{0,\Gamma_0} dt + 2\lambda \int_s^T |\theta|^2_{0,\partial\Gamma_0} dt, \quad (4.4.95)$$

where, we recall, the energy $E_\gamma(t)$ is defined by

$$E_\gamma(t) \equiv |\nabla z||^2_{0,\Omega} + d|z|^2_{0,\Gamma_1} + |z_t|^2_{0,\Omega} + |w_t|^2_{0,\Gamma_0} + \gamma|\nabla w_t|^2_{0,\Gamma_0} + a(w(t), w(t)) + |\theta|^2_{0,\Gamma_0}. \quad (4.4.96)$$

Proof. By applying the multipliers z_t on the wave equation, w_t on the elastic equation, and θ on the thermal equation and then integrating by parts, we obtain the above equality for smooth solutions. By using the regularization argument in section 3.3.2 (see also Remark 3.3.8) one can extend this equality to all solutions of finite energy. □

Our basic strategy is to obtain the estimates first for the thermoelastic equations on Γ_0 and then for the wave equation defined on Ω. A subtle point of the analysis is the treatment of the coupling and appropriate combination of the two estimates, in order to obtain the final estimate valid for the entire model. Notice that the coupling introduces terms that are not bounded by the energy norms—a notoriously difficult problem in studying stabilization. To cope with these we use trace regularity results presented in the previous subsection.

We also note that our decay rates are dependent on $\gamma \geq 0$. On the other hand, if $\gamma = 0$, then the thermoelastic system represents an analytic semigroup [137]. Instead, if $\gamma > 0$, the dynamics of the thermoelastic plate are hyperbolic-like with a finite speed of propagation. Thus, the dynamical properties of the overall system are very different in the cases of $\gamma = 0$ and $\gamma > 0$. In fact, the difference between the two types of dynamic will play a critical role through the analysis, particularly at the level of treating regularity of traces as well as the coupling between the two systems.

Thermoelastic equations. Let the plate energy $E_{w,\gamma}(t)$ be defined as

$$E_{w,\gamma}(t) = |w_t(t)|^2_{0,\Gamma_0} + \gamma|\nabla w_t(t)|^2_{0,\Gamma_0} + a(w(t), w(t)) + |\theta(t)|^2_{0,\Gamma_0}, \quad E_w(t) \equiv E_{w,0}(t). \tag{4.4.97}$$

The main result of this section is the following recovery estimate for the energy of thermoelastic structure.

Lemma 4.4.5. *With respect to the thermoelastic component of the model* (4.4.75), *with* $\gamma > 0$, *and* z *a solution to* (4.4.74), *the following inequality holds. For all* $\epsilon > 0, \exists C_{\gamma,T,\epsilon}$, $g > 0$ *such that*

$$\begin{aligned}\int_0^T [a(w,w) + |w_t|^2_{0,\Gamma_0} + \gamma|\nabla w_t|^2_{0,\Gamma_0)}]\, dt &\leq \int_0^T E_{w,\gamma}(t)\, dt \\ &\leq \epsilon C_\gamma \left[E_{w,\gamma}(0) + E_{w,\gamma}(T) + |z(0)|^2_{1,\Omega} + |z_t(0)|^2_{0,\Omega}\right] \\ &+ C_{T,\gamma,\epsilon}\left[\int_0^T |\theta|^2_{1,\Gamma_0} + |g(z_t)|^2_{0,\Gamma_1} dt + lot_\gamma(w,\theta) + lot(z)\right],\end{aligned} \tag{4.4.98}$$

where $lot(z)$ *are given below* (4.4.92) *and for* $\delta > 0$ *we denote*

$$lot_\gamma(w,\theta) \leq C_\delta \sup_{t\in[0,T]} [|w(t)|^2_{2-\delta,\Gamma_0} + \gamma|w_t(t)|^2_{1-\delta,\Gamma_0} + |w_t(t)|^2_{-\delta,\Gamma_0} + |\theta(t)|^2_{-1/4,\Gamma_0}],$$

With respect to the thermoelastic component of the model (4.4.75), *with* $\gamma = 0$, *and* w *satisfying* (4.4.74), *the following inequality holds. For all* $\epsilon > 0, \exists C_{T,\epsilon}, C > 0$ *such that*

$$\begin{aligned}\int_0^T \left[a(w,w) + |w_t|^2_{0,\Gamma_0}\right] dt &\leq \int_0^T E_w(t)\, dt \\ &\leq \epsilon C\left[E_w(0) + E_w(T) + |z(0)|^2_{1,\Omega} + |z_t(0)|^2_{0,\Omega}\right] \\ &+ C_{T,\epsilon}\left[\int_0^T [|\theta|^2_{1,\Gamma_0} + |g(z_t)|^2_{0,\Gamma_1} + |z_t|^2_{0,\Gamma_1}]\, dt + lot(w,\theta) + lot(z)\right],\end{aligned} \tag{4.4.99}$$

where

$$lot(w,\theta) \leq C_\delta \sup_{t\in[0,T]} [|w(t)|^2_{2-\delta,\Gamma_0} + |w_t(t)|^2_{-\delta,\Gamma_0} + |\theta(t)|^2_{-1/4,\Gamma_0}], \qquad \delta > 0.$$

The constants $C_{T,\gamma,\epsilon}$ are well defined for all $\gamma \geq 0$ but not necessarily uniformly bounded when $\gamma \to 0$.

Proof of Lemma 4.4.5 *(sketch).* We introduce some notation. Let $A_D u \equiv -\Delta u$; $D(A_D) = H^2(\Gamma_0) \times H_0^1(\Gamma_0)$. Let the Dirichlet map D be defined as follows:

$$D : L_2(\partial\Gamma_0) \to L_2(\Gamma_0), \quad \Delta Dg = 0 \text{ in } \Gamma_0, \quad Dg = g \text{ on } \partial\Gamma_0,$$

and let γ_0 denote a restriction (trace) operator to $\partial\Gamma_0$. With the above notation one can write

$$\begin{aligned} \Delta\theta = \Delta\theta - \Delta D\gamma_0\theta &= \Delta(\theta - D\gamma_0\theta) \\ = -A_D(\theta - D\gamma_0\theta) &\text{ in } [H_0^1(\Gamma_0)]', \quad \theta \in H^1(\Gamma_0), \\ \Delta w = -A_D(w - D\gamma_0 w) &\text{ in } L_2(\Gamma_0), \quad w \in H^2(\Gamma_0). \end{aligned} \tag{4.4.100}$$

We apply the technique introduced in [5, 6]. We multiply the first equation in (4.4.75) by $A_D^{-1}\theta$ and integrate from 0 to T to obtain

$$\int_0^T [(w_{tt} - \gamma\Delta w_{tt} + \Delta^2 w + \Delta\theta + z_t, A_D^{-1}\theta)]_{L^2(\Gamma_0)}\, dt = 0. \tag{4.4.101}$$

Rather tedious calculations based on several applications of Green's formula (details are given in [114]) yield the following.

Proposition 4.4.2. *For any $T > 0$ and ε, ϵ_1 small enough there exist positive constants $C, C_\epsilon, C_{\epsilon_i}$ so that for $\gamma > 0$,*

$$\begin{aligned} (1 - 2\varepsilon)\int_0^T [|w_t|_{0,\Gamma_0}^2 + \gamma|\nabla w_t|_{0,\Gamma_0}^2]\, dt \leq \epsilon_1 C[E_{w,\gamma}(0) + E_{w,\gamma}(T) \\ + |z(0)|_{1,\Omega}^2 + |z_t(0)|_{0,\Omega}^2] + C_{\epsilon,T}\int_0^T |\theta|_{1,\Gamma_0}^2\, dt + \varepsilon\int_0^T |w|_{2,\Gamma_0}^2\, dt \\ + C\int_0^T a(D\gamma_0 w, w)\, dt + C\int_0^T |g(z_t)|_{0,\Gamma_1}^2 + |z_t|_{0,\Gamma_1}^2 dt + C_{T,\epsilon_1,\gamma}[lot(z) + lot_\gamma(w,\theta)], \end{aligned} \tag{4.4.102}$$

and for $\gamma = 0$,

$$\begin{aligned} (1 - 2\varepsilon)\int_0^T |w_t|_{0,\Gamma_0}^2\, dt \leq \epsilon_1 C\left[E_w(0) + E_w(T) + |z(0|_{1,\Omega}^2 + |z_t(0)|_{0,\Omega}^2\right] \\ + C_{\epsilon,T}\int_0^T |\theta|_{1,\Gamma_0}^2\, dt + \varepsilon\int_0^T |w|_{2,\Gamma_0}^2\, dt + C\int_0^T a(D\gamma_0 w, w)\, dt \\ + C_{T,\epsilon_1}[lot(z) + lot(w,\theta) + \int_0^T |g(z_t)|_{0,\Gamma_1}^2 + |z_t|_{0,\Gamma_1}^2 dt]. \end{aligned} \tag{4.4.103}$$

The main technical issue to be tackled now is the estimate for the term $\int_0^T a(D\gamma_0 w, w)\, dt$ in (4.4.102). It is at this point where the sharp regularity for traces of solutions to plate equations is used.

Proposition 4.4.3. *Let $\gamma > 0$. For all $\epsilon, \epsilon_0 > 0$, there exist constants $C_\epsilon, C_{\epsilon_0,\epsilon}$ such that for $\gamma > 0$,*

$$\int_0^T a(D\gamma_0 w, w)\,dt \le \epsilon \int_0^T |w|^2_{2,\Omega} dt + C_{T,\gamma,\epsilon} \int_0^T |\theta|^2_{1,\Gamma_0} dt + \epsilon_0 C_{T,\gamma} E_{\gamma,w}(0) + C_{T,\gamma,\epsilon,\epsilon_0}[lot_\gamma(w,\theta) + lot(z)], \quad (4.4.104)$$

and for $\gamma = 0$,

$$\int_0^T a(D\gamma_0 w, w)\,dt$$
$$\le \epsilon \left[E_w(0) + E_w(T) + |z(0|^2_{1,\Omega} + |z_t(0)|^2_{0,\Omega} + \int_0^T |w_t|^2_{0,\Gamma_0} dt \right] + C_{T,\epsilon} \left[lot(w,\theta) + lot(z) + \int_0^T |g(z_t)|^2_{0,\Gamma_0} dt \right]. \quad (4.4.105)$$

Proof. The proof follows through several steps. We begin with the case $\gamma > 0$. We write $w \equiv w_1 + w_2$, where w_1 corresponds to the plate equation (4.4.75) with the zero initial data and without z_t term. The sharp regularity result in Lemma 4.4.2 applied to the Kirchhoff plate gives

$$|w_1(t)|^2_{2,\Gamma_0} + |w_{1t}|^2_{1,\Gamma_0} \le C_{T,\gamma} \int_0^T [|\Delta\theta|^2_{-1,\Gamma_0} + |\theta|^2_{1/2,\partial\Gamma_0}]\,dt \le C_{T,\gamma} \int_0^T |\theta|^2_{1,\Gamma_0} dt, \quad (4.4.106)$$

where we used the simple computations

$$|(\Delta\theta, \psi)_{\Gamma_0}| = \left| \left(\frac{\partial}{\partial\nu}\theta, \phi \right)_{\partial\Gamma_0} - (\nabla\theta, \nabla\phi)_{\Gamma_0} \right| \le C|\phi|_{1,\Gamma_0} |\theta|_{1,\Gamma_0}$$

together with the boundary conditions satisfied for θ variable and trace theorem.

Therefore, by regularity of the Dirichlet map $D \in \mathcal{L}(H^{3/2}(\partial\Gamma_0) \to H^2(\Gamma_0))$, and trace theory

$$\int_0^T a(D\gamma_0 w_1, w_1)\,dt \le C \int_0^T |w_1|^2_{2,\Gamma_0} dt \le C_{T,\gamma} \int_0^T |\theta|^2_{1,\Gamma_0} dt, \quad (4.4.107)$$

which gives the right estimate for the first component w_1. As for the second component w_2, we notice that w_2 satisfies the homogeneous Kirchhoff plate equation with the nonzero initial data and the forcing term $z_t|_{\Gamma_0}$. Thus, the sharp regularity result of Lemma 4.4.3 applies and gives

$$\int_0^T |w_2|^2_{3/2+\rho,\partial\Gamma_0} dt \le C_{T,\gamma} \left[|w(0)|^2_{2,\Gamma_0} + |w_t(0)|^2_{1,\Gamma_0} + \int_0^T |z|^2_{\rho,\Gamma_0} dt + \sup_{t\in[0,T]} |z(t)|^2_{-1+\rho,\Gamma_0} \right]$$
$$\le C_{T,\gamma} \left[|w(0)|^2_{2,\Gamma_0} + |w_t(0)|^2_{1,\Gamma_0} + \int_0^T |z|^2_{\rho+1/2,\Omega} dt + \sup_{t\in[0,T]} |z(t)|^2_{\rho+1/2,\Omega} \right]$$
$$\le C_{T,\gamma}[|w(0)|^2_{2,\Gamma_0} + |w_t(0)|^2_{1,\Gamma_0} + lot(z)]. \quad (4.4.108)$$

Trace theory, regularity of the Dirichlet map, the moment inequality, and the estimate in (4.4.108) above yield

$$\int_0^T a(D\gamma_0 w_2, w_2)\, dt \le \int_0^T |\gamma_0 w_2|_{3/2,\partial\Gamma_0} |w_2|_{2,\Gamma_0} dt$$
$$\le \epsilon \int_0^T |w_2|^2_{2,\Gamma_0} dt + C_\epsilon \int_0^T |w_2|^2_{3/2,\partial\Gamma_0} dt$$
$$\le \epsilon \int_0^T |w_2|^2_{2,\Gamma_0} dt + \epsilon_0 \int_0^T |w_2|^2_{3/2+\rho,\partial\Gamma_0} dt + C_{\epsilon,\epsilon_0} \int_0^T |w_2|^2_{1/2,\partial\Gamma_0} dt. \quad (4.4.109)$$

But, by (4.4.107),

$$\int_0^T |w_2|^2_{1/2,\partial\Gamma_0} dt \le C \int_0^T |w_2|^2_{1,\Gamma_0} \le C \int_0^T |w_1|^2_{1,\Gamma_0} + C \int_0^T |w|^2_{1,\Gamma_0}$$
$$\le C_{T,\gamma} \int_0^T |\theta|^2_{1,\Gamma_0} dt + C_T lot_\gamma(w,\theta). \quad (4.4.110)$$

Collecting these estimates

$$\int_0^T a(D\gamma_0 w_2, w_2)\, dt \le \epsilon \int_0^T |w_2|^2_{2,\Gamma_0} dt + \epsilon_0 C_{T,\gamma} E_{\gamma,w}(0)$$
$$+ C_{T,\epsilon,\epsilon_0} lot_\gamma(w,\theta) + C_{T,\gamma,\epsilon,\epsilon_0} \left[\int_0^T |\theta|^2_{1,\Gamma_0} dt + lot(z) \right]. \quad (4.4.111)$$

Combining (4.4.107) and (4.4.111) gives the desired result in Proposition 4.4.3 for $\gamma > 0$. The argument for $\gamma = 0$ is given next. We apply Lemma 4.4.5 and interpolation inequality

$$\int_0^T a(D\gamma_0 w, w)\, dt \le C \int_0^T |w|^2_{2,\Gamma_0} dt \le \epsilon \int_0^T |w|^2_{3,\Gamma_0} dt + C_\epsilon \int_0^T |w|^2_{1,\Gamma_0} dt$$
$$\le \epsilon C_T \left[E_w(0) + |z(0)|^2_{1,\Omega} + |z_t(0)|^2_{0,\Omega} + \int_0^T |z_t|^2_{-1,\Gamma_0} \right] + C_\epsilon \int_0^T |w|^2_{1,\Gamma_0} dt.$$
$$(4.4.112)$$

Applying next the second inequality in Lemma 4.4.1 to estimate z_t on the interface gives

$$\int_0^T |z_t|^2_{-1,\Gamma_0} dt \le C_T \left[|z(0)|^2_{1,\Omega} + |z_t(0)|^2_{0,\Omega} + \int_0^T \left| \frac{\partial}{\partial \nu} z \right|^2_{0,\Gamma} dt \right]$$

$$\le C_T[|z(0)|^2_{1,\Omega} + |z_t(0)|^2_{0,\Omega} + \int_0^T [|g(z_t)|^2_{0,\Gamma_1} + |w_t|^2_{0,\Gamma_0}]\, dt + lot(z)]. \tag{4.4.113}$$

Combining these two estimates gives

$$\int_0^T a(D\gamma_0 w, w)\, dt \le \epsilon C_T \left[E_w(0) + |z(0)|^2_{1,\Omega} + |z_t(0)|^2_{0,\Omega} + \int_0^T |w_t|^2_{0,\Gamma_0} dt \right]$$
$$+ C_{T,\epsilon} \left[\int_0^T |g(z_t)|^2_{0,\Gamma_1} dt + lot(w,\theta) + lot(z) \right]. \tag{4.4.114}$$

This gives the desired inequality for the case $\gamma = 0$. □

From Propositions 4.4.2 and 4.4.3 after taking ϵ_0, ϵ suitably small so that $\epsilon_0 C_{T,\gamma}, \epsilon C_T \le \epsilon_1 C_\gamma$ we obtain the following.

Proposition 4.4.4. *Let $\gamma > 0$. For all sufficiently small $\epsilon, \epsilon_1 \exists C_{\gamma,T,\epsilon,\epsilon_1}$ such that*

$$\int_0^T [|w_t|^2_{0,\Gamma_0} + \gamma |\nabla w_t|^2_{0,\Gamma_0}]\, dt$$
$$\le \epsilon_1 C_\gamma [E_{w,\gamma}(0) + E_{w,\gamma}(T) + |z(0)|^2_{1,\Omega} + |z_t(0)|^2_{0,\Omega}] + \epsilon \int_0^T |w|^2_{2,\Gamma_0}$$
$$+ C_{T,\gamma,\epsilon,\epsilon_1} \left[\int_0^T |\theta|^2_{1,\Gamma_0} + |g(z_t)|^2_{0,\Gamma_1} dt + lot(w,\theta) + lot(z) \right]. \tag{4.4.115}$$

Let $\gamma = 0$. For all sufficiently small $\epsilon, \epsilon_1, \exists C_{T,\epsilon,\epsilon_1}$ such that

$$\int_0^T |w_t|^2_{0,\Gamma_0} dt \le \epsilon_1 [E_w(0) + E_w(T)$$
$$+ |z(0)|^2_{1,\Omega} + |z_t(0)|^2_{0,\Omega}] + \epsilon \int_0^T |w|^2_{2,\Gamma_0}$$
$$+ C_{T,\epsilon,\epsilon_1} \left[\int_0^T [|\theta|^2_{1,\Gamma_0} + |g(z_t)|^2_{0,\Gamma_1} + |z_t|^2_{0,\Gamma_1}] dt + lot(w,\theta) + lot(z) \right]. \tag{4.4.116}$$

Next, we multiply the same equation of (4.4.75) by w and integrate from 0 to T to obtain

$$(w_t, w)_{L^2(\Gamma_0)}|_0^T + \gamma (\nabla w_t, \nabla w)_{L^2(\Gamma_0)}|_0^T - \left[\int_0^T [|w_t|^2_{0,\Gamma_0} + \gamma |\nabla w_t|^2_{0,\Gamma_0}]\, dt \right]$$
$$= -\int_0^T a(w,w)\, dt + \int_0^T (\nabla\theta, \nabla w)_{L^2(\Gamma_0)}\, dt - \int_0^T \left(\theta, \frac{\partial}{\partial \nu} w \right)_{\partial\Gamma_0} dt$$
$$- \int_0^T (z_t, w)_{L^2(\Gamma_0)}\, dt.$$

Taking norms and using the trace theorem gives that

$$\int_0^T a(w,w)\,dt$$

$$\leq \left| (w_t, w)_{L^2(\Gamma_0)}|_0^T + \gamma(\nabla w_t, \nabla w)_{L^2(\Gamma_0)}|_0^T - \left[\int_0^T [|w_t|_{0,\Gamma_0}^2 + \gamma|\nabla w_t|_{0,\Gamma_0}^2]\,dt \right] \right|$$
$$+ \left| \int_0^T (\nabla\theta, \nabla w)_{L^2(\Gamma_0)}\,dt \right| + \left| \int_0^T \left(\theta, \frac{\partial}{\partial\nu} w\right)_{L_2(\partial\Gamma_0)} dt \right| - \left| \int_0^T (z_t, w)_{L^2(\Gamma_0)}\,dt \right|.$$

Direct computations yield

$$|(w_t, w)_{L^2(\Gamma_0)}|_0^T + \gamma(\nabla w_t, \nabla w)_{L^2(\Gamma_0)}|_0^T \leq \epsilon[|w_t|_{0,\Gamma_0}^2 + \gamma|\nabla w_t|_{0,\Gamma_0}^2]|_0^T$$
$$+ C_\epsilon[|w|_{0,\Gamma_0}^2 + \gamma|\nabla w|_{0,\Gamma_0}^2]|_0^T \leq \epsilon[E_{w,\gamma}(0) + E_{w,\gamma}(T)] + C_\epsilon lot(w,\theta),$$

$$\int_0^T (z_t, w)_{\Gamma_0} = (z, w)_{\Gamma_0}|_0^T - \int_0^T (z, w_t)_{\Gamma_0} dt$$
$$\leq |z(t)|_{1-\delta,\Omega}|w(t)|_{0,\Gamma_0}|_0^T + C\int_0^T |z|_{1-\delta,\Omega}|w_t|_{0,\Gamma_0} dt$$
$$\leq C_{T,\delta}[lot(z) + lot(w,\theta)] + C\int_0^T |w_t|_{0,\Gamma_0}^2 dt, \tag{4.4.117}$$

where the value of δ is less than $1/2$.

Hence, for all $\gamma \geq 0$,

$$\int_0^T a(w,w)\,dt \leq \epsilon_0[E_{w,\gamma}(0) + E_{w,\gamma}(T)] + C_1\int_0^T [\|w_t\|_{L^2(\Gamma_0)}^2 + \gamma\|\nabla w_t\|_{L^2(\Gamma_0)}^2]\,dt$$
$$+ \varepsilon\int_0^T |w|_{2,\Gamma_0}^2\,dt + C_\varepsilon\int_0^T |\theta|_{1,\Gamma_0}^2 dt + C_{T,\epsilon_0} lot(w,\theta) + C_{T,\epsilon_0} lot(z).$$

Thus, we have that for all $\epsilon, \epsilon_1 > 0$ there exist suitable constants $C_{\epsilon,\epsilon_1} > 0$ such that for all $\gamma \geq 0$,

$$(1-\varepsilon)\int_0^T a(w,w)\,dt \leq \epsilon_1 C\big[E_{w,\gamma}(0) + E_{w,\gamma}(T)\big]$$
$$+ C_1\int_0^T [|w_t|_{0,\Gamma_0}^2 + \gamma|\nabla w_t|_{0,\Gamma_0}^2]\,dt$$
$$+ C_\epsilon\int_0^T |\theta|_{1,\Gamma_0}^2\,dt + C_{T,\epsilon_1}[lot(w,\theta) + lot(z)]. \tag{4.4.118}$$

If the ε of (4.4.115) and (4.4.118) are small enough, they can be combined to produce the inequality

$$\int_0^T [|w|_{2,\Gamma_0}^2 + |w_t|_{0,\Gamma_0}^2 + \gamma|\nabla w_t|_{0,\Gamma_0}^2]\,dt \leq \epsilon_1 C_\gamma[E_{w,\gamma}(0) + E_{w,\gamma}(T)]$$
$$+ C_{T,\gamma,\epsilon_1}\left[\int_0^T |\theta|_{1,\Gamma_0}^2 + |g(z_t)|_{0,\Gamma_1}^2\,dt + lot_\gamma(w,\theta) + lot(z)\right],$$

and for $\gamma = 0$,

$$\int_0^T \left[|w|^2_{2,\Gamma_0} + |w_t|^2_{0,\Gamma_0}\right] dt \leq \epsilon_1 C \left[E_{w,\gamma}(0) + E_{w,\gamma}(T) + |z(0)|^2_{1,\Omega} + |z_t(0)|^2_{0,\Omega}\right]$$
$$+ C_{T,\epsilon_1} \left[\int_0^T [|\theta|^2_{1,\Gamma_0} + |g(z_t)|^2_{0,\Gamma_1}\, dt + lot(w,\theta) + lot(z)\right],$$

which is the desired result in Lemma 4.4.5 for $\gamma \geq 0$. □

4.4.5 Wave equation

Let $E_z(t)$ be the energy defined by

$$E_z(t) = |z_t(t)|_{L^2(\Omega)} + |\nabla z(t)|_{L^2(\Omega)} + d|z(t)|^2_{0,\Gamma_2}. \tag{4.4.119}$$

The recovery estimate for the wave equation is given below.

Lemma 4.4.6. *Consider the wave equation with the boundary conditions*

$$\begin{aligned} z_{tt} &= \Delta z \quad in \quad \Omega \times (0,\infty),\\ \frac{\partial}{\partial\nu} z + dz &= -g(z_t) \quad on \quad \Gamma_1 \times (0,\infty),\\ \frac{\partial}{\partial\nu} z &= w_t \quad on \quad \Gamma_0 \times (0,\infty), \end{aligned} \tag{4.4.120}$$

where w is a solution to (4.4.75). Then, for any $\alpha < T/2$ there exist positive constants C, possibly depending on α, such that for $\gamma > 0$,

$$\begin{aligned}\int_\alpha^{T-\alpha} E_z(t)\, dt \leq C\left[E_z(\alpha) + E_z(T-\alpha) + E_z(0)\right]\\ + C_T \int_0^T [|z_t|^2_{0,\Gamma_1} + |g(z_t)|^2_{0,\Gamma_1}]\, dt\\ + C_{\gamma,T}[lot(z) + lot_\gamma(w,\theta)],\end{aligned} \tag{4.4.121}$$

and for $\gamma = 0$,

$$\begin{aligned}\int_\alpha^{T-\alpha} E_z(t)\, dt \leq C\left[E_z(\alpha) + E_z(T-\alpha) + E_z(0) + E_w(0)\right]\\ + C_T \int_0^T [|z_t|^2_{0,\Gamma_1} + |g(z_t)|^2_{0,\Gamma_1} + |w_t|^2_{0,\Gamma_0}]\, dt\\ + C_T[lot(z) + lot(w,\theta)].\end{aligned} \tag{4.4.122}$$

Proof. In the first step of the proof we use the multipliers method. Applicability of this method rests on the assumption, as usual, that the solutions are sufficiently regular to perform standard differential calculus. In the nonlinear case, such solutions may not display enough regularity (although the initial data are taken sufficiently smooth). However, a regularization argument proposed in [120] allows us to circumvent the problem by constructing an appropriate regularization for the wave equation to which standard calculus applies. We do not repeat this argument here but only mention that the calculus performed below should be applied to the regularized wave equation for which the estimates in (4.4.121) and (4.4.122) are obtained. Passage through the limit (with the regularization parameter) allows us to reconstruct (see section 2.3.2) these inequalities for the original problem. The details of this limiting argument are given in [114].

As a first step we use the standard multipliers method. We take $(x - x_0) \cdot \nabla z$, with $x_0 \in \Gamma_0$ and z, as the multipliers in (4.4.120). Noting that $(x - x_0) \cdot \nu = 0$, on Γ_0, and performing standard, by now, calculations lead to

$$\begin{aligned} \int_s^T E_z(t)\,dt &\le C[E_z(s) + E_z(T)] + C\int_s^T \int_{\Gamma_1} [z_t^2 + g^2(z_t) + z^2] d\Gamma_1 dt \\ &+ C\int_s^T |w_t|^2_{0,\Gamma_0} dt + C\int_s^T \left|\frac{\partial}{\partial \tau} z\right|^2_{0,\Gamma_1} dt + C\int_s^T \left|\left(w_t, \frac{\partial}{\partial \tau} z\right)_{\Gamma_0}\right| dt. \end{aligned} \tag{4.4.123}$$

The terms that need to be further estimated are the two last boundary terms in (4.4.123), which involve tangential derivatives and are not controlled by the energy norms. To accomplish this we use the result of Lemma 4.4.1. This gives

$$\begin{aligned} \int_\alpha^{T-\alpha} \left|\frac{\partial}{\partial \tau} z\right|^2_{0,\Gamma_1} dt &\le C_T \int_0^T \left[|z_t|^2_{0,\Gamma_1} + \left|\frac{\partial}{\partial \nu} z\right|^2_{0,\Gamma}\right] dt + C_T lot(z) \\ &\le C_T \left[\int_0^T [|z_t|^2_{0,\Gamma_1} + |g(z_t)|^2_{0,\Gamma_1} + |w_t|^2_{0,\Gamma_0}]\,dt + lot(z)\right], \end{aligned} \tag{4.4.124}$$

which gives the estimate for the first tangential term in (4.4.123). The estimate for the second one is

$$\begin{aligned} \int_0^T \left(w_t, \frac{\partial}{\partial \tau} z\right)_{\Gamma_0} dt &\le \epsilon \int_0^T \left|\frac{\partial}{\partial \tau} z\right|^2_{-1/2,\Gamma_0} dt + C_\epsilon \int_0^T |w_t|^2_{1/2,\Gamma_0} dt \\ &\le \epsilon \int_0^T |z(t)|^2_{1,\Omega} dt + C_\epsilon \int_0^T |w_t|^2_{1/2,\Gamma_0}. \end{aligned} \tag{4.4.125}$$

Hence

$$\int_0^T \left(w_t, \frac{\partial}{\partial \tau} z\right)_{\Gamma_0} dt \le \epsilon \int_0^T E_z(t)\,dt + C_{\epsilon,\gamma} lot_\gamma(w, \theta), \tag{4.4.126}$$

which gives the estimate for $\gamma > 0$. For $\gamma = 0$, the main point is to estimate the last integral on the right-hand side of (4.4.125), which no longer contributes the lower-order terms. By the second statement in Lemma 4.4.4 applied with $\delta = 1/2$ we obtain

$$\int_0^T |w_t|^2_{1/2,\Gamma_0} dt \le \epsilon_0[E_w(0) + E_z(0)] + C_{\epsilon_0,T}[lot(w,\theta) + lot(z)] + C\int_0^T |g(z_t)|^2_{0,\Gamma_1} dt. \tag{4.4.127}$$

Hence

$$C_\epsilon \int_0^T |w_t|^2_{1/2,\Gamma_0} dt \le \epsilon_0 C_\epsilon[E_w(0) + E_z(0)] + C_{\epsilon_0,\epsilon,T}\left[lot(w,\theta) + lot(z) + \int_0^T |g(z_t)|^2_{0,\Gamma_1} dt\right]$$
$$\le C[E_w(0) + E_z(0)] + C_{T,\epsilon}\left[lot(w,\theta) + lot(z) + \int_0^T |g(z_t)|^2_{0,\Gamma_1} dt\right], \tag{4.4.128}$$

where we have made a suitable choice of $\epsilon_0 = \epsilon_0(T, \epsilon(T)) = \epsilon_0(T)$. From (4.4.125) and the above inequality we obtain for $\gamma = 0$

$$\int_0^T \left(w_t, \frac{\partial}{\partial\tau} z\right)_{\Gamma_0} dt \le C[E_z(0) + E_w(0)] + \epsilon\int_0^T E_z(t)\, dt + C_{T,\epsilon}\left[lot(w,\theta) + lot(z) + \int_0^T |g(z_t)|^2_{0,\Gamma_1} dt\right]. \tag{4.4.129}$$

Collecting (4.4.126) (for $\gamma > 0$) and (4.4.129) (for $\gamma = 0$) and combining with (4.4.123) applied with s replaced by α and T replaced by $T - \alpha$ and using classical trace theory to absorb $\int_{\Gamma_1} z^2 dx$ yields, after taking ϵ sufficiently small, for all $\gamma > 0$,

$$\int_\alpha^{T-\alpha} E_z(t)\, dt \le C[E_z(\alpha) + E_z(T-\alpha) + E_z(0)]$$
$$\le C_T\left[\int_0^T [|z_t|^2_{0,\Gamma_1} + |g(z_t)|^2_{0,\Gamma_1} + |w_t|^2_{0,\Gamma_0}]\, dt + C_{\gamma,T}[lot(z) + lot_\gamma(w,\theta)\right], \tag{4.4.130}$$

and for $\gamma = 0$,

$$\int_\alpha^{T-\alpha} E_z(t)\, dt \le C[E_z(\alpha) + E_z(T-\alpha) + E_z(0) + E_w(0)]$$
$$\le C_T \int_0^T [|z_t|^2_{0,\Gamma_1} + |g(z_t)|^2_{0,\Gamma_1} + |w_t|^2_{0,\Gamma_0}]\, dt + C_T[lot(z) + lot(w,\theta)]. \tag{4.4.131}$$

This provides the desired conclusion in Lemma 4.4.6. □

4.4.6 Uniform stability analysis for the coupled system

In the final analysis, we combine the energy estimates obtained for the plate and wave equations and then absorb the lower terms by means of a standard compactness-uniqueness argument.

Proposition 4.4.5. *Let $T > 0$ be sufficiently large. The following estimate holds for the solution to (4.4.74) and (4.4.75). For $\gamma > 0$,*

$$\int_0^T E_\gamma(t)\,dt + E_\gamma(T) \le C_{\gamma,T} \int_0^T [|\theta|^2_{1,\Gamma_0} + |z_t|^2_{0,\Gamma_1} + |g(z_t)|^2_{0,\Gamma_1}]\,dt + C_{\gamma,T}[lot_\gamma(w,\theta) + lot(z)], \tag{4.4.132}$$

and for $\gamma = 0$,

$$\int_0^T E_\gamma(t)\,dt + E_\gamma(T) \le C_T \int_0^T [|\theta|^2_{1,\Gamma_0} + |z_t|^2_{0,\Gamma_1} + |g(z_t)|^2_{0,\Gamma_1}]\,dt + C_T[lot(w,\theta) + lot(z)]. \tag{4.4.133}$$

Proof. Here the argument is the same for $\gamma > 0$ and $\gamma = 0$. So, it suffices to provide the argument for $\gamma > 0$. We multiply (4.4.98) by a suitably large constant A_T (possibly depending on T) and add the result to (4.4.121). This gives

$$\begin{aligned} A_T \int_0^T E_{w,\gamma}(t)\,dt + \int_\alpha^{T-\alpha} E_z(t)\,dt &\le \epsilon C_\gamma A_T[E_{w,\gamma}(0) + E_{w,\gamma}(T) + E_z(0)] \\ &+ C[E_z\alpha + E_z(T-\alpha) + E_z(0) + E_w(0)] + A_T C_{T,\gamma,\epsilon} \int_0^T [|\theta|^2_{1,\Gamma_0}]\,dt \\ + C_T \int_0^T [|z_t|^2_{0,\Gamma_1} + |g(z_t)|^2_{0,\Gamma_0} &+ |w_t|^2_{0,\Gamma_0}]\,dt + A_T C_{T,\gamma,\epsilon} lot_\gamma(w,\theta) + C_T lot(z). \end{aligned} \tag{4.4.134}$$

We take $A_T > 2C_T$, which allows us to eliminate the term with $|w_t|^2_{0,\Gamma_0}$ from the right-hand side of the inequality in (4.4.134). We also select small $\epsilon = \epsilon(T,\gamma)$ so that $\epsilon C_\gamma A_T \le C_\gamma$. This gives

$$\begin{aligned} &\int_0^T E_\gamma(t)\,dt \\ &+ \int_\alpha^{T-\alpha} E_z(t)\,dt \le C_\gamma\big[E_{w,\gamma}(0) + E_{w,\gamma}(T) + E_z(\alpha) + E_z(T-\alpha) + E_z(0)\big] \\ &+ C_{\gamma,T} \int_0^T (|\theta|^2_{H^1(\Gamma_0)} + |z_t|^2_{L^2(\Gamma_1)} + |g(z_t)|^2_{0,\Gamma_1})\,dt + C_{T,\gamma}[lot_\gamma(w,\theta) + lot(z)]. \end{aligned}$$

Hence

$$\int_\alpha^{T-\alpha} E_\gamma(t)\,dt \le C_\gamma\big[E_\gamma(0) + E_\gamma(T) + E_\gamma(\alpha) + E_\gamma(T-\alpha)\big] \tag{4.4.135}$$

$$+ C_{\gamma,T} \int_0^T (|\theta|^2_{H^1(\Gamma_0)} + |z_t|^2_{L^2(\Gamma_1)} + |g(z_t)|^2_{0,\Gamma_1})\,dt + C_{T,\gamma}[lot_\gamma(w,\theta) + lot(z)]. \tag{4.4.136}$$

Here, we recall, the energy $E_\gamma(t)$ is defined as in (4.4.96). Our next step is to use dissipativity of the energy to eliminate the terms with α. By using the energy identity (4.4.95) in Proposition 4.4.1 and the simple inequality

$$\left(\int_0^\alpha + \int_{T-\alpha}^T\right) E_\gamma(t)\,dt \le 2\alpha E_\gamma(0),$$

we obtain

$$\int_0^T E_\gamma(t)\,dt \le C_\gamma\big[E_\gamma(0)+E_\gamma(T)\big]$$
$$+\,C_{T,\gamma}\int_0^T (|\theta|^2_{H^1(\Gamma_0)}+|z_t|^2_{L^2(\Gamma_1)}+|g(z_t)|^2_{0,\Gamma_1})\,dt$$
$$+\,C_{T,\gamma}[lot_\gamma(w,\theta)+lot(z)]. \tag{4.4.137}$$

Using once more the energy relation gives for $t \le T$

$$E_\gamma(t) \ge E_\gamma(T) - C_{T,\gamma}\int_0^T (|\theta|^2_{H^1(\Gamma_0)}+|z_t|^2_{L^2(\Gamma_1)}+|g(z_t)|^2_{0,\Gamma_1})\,dt.$$

Hence

$$TE_\gamma(T)\,dt \le C_\gamma E_\gamma(T)$$
$$+\,C_{T,\gamma}\int_0^T (|\theta|^2_{H^1(\Gamma_0)}+|z_t|^2_{L^2(\Gamma_1)}+|g(z_t)|^2_{\Gamma_1})\,dt$$
$$+\,C_{T,\gamma}[lot_\gamma(w,\theta)+lot(z)]. \tag{4.4.138}$$

Combining (4.4.138) and (4.4.137) and taking $T > C_\gamma$ leads to the desired conclusion in Proposition 4.4.5. □

Our next step is to eliminate the lower-order terms from the inequality in Proposition 4.4.5. This is done via the usual compactness-uniqueness argument.

Proposition 4.4.6. *Let T be large enough (with a value depending on γ). With respect to the coupled PDE system* (4.4.74), (4.4.75) *there exists a constant $C_{\gamma,T} > 0$ (resp., C_T) such that for $\gamma > 0$,*

$$lot_\gamma(w) + lot(z) \le C_{\gamma,T}\int_0^T (|z_t|^2_{0,\Gamma_1}+|g(z_t)|^2_{0,\Gamma_1}+|\theta|^2_{1,\Gamma_0})\,dt, \tag{4.4.139}$$

and for $\gamma = 0$,

$$lot(w) + lot(z) \le C_T\int_0^T (|z_t|^2_{0,\Gamma_1}+|g(z_t)|^2_{0,\Gamma_1}+|\theta|^2_{1,\Gamma_0})\,dt. \tag{4.4.140}$$

Proof. The conclusion follows by a contradiction from the usual compactness-uniqueness argument. Since this argument is standard, we do not report all the details; we will only point out the main steps. The compactness of $lot_\gamma(w,\theta)+lot(z)$ (resp., $lot(w,\theta)+lot(z)$), with respect to the topology induced by the energy E_γ, $\gamma \ge 0$, follows from the compact embeddings

$$H^{2-\epsilon}(\Gamma_0)\times H^{1-\epsilon}(\Gamma_0)\times H^{1-\epsilon}(\Omega)\times H^{-\epsilon}(\Omega)$$
$$\subset H^2(\Gamma_0)\times H^1(\Gamma_0)\times H^1(\Omega)\times L_2(\Omega), \qquad \epsilon > 0.$$

As for the uniqueness part, we deal with the following overdetermined system (here we consider only the more difficult case $\gamma > 0$):

$$\begin{aligned}
&\tilde z_{tt} = \Delta \tilde z && \text{on } [0,T]\times\Omega,\\
&\tilde z_t = 0 && \text{on } [0,T]\times\Gamma_1,\\
&\tfrac{\partial \tilde z}{\partial\nu} = \tilde w_t,\ \tilde\theta = 0 && \text{on } [0,T]\times\Gamma_0,\\
&\tfrac{\partial}{\partial\nu}\tilde z + d\tilde z = 0 && \text{on } [0,T]\times\Gamma_1.
\end{aligned} \tag{4.4.141}$$

Since $\tilde{z}_t = 0$, $\frac{\partial}{\partial\nu}\tilde{z}_t + d\tilde{z}_t = 0$, on $\Gamma_1 \times (0,T)$, a version of the Holmgren uniqueness theorem applies (see Theorem 3.5 in [86]) to conclude $\tilde{z}_t \equiv 0$. Feeding back this information into the plate equation yields the following overdetermined system for the variable $\tilde{w}$:

$$\begin{array}{ll} \tilde{w}_{tt} - \gamma\Delta\tilde{w}_{tt} + \Delta^2\tilde{w} = 0 & \text{on } [0,T]\times\Gamma_0, \\ \tilde{\theta} \equiv 0, \ \Delta\tilde{w}_t = 0 & \text{on } [0,T]\times\Gamma_0, \\ \Delta\tilde{w} + (1-\mu)B_1\tilde{w} = 0 & \text{on } [0,T]\times\partial\Gamma_0, \\ \frac{\partial}{\partial\nu}\Delta\tilde{w} + (1-\mu)B_2\tilde{w} - \gamma\frac{\partial}{\partial\nu}\tilde{w}_{tt} = 0 & \text{on } [0,T]\times\partial\Gamma_0, \end{array}$$

$$\{\tilde{z}(0), \tilde{z}_t(0), \tilde{w}(0), \tilde{w}_t(0), \tilde{\theta}(0)\} = \{\tilde{z}_0, \tilde{z}_1, \tilde{w}_0, \tilde{w}_1, \tilde{\theta}_0\} \in Y.$$

Our aim is to show that the above overdetermined system admits only zero solution. Since also

$$\tilde{w}_{ttt} = \gamma\Delta\tilde{w}_{ttt} - \Delta^2\tilde{w}_t \equiv 0, \quad \text{on } [0,T]\times\Gamma_0,$$

we obtain

$$\begin{array}{ll} \Delta^2\tilde{w}_t = 0 & \text{on } [0,T]\times\Gamma_0, \\ \tilde{\theta} \equiv 0 & \text{on } [0,T]\times\Gamma_0, \\ \Delta\tilde{w}_t + (1-\mu)B_1\tilde{w}_t = 0 & \text{on } [0,T]\times\partial\Gamma_0, \\ \frac{\partial}{\partial\nu}\Delta\tilde{w}_t + (1-\mu)B_2\tilde{w}_t = 0 & \text{on } [0,T]\times\partial\Gamma_0. \end{array}$$

Therefore, by the uniqueness of solutions to elliptic equations, $\tilde{w}_t \equiv 0$. This reduces the entire problem to the following static equations:

$$\begin{array}{ll} \Delta^2\tilde{w} = 0 & \text{on } [0,T]\times\Gamma_0, \\ \tilde{\theta} \equiv 0 & \text{on } [0,T]\times\Gamma_0, \\ \Delta\tilde{w} + (1-\mu)B_1\tilde{w} = 0 & \text{on } [0,T]\times\partial\Gamma_0, \\ \frac{\partial}{\partial\nu}\Delta\tilde{w} + (1-\mu)B_2\tilde{w} = 0 & \text{on } [0,T]\times\partial\Gamma_0, \end{array} \tag{4.4.142}$$

$$\begin{array}{ll} \Delta\tilde{z} = 0 & \text{on } [0,T]\times\Omega, \\ \frac{\partial}{\partial\nu}\tilde{z} = 0 & \text{on } [0,T]\times\Gamma_0, \\ \frac{\partial}{\partial\nu}\tilde{z} + d\tilde{z} = 0 & \text{on } [0,T]\times\Gamma_1. \end{array} \tag{4.4.143}$$

By the uniqueness of elliptic solutions (note that $l, d > 0$) we conclude that

$$\tilde{z} = \tilde{w} = \tilde{\theta} = 0$$

as desired. □

Completion of the proof of Theorem 4.4.2. Combining the results of Propositions 4.4.5 and Proposition 4.4.6 we obtain

$$E_\gamma(T) \le C_{T,\gamma}\int_0^T [|z_t|^2_{\Gamma_1} + |g(z_t)|^2_{\Gamma_1} + |\theta|^2_{\Gamma_0}]\,dt, \qquad \gamma > 0, \tag{4.4.144}$$

$$E_0(T) \le C_T\int_0^T [|z_t|^2_{\Gamma_1} + |g(z_t)|^2_{\Gamma_1} + |\theta|^2_{\Gamma_0}]\,dt, \qquad \gamma = 0. \tag{4.4.145}$$

By using the assumptions imposed on the nonlinear function g and splitting the region of integration into two—$z_t \le 1$ and $z_t > 1$ (as done before)—we also obtain

$$\int_0^T [|z_t|^2_{\Gamma_1} + |g(z_t)|^2_{\Gamma_1} + |\theta|^2_{\Gamma_0}]\,dt \le C_{T,m,M}[I + h_0]\int_0^T\int_{\Gamma_1} g(z_t)z_t\,dx\,dt, \tag{4.4.146}$$

where we have used Jensen's inequality. Combining (4.4.144) and (4.4.146) and recalling monotonicity of h_0 we obtain

$$E_\gamma(T) \le C_{T,\gamma,m,M}[I + h_0] \int_0^T \int_{\Gamma_1} g(z_t) z_t dx + |\nabla\theta|^2_{1,\Gamma_0} dt$$
$$= C_{T,\gamma,m,M}[I + h_0][E_\gamma(0) - E_\gamma(T)], \qquad (4.4.147)$$

where in the last step we have used the energy relation. Since $[I + h_0]$ is invertible, this gives

$$[I + h_0]^{-1}(C^{-1}_{\gamma,T,m,M} E_\gamma(T)) \le E_\gamma(0) - E_\gamma(T); \qquad (4.4.148)$$

this gives

$$p(E_\gamma(T)) + E_\gamma(T) \le E_\gamma(0)$$

with p defined by Theorem 4.4.2. The final conclusion of Theorem 4.4.2 follows now from the application of Theorem 3.2.1. The argument for $\gamma = 0$ is identical. We apply (4.4.145) instead of (4.4.144). □

4.5 Comments and Open Problems

- One may generalize the results obtained in this chapter by considering nonlinear plate equations modeling structural walls of the acoustic chamber. This will involve compilation of techniques developed in the context of nonlinear plates (see section 2.4) with those presented in the present chapter. For thermoelastic plates this was carried out in [115].
- All the results presented in this chapter remain unchanged if some internal viscous damping is added to the plate models. In fact, the goal of our analysis was not to account for potential beneficial effects of this damping. However, if such a damping is present, then only cosmetic alterations of proofs are needed to adjust the arguments that would then lead to the same decay rates in the presence of some unaccounted viscous damping.
- Similar comments apply to a potential presence of structural (Kelvin–Voight) damping added to plate models with viscous, boundary, or thermal damping. This is to say that mechanically or thermally damped plates contain an additional term, say, $\alpha(x)\Delta^2 v_t$, where $\alpha(x) \ge 0$ in Γ_0. In such cases the goal is to show that the uniform decay rates valid for the energy function are independent on $\alpha(x)$ for $\alpha \in C^1(\Omega)$. While such a result has been obtained in the case of thermal plates with hinged or clamped boundary conditions [147], the same result for thermal plates with free boundary conditions or for the plates with boundary damping is more challenging to obtain.
- An interesting aspect of the analysis is the location of the absorbing boundary conditions on Γ. Two extreme cases are of interest. The boundary damping $g(z_t)$ is placed on the interface only, i.e., flexible wall Γ_0 and not on the hard walls Γ_1. Or, the damping is placed on the hard walls only and not on the interface Γ_0.

  In the case of thermal damping affecting the plate, these two scenarios were successfully addressed in [112, 114] under the convexity assumption imposed on the undamped wall Γ_1. (If the damping is active on the entire Γ_1, there is no need for any geometrical constraints.) In fact, the analysis in [112, 114] reveals that the first scenario leads to an interesting microlocal analysis problem, while the second scenario gives rise to new geometric problems.

- In a similar vein one may ask whether it is possible to stabilize a structural acoustic model by damping only a portion of hard walls on Γ_1 (and no damping on the interface Γ_0). The first insight into this problem suggests that one needs to solve both: the geometric problem addressed in [112] and the microanalytic problem addressed in [114]. In this case, the final result would predict that slightly more than a quarter of the entire boundary Γ_1 is necessary for stabilization. This should be contrasted with a popular statement made in the context of boundary stabilization that slightly more than half the boundary is needed for stabilization [112].
- Another interesting problem, both physically and mathematically, is the following: Can we stabilize the structural acoustic model with zero Dirichlet data prescribed on a (sufficiently small) portion of the boundary Γ? The main issue here is that of regularity of solutions corresponding to appropriate static problems. As is well known, these solutions may develop singularities that prevent proper justification of the PDE estimates needed to establish the stability. While the methods developed by Grisvard are successful in dealing with a pure wave equation, it is far from being clear that the same method can be applied to the entire structural acoustic problem. In fact, so far, we are not aware of any rigorous justification of these estimates, even in a simple case of rectangular domains; see [58].

Chapter 5

Structural Acoustic Control Problems: Semigroup and PDE Models

5.1 Orientation

The next three chapters deal with the mathematical optimal control theory of linear structural acoustic problems. Physical motivation for studying this kind of problem comes from a variety of engineering applications that arise, for example, in the context of controlling the pressure in a helicopter's cabin or in reducing unwanted cabin noise generated by some exterior field. There is substantial engineering literature dealing with the practical aspects of structural acoustic problems (see [149, 52, 53, 56] and the references therein). The PDE equations describing acoustic interactions have been known for a long time (e.g., [166, 24]). However, the first mathematical (PDE) model for the underlying control problem was provided in [16, 17, 15], where the structural acoustic problem is formulated in the setting of C_0 semigroups with unbounded control operators. References [16, 15, 18] also provide several finite-dimensional numerical approximations. However, until recently, relatively little was known about the mathematical properties of these models. In particular, a PDE theory giving useful qualitative and quantitative information about these problems was lacking. This is hardly surprising, inasmuch as the mathematical control theory of PDE models under the influence of unbounded control operators (which model boundary and point control actions) was developed only recently (see, e.g., the books [129, 28, 140] and references therein).

Since the stated structural acoustic problem is described by a coupled system of PDEs composed of a parabolic equation coupled with a hyperbolic equation, it is reasonable to speculate that the already-available control theoretic results for (disparate) hyperbolic and parabolic dynamics should be of use in the course of the analysis. Indeed, the now well-developed theory for single equations [129, 28, 140] has been instrumental, particularly in the last several years, in spurring interest and initiating progress toward the development of a mathematical control theory for interactive structures. However, what distinguishes the analysis of interactive structures from that of uncoupled PDEs is the need to reconcile and understand the coupling effects in the former: distinctly new mathematical phenomena arise from coupled structures, such as the structural acoustic PDE, because the particular mode of coupling creates a situation where the behavior of one component, say, the hyperbolic component, is influenced by the other component, say, the parabolic component, and vice versa. To understand thoroughly the reciprocal effects of these component equations, one influencing the behavior of the other, techniques such as pseudodifferential operators, microlocal analysis, propagation of singularities, and analyticity must be critically invoked.

The purpose of these lectures is to expose some of the major mathematical problems and to present several new results in this area. In particular, regarding the control theory of the canonical prototype structural acoustic model, we attempt to provide an answer to the question, What is possible and achievable, and what is not? As we will see, a central role in the arguments will necessarily be played by the sharp regularity theory—particularly that of trace regularity—of the underlying boundary value problems, as well as the theory of propagation of singularities.

Two greatly distinct models—with vastly and necessarily different analyses—are considered: (i) a system with parabolic-hyperbolic coupling, and (ii) a system with hyperbolic-hyperbolic coupling. In particular, corresponding regularity results, including those of the so-called gain operators, will be drastically different for these two cases. Indeed, while in the case of parabolic-hyperbolic coupling there is a transfer of regularity properties from the parabolic component onto the entire structure, which ultimately yields a boundedness for the gain operators, such is not the case for the hyperbolic-hyperbolic coupling. This marked difference in the corresponding theories between these two cases has implications on the practical implementation of control algorithms and related numerical schemes devised for an effective computation of control laws.

The main goal of the present chapter is to provide an introduction to the study of optimal control problems, which will be extensively studied in Chapters 6 and 7. Thus, we begin here with a description of general structural acoustic control models. These include systems of coupled PDEs describing an acoustic chamber subject to an unwanted noise (acoustic pressure), with a flat elastic wall, modeled by plate equations, or curved walls, modeled by shell equations, (see [118, 116, 108, 141]). The elastic wall is assumed to be described either by (i) an Euler–Bernoulli equation subject to structural damping, e.g., Kelvin–Voight type, or (ii) a Kirchhoff plate equation, which is undamped but does account for rotational forces. While realistic physical models surely can be derived that contain combinations of various types of damping, we restrict our focus here on the two distinct situations (i) and (ii) above, to clearly illuminate the drastically different characteristics and results corresponding to each of these two cases.

In section 5.2 we provide an abstract semigroup framework that will unify the various models considered. In section 5.3 we specialize this abstract framework to the concrete PDE systems subject to specified boundary conditions. In section 5.4 we collect several uniform stability results formulated in the context of linear structural acoustic models introduced before. These results are critically used in the subsequent Chapters 6 and 7, which deal with optimal control problems for structural acoustic models.

The material reported in these notes pertains only to theoretical results associated with control theory of structural acoustic models. For an exposition of numerical methods and numerical treatment of various control algorithms we refer the reader to [16, 15, 17, 70] and the references therein.

5.2 Abstract Setting: Semigroup Formulation

A structural acoustic interaction can be described by a coupled system of equations which describe (i) the acoustic medium in a given two- or three-dimensional chamber (wave equation) and (ii) the structure (beam, plate, or shell equation) representing a flexible wall of the chamber. The interaction takes place on the boundary between the acoustic medium and the structure (flexible wall). This leads to a mathematical model of a coupled hyperbolic (wave)-parabolic (wall) equation. In this chapter we are concerned with linear models. In some cases, when the structure (wall) is modeled by higher-order approximations such as a Kirchhoff model without accounting for structural damping, the coupling is between two hyperbolic systems. It will be seen that the properties of the system and of the resulting control laws depend heavily on the type of coupling.

To give the flavor of the problem, we begin with a model that is fairly general and that encompasses a variety of more specific models discussed in detail in section 5.3.

Let $\Omega \in R^n$, $n = 2$ or 3, be an open bounded domain with boundary Γ. The boundary $\Gamma = \overline{\Gamma_0 \cup \Gamma_1}$ is connected and consists of two simply connected and nonintersecting open regions Γ_1, Γ_0. Thus, Γ_0 is a connected, bounded manifold in R^{n-1}; see Figure 1. In what follows we assume that either Γ is sufficiently smooth or Ω is convex. This assumption guarantees that solutions to classical elliptic equations with $L_2(\Omega)$ forcing terms are in $H^2(\Omega)$ [68]. The acoustic medium is described by the wave equation in the variable z, which describes the velocity potential, while the quantity ρz_t represents the acoustic pressure, where ρ denotes the density of the fluid. The structural vibrations of the elastic wall (structure) are described by the variable $\mathbf{v} = [v_1, v_2, v]$, where v_1, v_2 denote in-plane displacements and the last component of the vector $\mathbf{v}$, denoted by v, represents vertical displacement of the beam, plate, or shell. The variable v will then satisfy a suitable beam, plate, or shell equation defined on the manifold Γ_0. In most applications (beams, plates) we consider only vertical displacements, in which case $\mathbf{v}$ will coincide with scalar v.

Acoustic medium. The following PDE describes an acoustic pressure field for the model under consideration:

$$z_{tt} = c^2 \Delta z - d_1 z_t + f \text{ in } \Omega \times (0,T)$$

with the boundary conditions on Γ,

$$\frac{\partial}{\partial \nu} z + d_2 z_t = 0 \text{ on } \Gamma_1 \times (0,T),$$

$$\frac{\partial}{\partial \nu} z + d_3 z_t = v_t \text{ on } \Gamma_0 \times (0,T),$$

and the initial conditions given by

$$z(t=0) = z_0 \in H^1(\Omega), \quad z_t(t=0) = z_1 \in L_2(\Omega). \tag{5.2.1}$$

The system (5.2.1) couples variable z with variable v; the latter satisfies (5.2.2) below. The constant c^2 denotes, as usual, the speed of sound in the fluid.

The function $f \in L_2(0, T \times \Omega)$ represents a deterministic external noise entering the acoustic environment. The time T may be finite or infinite. The constants $d_i \geq 0$ introduce potential interior or boundary damping in the model. In the case of the finite horizon control problem ($T < \infty$), we may take these constants equal to zero. However, in the infinite horizon case, where stability becomes an issue, we assume that some of these constants are strictly positive.

Remark 5.2.1. *Instead of prescribing the Neumann data on Γ_1, one could assume that the boundary data on Γ_1 are Dirichlet data. However, in this case, attention must be paid to the regularity issues that arise from considering elliptic problems with mixed boundary conditions (Dirichlet and Neumann) defined on the same portion of the boundary (see [69]). To avoid this technical issue, one may assume further that $\overline{\Gamma_0} \cap \overline{\Gamma_1} = \emptyset$; the issue is nonexistent. This type of assumption is not very attractive in the context of structural acoustic models, however, inasmuch as $\overline{\Gamma_0} \cap \overline{\Gamma_1} = \emptyset$ corresponds physically to an acoustic chamber that is doughnut-like, that is, with a hole, where the external wall is rigid and the internal wall is flexible. (See the comments at the end of this chapter.)*

At this point, we represent the structural vibrations of the elastic wall by an abstract second-order equation—to govern a beam, plate or, shell—and specify this quantity later. This operator theoretic model will allow the consideration of undamped or internally damped, in particular structurally damped, plate, beam, or shell equations. The latter case arises in the modeling of Kelvin–Voight damping, which is present in structurally damped plates or beams. Kelvin–Voight damping may be achieved by selecting material with appropriate elastic properties.

Elastic wall structure. With $\mathcal{H} = [L_2(\Gamma_0)]^s$, where the values of parameters $s = 1, 2$ or 3, the general form of the equation describing the wall is

$$\begin{aligned} &\mathcal{M}_\gamma \mathbf{v}_{tt} + \alpha \mathcal{A}^{2\theta} \mathbf{v}_t + \mathcal{A}\mathbf{v} + \mathcal{C}(\rho z_t|_{\Gamma_0}) + \mathcal{A} g \mathcal{G} \mathcal{G}^* \mathcal{A} \mathbf{v}_t \\ &= \mathcal{B}u \text{ in } \mathcal{D}(\mathcal{A}^{1/2})' \times (0, T) \qquad \text{with the initial conditions} \\ &\mathbf{v}(t=0) = \mathbf{v}_0 \in \mathcal{D}(\mathcal{A}^{1/2}), \quad \mathbf{v}_t(t=0) = \mathbf{v}_1 \in \mathcal{D}({\mathcal{M}_\gamma}^{1/2}). \end{aligned} \tag{5.2.2}$$

Above, the linear operators

$$\mathcal{A} : \mathcal{D}(\mathcal{A}) \subset \mathcal{H} \to \mathcal{H}, \quad \mathcal{M}_\gamma : \mathcal{D}(\mathcal{M}_\gamma) \subseteq \mathcal{H} \to \mathcal{H}, \qquad \mathcal{D}(\mathcal{A}^{1/2}) \subseteq \mathcal{D}({\mathcal{M}_\gamma}^{1/2})$$

are each positive, self-adjoint, and densely defined operators. Here the parameter $\alpha \geq 0$ is the damping parameter and $0 \leq 2\theta \leq 1$. In what follows, we are mostly concerned with the case $\theta = 1/2$.

The operator $\mathcal{M}_\gamma$ is typically labeled as the stiffness operator, and $\mathcal{A}$ is the elastic operator. We use the notation $\mathcal{M}_\gamma$ to indicate the dependence of this operator on the parameter $\gamma \geq 0$, which typically represents rotational inertia in the model. The variable $\mathbf{v}$ is more general than v (appearing in (5.2.1)) and will contain v as its last component. Indeed, $\mathbf{v} = [v_1, v_2, v]$ may denote all the displacements of the vibrating structure (vertical and horizontal). In the case of plates and beams, we have $s = 1$ and $\mathbf{v} = v$. In the case of shells, the variable $\mathbf{v}$ represents horizontal and vertical displacements, and so $s = 3$.

The coupling between the structure and the acoustic medium is represented by the operator $\mathcal{C} : L_2(\Gamma_0) \to [L_2(\Gamma_0)]^s$, which acts on the trace of z_t, i.e., the acoustic pressure. The quantity $\rho z_t|_{\Gamma_0}$ then represents the back pressure against the moving wall.

The vector $g = (g_1, g_2), g_i \geq 0$, represents potential boundary damping on the wall. The operators

$$\mathcal{G} = (\mathcal{G}_1, \mathcal{G}_2) : U(\partial\Gamma_0) = U_1(\partial\Gamma_0) \times U_2(\partial\Gamma_0) \to \mathcal{H} \times \mathcal{H}$$

are suitable Green's operators modeling the two boundary conditions associated with the plates. The above notation means

$$g\mathcal{G} = g_1\mathcal{G}_1 + g_2\mathcal{G}_2.$$

The adjoints of $\mathcal{G}_i$, $i = 1, 2$, are computed with respect to the $U_i(\partial\Gamma_0) \to \mathcal{H}$ topology.

To have a well-posed plate system, in line with Assumption 2.2.1 (see Corollary 2.3.3, where $g\mathcal{G} = G$), we make the following standing assumption.

Assumption 5.2.1. *$\mathcal{A}^{1/2}\mathcal{G}_i : U_i(\partial\Gamma_0) \to \mathcal{H}$ is bounded for each $i = 1, 2$.*

The control operator $\mathcal{B}$ describes the effect of the control u acting on the moving wall. Typically, it will be an unbounded operator, such as point or boundary control. We let $\mathcal{U}$ denote the space of controls u. In applications to smart materials, where the control $u(t)$ represents a voltage applied to patches bonded on the wall, the space $\mathcal{U}$ may be taken to be R^k.

We make the following abstract assumption on the control operator $\mathcal{B}$.

Assumption 5.2.2. *$\mathcal{A}^{-r}\mathcal{B} \in \mathcal{L}(\mathcal{U}; [L_2(\Gamma_0)]^s)$ for some $r < 1/2$.*

The important property of the $\mathbf{v}$-equation (5.2.2) is that, when it is uncoupled from the wave component, i.e., when $\rho = 0$, then the map $\{\mathbf{v}_0, \mathbf{v}_1\} \to \{\mathbf{v}(t), \mathbf{v}_t(t)\}$ generates a strongly continuous semigroup on the space $D(\mathcal{A}^{1/2}) \times D({\mathcal{M}_\gamma}^{1/2})$, which, moreover, is analytic when $\alpha > 0$, $\theta \geq 1/4$, and $\mathcal{M}_\gamma = I$ [48, 185, 47]. This property will critically influence the results of the corresponding optimal control problem as well as the techniques employed to prove them. In this case problem (5.2.1), (5.2.2) with $\alpha > 0$ couples a parabolic dynamics in $\mathbf{v}$ with a hyperbolic dynamics in z. The coupling involves the traces $v_t|_{\Gamma_0}$ in (5.2.1) for z and $z_t|_{\Gamma_0}$ in (5.2.2) for v and thus is given by boundary trace operators (unbounded and unclosable).

These are the basic features of the problem concerning the general model (5.2.1), (5.2.2): we have a hyperbolic equation (5.2.1) coupled with another PDE (often parabolic-like) (5.2.2) defined on a manifold Γ_0. The coupling takes place on the boundary Γ_0 and the dynamics describing the structure are associated with an analytic semigroup, if $\alpha > 0$. If $\alpha > 0$, we are then dealing with a hyperbolic-parabolic coupling. A few words about the coupling: since the control action is concentrated on the wall only, the coupling must be strong enough to transfer the effects of the control onto the entire system. On the other hand, the coupling is represented by a boundary-trace type of operator, which is unbounded and, in fact, unclosable. Thus, as the coupling is accomplished by the action of unbounded operators, it should be expected that the mathematical analysis for this model will be nontrivial and will partly depend on the analytic properties of the underlying solutions of (5.2.2) with $\rho = 0$. Our next step is to provide a semigroup setting for the model.

Semigroup framework for (5.2.1), (5.2.2). We put problems (5.2.1) and (5.2.2) into a semigroup framework. To accomplish this, we define the operators A_N : $\mathcal{D}(A_N) \subset L_2(\Omega) \to L_2(\Omega)$,

$$A_N = -c^2 \Delta z \text{ for } z \in \mathcal{D}(A_N) = \left\{ z \in H^2(\Omega) : \frac{\partial z}{\partial \nu} = 0 \text{ on } \Gamma \right\}, \qquad (5.2.3)$$

and the operators $N, \hat{N} : L_2(\Gamma) \to L_2(\Omega)$ by[1]

$$(c^2 \Delta - I) N g = 0 \text{ in } \Omega, \; \frac{\partial}{\partial \nu} N g = \begin{pmatrix} 0 \text{ on } \Gamma_1 \\ g \text{ on } \Gamma_0 \end{pmatrix},$$

$$(c^2 \Delta - I) \hat{N} g = 0 \text{ in } \Omega, \; \frac{\partial}{\partial \nu} \hat{N} g = \begin{pmatrix} g \text{ on } \Gamma_1 \\ 0 \text{ on } \Gamma_0 \end{pmatrix}.$$

It is well known [151] that $N, \hat{N} \in \mathcal{L}(L_2(\Gamma); H^{3/2}(\Omega))$ and, moreover, by Green's formula [129, 140] the operator $N^*(A_N + I)$ coincides with the trace operator, i.e. [67],

$$\begin{aligned} N^*(A_N + I) z &\equiv c^2 z|_{\Gamma_0}, \; \hat{N}^*(A_N + I) z \\ &\equiv c^2 z|_{\Gamma_1} \text{ for } z \in \mathcal{D}({A_N}^{1/2}) \sim H^1(\Omega). \end{aligned} \qquad (5.2.4)$$

With the above notation, the equivalent form of (5.2.1) and (5.2.2) is

$$\begin{cases} z_{tt} + A_N z - (A_N + I) N v_t + d_1 z_t + c^{-2} d_3 (A_N + I) N N^* (A_N + I) z_t \\ \quad + c^{-2} d_2 (A_N + I) \hat{N} \hat{N}^* (A_N + I) z_t = f \text{ on } \mathcal{D}(A_N^{1/2})' \times (0, T), \end{cases}$$

$$\begin{cases} \mathcal{M}_\gamma \mathbf{v}_{tt} + \mathcal{A} \mathbf{v} + \alpha \mathcal{A}^{2\theta} \mathbf{v}_t + c^{-2} \rho \mathcal{C} N^* (A_N + I) z_t \\ \quad + \mathcal{A} g \mathcal{G} \mathcal{G}^* \mathcal{A} \mathbf{v}_t = \mathcal{B} u \text{ on } \mathcal{D}(\mathcal{A}^{1/2})' \times (0, T), \end{cases}$$

$$(z(0), z_t(0), \mathbf{v}(0), \mathbf{v}_t(0)) \in H^1(\Omega) \times L_2(\Omega) \times \mathcal{D}(\mathcal{A}^{1/2}) \times \mathcal{D}({\mathcal{M}_\gamma}^{1/2}), \qquad (5.2.5)$$

where

$$\mathcal{C} z \equiv [\underbrace{0, 0, \ldots, 0,}_{(s-1) \text{ times}} z]^T.$$

Note that the operator $\mathcal{C}$ projects on the vertical component of the displacement $\mathbf{v}$. (So $\mathcal{C} = I$ for plates and beams, $s = 1$, and for shells we have $s = 3$.) We recall that the control operator $\mathcal{B}$ is subject to Assumption 5.2.2.

We define next the following spaces:

$$H_z \equiv H^1(\Omega) \times L_2(\Omega), \quad H_v \equiv D(\mathcal{A}^{1/2}) \times D({\mathcal{M}_\gamma}^{1/2}), \quad X \equiv H_z \times H_v. \qquad (5.2.6)$$

[1] The Neumann map is defined with respect to $c^2 \Delta - I$, rather than just with respect to $c^2 \Delta$. This is to avoid potential nonuniqueness of harmonic extensions corresponding to the Neumann boundary data. Another way to cope with the issue is to define the Neumann map as a classical harmonic extension with the range in the factor space $L_2(\Omega)/\ker(N)$.

Define the (unbounded) operator $C : \mathcal{D}(C) \subset H_z \to H_v$ by

$$C\begin{pmatrix} z_1 \\ z_2 \end{pmatrix} \equiv \begin{pmatrix} \underbrace{[0, 0, \ldots, 0]^T}_{s \text{ times}} \\ c^{-2}\rho\mathcal{M}_\gamma{}^{-1}\mathcal{C}N^*(A_N + I)z_2 \end{pmatrix},$$

where H_z and H_v are given by (5.2.6). As usual, the domain of the operator C is defined maximally, i.e.,

$$D(C) = \{(z_1, z_2) \in H_z;\ \mathcal{C}N^*A_N z_2 \in [D(\mathcal{M}_\gamma{}^{1/2})]'\},$$

where the duality is understood with respect to duality pairing with $[L_2(\Gamma_0)]^s$. Clearly, by (5.2.4) $H^1(\Omega) \times H^1(\Omega) \subset D(C)$. Thus, C is densely defined. Note, however, that the operator C, being a trace operator, is both unbounded and unclosable on the state space H_z. Define next

$$A_z : H_z \to H_z;\quad A_v : H_v \to H_v,$$

$$A_z \equiv \begin{pmatrix} 0 & -I \\ A_N & d_1 I + c^{-2}(A_N + I)[d_2\hat{N}\hat{N}^* + d_3 NN^*](A_N + I)] \end{pmatrix},$$

$$A_v \equiv \begin{pmatrix} 0 & -I \\ \mathcal{M}_\gamma{}^{-1}\mathcal{A} & \alpha\mathcal{M}_\gamma{}^{-1}\mathcal{A}^{2\theta} + \mathcal{M}_\gamma{}^{-1}\mathcal{A}g\mathcal{G}\mathcal{G}^*\mathcal{A} \end{pmatrix}.$$

We note that $-A_z$ corresponds to the generator of the wave equation, while $-A_v$ corresponds to the generator of plate, beam, or shell model accounting for moments of inertia. It can be easily verified, after accounting for Assumption 5.2.1, that both generators are maximally dissipative on the respective topologies of H_z and H_v (we topologize $H^1(\Omega)$ by the graph norm of $A_N{}^{1/2}$) (see Corollary 2.3.3). Moreover, if $\alpha > 0$, $\theta \geq 1/4$, $g \equiv 0$, $-A_v$ generates an analytic semigroup on H_v [48]. (If $g \neq 0$, we need to take $\theta = 1/2$.)

With the above notations, the generator $A : X \to X$, describing the entire structure (5.2.5) is given by

$$-A \equiv \begin{pmatrix} A_z & -c^2\rho^{-1}C^* \\ C & A_v \end{pmatrix}, \tag{5.2.7}$$

where the adjoint of C is defined via duality

$$\left(C\begin{pmatrix} z_1 \\ z_2 \end{pmatrix}, \begin{pmatrix} \mathbf{v}_1 \\ \mathbf{v}_2 \end{pmatrix}\right)_{H_v} = \left(\begin{pmatrix} z_1 \\ z_2 \end{pmatrix}, C^*\begin{pmatrix} \mathbf{v}_1 \\ \mathbf{v}_2 \end{pmatrix}\right)_{H_z}$$

for $\binom{z_1}{z_2} \in D(C)$, $\binom{\mathbf{v}_1}{\mathbf{v}_2} \in H_v$. Specifically, $C^* : H_v \to \mathcal{D}(A_z)'$ is given by

$$C^*\begin{pmatrix} \mathbf{v}_1 \\ \mathbf{v}_2 \end{pmatrix} = \begin{pmatrix} 0 \\ c^{-2}\rho(A_N + I)Nv_2 \end{pmatrix}$$

and the domain of A is defined maximally, i.e.,

$$\begin{aligned} D(A) = \{x \equiv (z_1, z_2, \mathbf{v}_1, \mathbf{v}_2) \in X; z_2 \in H^1(\Omega); \\ z_1 - c^{-2}d_2\tilde{N}\tilde{N}^*(A_N + I)z_2 - c^{-2}d_3 NN^*(A_N + I)z_2 - Nv_2 \in D(A_N); \\ \mathbf{v}_2 \in D(\mathcal{A}^{1/2});\ \mathcal{A}(\mathbf{v}_1 - \alpha\mathcal{A}^{2\theta-1}\mathbf{v}_2 - g\mathcal{G}G^*\mathcal{A}\mathbf{v}_2) \in D(\mathcal{M}_\gamma{}^{1/2})'\}. \end{aligned} \tag{5.2.8}$$

Now we can rewrite (5.2.5) as a first-order system,

$$\frac{d}{dt}\begin{pmatrix} z \\ z_t \\ \mathbf{v} \\ \mathbf{v}_t \end{pmatrix} - A\begin{pmatrix} z \\ z_t \\ \mathbf{v} \\ \mathbf{v}_t \end{pmatrix} = \begin{pmatrix} 0 \\ 0 \\ 0 \\ \mathcal{M}_\gamma{}^{-1}\mathcal{B}u \end{pmatrix} + \begin{pmatrix} 0 \\ f \\ 0 \\ 0 \end{pmatrix} \equiv Bu + F. \tag{5.2.9}$$

It can be easily verified that both A and A^* are dissipative (after rescaling the inner product on H_v to account for the factor $c^{-2}\rho$); hence by the standard semigroup result [175], A is the generator of a contraction semigroup on X. In what follows, we use the adjoint of B defined via duality as

$$(B^*x, u)_{\mathcal{U}} = (x, Bu)_X \text{ for all } x \in D(B^*) \supseteq D(A^*),$$

where $D(B^*)$ is defined as a maximal extension by density of $D(A^*) \subset D(B^*)$. By Assumption 5.2.2 we always have

$$|B^*x|_{\mathcal{U}} \le C|\mathcal{A}^r \mathbf{v}_2|_{\mathcal{H}} \le C|x|_{D(A)}.$$

The dynamics described by (5.2.9) are an abstract semigroup formulation for the structural acoustic model. We note that while for $\alpha > 0, \theta \ge 1/4, g \equiv 0$, $-A_v$ generates an analytic semigroup [48], this is not the case for the big operator A. The operator A has a substantial hyperbolic component.

5.3 PDE Models Illustrating the Abstract Wall Equation (5.2.2)

In the following two subsections we provide several concrete models for the structure (wall), including the cases of structurally damped beams, plates, and shells. To simplify matters and motivated as we are by physical relevance and importance, we assume that the value of the parameter $\theta = 1/2$. In the case of beams or plates we assume that the boundary Γ_0 is flat. In a more general case when Γ_0 is curved, we need to resort to shell models. These are discussed later.

5.3.1 Plates and beams—Flat Γ_0

As mentioned above, for plates and beams we have $s = 1$ and the general model (5.2.2) becomes

$$\begin{aligned} \mathcal{M}_\gamma v_{tt}(t) + \alpha\mathcal{A}v_t(t) + \mathcal{A}v(t) + \rho z_t(t)|_{\Gamma_0} + \mathcal{A}g\mathcal{G}\mathcal{G}^*\mathcal{A}v_t(t) \\ = \mathcal{B}u(t) \text{ on } \mathcal{D}(\mathcal{A}^{1/2})' \times (0,T), \\ v(0) = v_0 \in \mathcal{D}(\mathcal{A}^{1/2}), \quad v_t(0) = v_1 \in \mathcal{D}(\mathcal{M}_\gamma{}^{1/2}), \end{aligned} \tag{5.3.10}$$

so that operator $\mathcal{C}$ in (5.2.2) is simply the identity operator. Different sets of boundary conditions will yield different realizations of $\mathcal{A}$ and $\mathcal{M}_\gamma$ and $\mathcal{G}$.

Choices of the operators $\mathcal{A}$, $\mathcal{G}$, and $\mathcal{M}_\gamma$. The following elliptic operators are used frequently: $\mathcal{A}_N, \mathcal{A}_D : L_2(\Gamma_0) \to L_2(\Gamma_0)$ defined by

$$\begin{aligned} \mathcal{A}_N v = -\Delta v, \quad v \in \mathcal{D}(\mathcal{A}_N) = \left\{ v \in H^2(\Gamma_0); \frac{\partial}{\partial \nu} v = 0 \text{ on } \partial\Gamma_0 \right\} \\ \mathcal{A}_D v = -\Delta v, \quad v \in \mathcal{D}(\mathcal{A}_D) = H^2(\Gamma_0) \cap H_0^1(\Gamma_0). \end{aligned} \tag{5.3.11}$$

Remark 5.3.1. *For simplicity of exposition we restrict ourselves to elliptic operators with constant coefficients. However, our analysis will not be affected if one takes variable coefficients (space dependent) subject to suitable ellipticity/regularity conditions.*

Clamped plates and beams. Here we take

$$\mathcal{A} = \Delta^2, \ \mathcal{D}(\mathcal{A}) = \left\{ v \in H^4(\Gamma_0); v = \frac{\partial}{\partial \nu} v = 0 \text{ on } \partial\Gamma_0 \right\} \\ = H^4(\Gamma_0) \cap H_0^2(\Gamma_0), \tag{5.3.12}$$

$$\mathcal{G} = [\mathcal{G}_1, \mathcal{G}_2] : L_2(\partial\Gamma_0) \times L_2(\partial\Gamma_0) \to L_2(\Gamma_0) \times L_2(\Gamma_0),$$

$$\begin{aligned} &\Delta^2 \mathcal{G}_1 g = 0 \text{ in } \Gamma_0, \qquad \mathcal{G}_1 g = g \text{ on } \partial\Gamma_0, \qquad \frac{\partial}{\partial \nu} \mathcal{G}_1 g = 0 \text{ on } \partial\Gamma_0, \\ &\Delta^2 \mathcal{G}_2 g = 0 \text{ in } \Gamma_0, \qquad \mathcal{G}_2 g = 0 \text{ on } \partial\Gamma_0, \qquad \frac{\partial}{\partial \nu} \mathcal{G}_2 g = g \text{ on } \partial\Gamma_0. \end{aligned} \tag{5.3.13}$$

By elliptic regularity theory [151] we have that the maps

$$[\mathcal{G}_1, \mathcal{G}_2] : L_2(\partial\Gamma_0) \times L_2(\partial\Gamma_0) \to H^{1/2}(\Gamma_0) \times H^{3/2}(\Gamma_0)$$

are bounded.

The operator $\mathcal{M}_\gamma : L_2(\Gamma_0) \to L_2(\Gamma_0)$ is given by

$$\mathcal{M}_\gamma \equiv I + \gamma \mathcal{A}_D.$$

Hinged plates and beams. Here we take

$$\mathcal{A} = \Delta^2, \ \mathcal{D}(\mathcal{A}) = \{ v \in H^4(\Gamma_0); v = 0, \Delta v = 0 \text{ on } \partial\Gamma_0 \}, \tag{5.3.14}$$

$$\mathcal{G} = [\mathcal{G}_1, \mathcal{G}_2] : L_2(\partial\Gamma_0) \times L_2(\partial\Gamma_0) \to L_2(\Gamma_0) \times L_2(\Gamma_0)$$

defined by

$$\begin{aligned} &\Delta^2 \mathcal{G}_1 g = 0 \text{ in } \Gamma_0, \qquad \mathcal{G}_1 g = g \text{ on } \partial\Gamma_0 \qquad \Delta \mathcal{G}_1 g = 0 \text{ on } \partial\Gamma_0, \\ &\Delta^2 \mathcal{G}_2 g = 0 \text{ in } \Gamma_0, \qquad \mathcal{G}_2 g = 0 \text{ on } \partial\Gamma_0, \qquad \Delta \mathcal{G}_2 g = g \text{ on } \partial\Gamma_0. \end{aligned} \tag{5.3.15}$$

By elliptic regularity theory [151] we have that

$$[\mathcal{G}_1, \mathcal{G}_2] : L_2(\partial\Gamma_0) \times L_2(\partial\Gamma_0) \to H^{1/2}(\Gamma_0) \times H^{5/2}(\Gamma_0)$$

is bounded.

The operator $\mathcal{M}_\gamma : L_2(\Gamma_0) \to L_2(\Gamma_0)$ is given by

$$\mathcal{M}_\gamma \equiv I + \gamma \mathcal{A}_D = I + \gamma \mathcal{A}^{1/2}.$$

Free plates and beams. Here we take

$$\mathcal{A} = \Delta^2, \ \mathcal{D}(\mathcal{A}) = \{ v \in H^4(\Gamma_0); \Delta v + B_1 v = 0, \\ \frac{\partial}{\partial \nu} \Delta v + B_2 v = 0 \text{ on } \partial\Gamma_0 \}, \tag{5.3.16}$$

where B_1, B_2 are boundary operators given by (see [99, 101])

$$B_1 w = (1-\mu)[2\nu_1\nu_2 w_{xy} - \nu_1^2 w_{yy} - \nu_2^2 w_{xx}],$$
$$B_2 w = (1-\mu)\frac{\partial}{\partial \tau}[(\nu_1^2 - \nu_2^2) w_{xy} + \nu_1\nu_2(w_{yy} - w_{xx})] + lw, \tag{5.3.17}$$

and μ is Poisson's modulus, parameter $l \geq 0$. Vectors $\nu \equiv (\nu_1, \nu_2)$ and τ denote, respectively, the positively oriented outward unit normal and tangential vectors to the boundary $\partial\Gamma_0$.

The formulas for the beam are simpler; it suffices to take $B_i \equiv 0$. We note that the presence of boundary operators B_i in the two-dimensional case is necessary. Indeed, this is to avoid the infinite dimensionality of the null space of the elliptic operator $\mathcal{A}$ (otherwise all the harmonic functions would belong to the nullspace of $\mathcal{A}$).

$$\mathcal{G} = [\mathcal{G}_1, \mathcal{G}_2] : L_2(\partial\Gamma_0) \times L_2(\partial\Gamma_0) \to L_2(\Gamma_0) \times L_2(\Gamma_0),$$

$$\Delta^2 \mathcal{G}_1 g = 0 \text{ in } \Gamma_0,\ \Delta \mathcal{G}_1 g + B_1 \mathcal{G}_1 g = g \text{ on } \partial\Gamma_0,$$
$$\frac{\partial}{\partial \nu}\Delta \mathcal{G}_1 g + B_2 \mathcal{G}_1 g = 0 \text{ on } \partial\Gamma_0,$$
$$\Delta^2 \mathcal{G}_2 g = 0 \text{ in } \Gamma_0,\ \Delta \mathcal{G}_2 g + B_1 \mathcal{G}_2 g = 0 \text{ on } \partial\Gamma_0,$$
$$\frac{\partial}{\partial \nu}\Delta \mathcal{G}_2 g + B_2 \mathcal{G}_2 g = g \text{ on } \partial\Gamma_0. \tag{5.3.18}$$

By elliptic regularity we have that

$$[\mathcal{G}_1, \mathcal{G}_2] : L_2(\partial\Gamma_0) \times L_2(\partial\Gamma_0) \to H^{5/2}(\Gamma_0) \times H^{7/2}(\Gamma_0)$$

is bounded.

The operator $\mathcal{M}_\gamma : L_2(\Gamma_0) \to L_2(\Gamma_0)$ is given by [128, 34]

$$\mathcal{M}_\gamma \equiv I + \gamma \mathcal{A}_N.$$

Remark 5.3.2. *When moments of inertia (rotational forces) are accounted for in the models, then additional terms involving the second-order derivatives are added to the equation. This corresponds to the so-called Kirchhoff equation and is modeled by the operator $\mathcal{M}_\gamma$ with $\gamma > 0$, where γ is proportional to the square of the thickness of the plate. While the differential form of the operator $\mathcal{M}_\gamma$ is always $I - \gamma\Delta$, its operator's form depends, as seen above, on the boundary conditions imposed on the plate.*

Choice for the control operator $\mathcal{B}$. Finally, we return to the issue of the control acting on the system (5.2.1), (5.2.2), governed by a control operator $\mathcal{B}$. As mentioned, these control operators are motivated by smart structures technology and typically represent some sort of delta functions. In what follows, we provide a more precise definition of these operators.

In the case of plates or beams, a common choice for the control operator is

$$\mathcal{B}u = \sum_{i=1}^{k} \alpha_i u_i \delta'_{\xi_i}, \qquad u = [u_1, u_2 \cdots u_k] \in R^k = \mathcal{U}, \tag{5.3.19}$$

where ξ_i are either points ($\dim \Gamma_0 = 1$) or closed curves ($\dim \Gamma_0 = 2$) in Γ_0 and the corresponding α_i are either constants ($\dim \Gamma_0 = 1$) or sufficiently smooth functions ($\dim \Gamma_0 = 2$). The symbol δ'_{ξ_i} denotes the derivative of the delta distribution supported on ξ_i (when $\dim \Gamma_0 = 2$, we take normal derivatives to the curves ξ_i). Thus, for the one-dimensional Γ_0 we have

$$(\delta'_{\xi_i}, \phi)_{L_2(\Gamma_0)} \equiv -\phi'(\xi_i) \qquad \text{for all } \phi \in H^{3/2+\epsilon}(\Gamma_0).$$

For the two-dimensional Γ_0 we have

$$(\delta'_{\xi_i}, \phi)_{L_2(\Gamma_0)} \equiv -\int_{\xi_i} \nabla\phi \cdot \nu d\xi_i \qquad \text{for all } \phi \in H^{3/2+\epsilon}(\Gamma_0),$$

where the outer vector ν is normal to the curve ξ_i and ϵ is an arbitrary small, positive constant.

This choice of the boundary operator is typically seen in smart materials technology, where control action is implemented via piezoceramic or piezoelectric patches bonded to the wall (see [18, 15, 72, 71]). The voltage $u_i(t)$ applied to a patch creates a bending moment, which, in turn, causes the bending of the wall. This has the effect of reducing the vibrations.

We next verify that the operator $\mathcal{B}$ in (5.3.19) satisfies the Assumption 5.2.2. Indeed, this follows from the embeddings $D(\mathcal{A}^\theta) \subset H^{4\theta}(\Gamma_0)$, $0 \leq \theta \leq 1$ [67], by which we conclude via (5.3.19) that Assumption 5.2.2 is satisfied with $r = 3/8 + \frac{1}{4}\epsilon < 1/2$ for some small ϵ.

In what follows we discuss in detail how various choices of the boundary conditions affect the PDE structure of the models considered. While the situation is rather simple in the case of undamped boundary conditions, the structure and the functional setup are more complicated when boundary damping is accounted for. We begin with the simplest case when the boundary conditions are homogeneous, so there is no boundary damping and $g \equiv 0$.

5.3.2 "Undamped" boundary conditions: $g \equiv 0$ in (5.3.10)

In this case the operators $\mathcal{A}, \mathcal{M}_\gamma$ are defined for each respective set of boundary conditions in section 5.3.1, and $\mathcal{H} \equiv L_2(\Gamma_0)$. The abstract equation (5.3.10) takes the following forms.

Hinged boundary conditions.

$$\begin{aligned} v_{tt}(t) + \gamma\mathcal{A}^{1/2}v_{tt}(t) + \alpha\mathcal{A}v_t(t) + \mathcal{A}v(t) + \rho z_t(t)|_{\Gamma_0} &= \mathcal{B}u(t), \\ &\text{on } \mathcal{D}(\mathcal{A}^{1/2})' \times (0,T), \\ v(0) = v_0 \in \mathcal{D}(\mathcal{A}^{1/2}) &\sim H^2(\Gamma_0) \cap H_0^1(\Gamma_0), \\ v_t(0) = v_1 \in \mathcal{D}(\mathcal{A}^{1/4}) &\sim H_0^1(\Gamma_0), \end{aligned} \tag{5.3.20}$$

where the constant γ is positive and proportional to the square of the thickness of the plate and $\mathcal{A}$ is given by (5.3.14). In this case the stiffness operator $\mathcal{M}_\gamma$ appearing in the abstract model (5.2.2) coincides with $I + \gamma\mathcal{A}^{1/2} = I - \gamma\Delta$ and $\mathcal{D}({\mathcal{M}_\gamma}^{1/2}) = D(\mathcal{A}^{1/4}) \sim H_0^1(\Gamma_0)$. When the constant $\gamma = 0$, then we have $D({\mathcal{M}_\gamma}^{1/2}) = L_2(\Gamma_0)$ and $v_1 \in L_2(\Gamma_0)$.

The PDE version of this equation becomes (see [140, p. 209])

$$\begin{aligned} v_{tt}(t) - \gamma\Delta v_{tt}(t) + \alpha\Delta^2 v_t(t) + \Delta^2 v(t) + \rho z_t(t)|_{\Gamma_0} &= \mathcal{B}u(t) \\ &\text{on } \mathcal{D}(\mathcal{A}^{1/2})' \times (0,T), \\ v = 0, \qquad \Delta(v + \alpha v_t) &= 0 \text{ on } \partial\Gamma_0 \times (0,T), \\ v(0) = v_0 \in H^2(\Gamma_0) \cap H_0^1(\Gamma_0); \quad v_t(0) &= v_1 \in H^1_{\gamma,0}(\Gamma_0), \end{aligned} \tag{5.3.21}$$

where $H^1_{\gamma,0}(\Gamma_0)$ is defined by (4.4.73).

Clamped boundary conditions.

$$\begin{aligned} v_{tt}(t) + \gamma \mathcal{A}_D v_{tt}(t) + \alpha \mathcal{A} v_t(t) + \mathcal{A} v(t) + \rho z_t(t)|_{\Gamma_0} &= \mathcal{B} u(t) \\ \text{on } \mathcal{D}(\mathcal{A}^{1/2})' \times (0,T), \\ v(0) = v_0 \in H_0^2(\Gamma_0) = \mathcal{D}(\mathcal{A}^{1/2}),\ v_t(0) = v_1 \in H_0^1(\Gamma_0) &= \mathcal{D}(\mathcal{A}^{1/4}), \end{aligned} \tag{5.3.22}$$

where $\mathcal{A}$ is given by (5.3.12). The operator $I + \gamma \mathcal{A}_D$ coincides with the stiffness operator $\mathcal{M}_\gamma$ in the abstract model and $\mathcal{D}({\mathcal{M}_\gamma}^{1/2}) = \mathcal{D}(\mathcal{A}^{1/4}) \sim H_0^1(\Gamma_0)$ when $\gamma > 0$. For $\gamma = 0$, we have $\mathcal{D}({\mathcal{M}_\gamma}^{1/2}) = L_2(\Gamma_0)$ and $v_1 \in L_2(\Gamma_0)$.

The PDE version of this equation becomes

$$\begin{aligned} v_{tt}(t) - \gamma \Delta v_{tt}(t) + \alpha \Delta^2 v_t(t) + \Delta^2 v(t) + \rho z_t(t)|_{\Gamma_0} &= \mathcal{B} u(t) \\ \text{on } \mathcal{D}(\mathcal{A}^{1/2})' \times (0,T), \\ v = 0, \quad \frac{\partial}{\partial \nu} v &= 0 \text{ on } \partial\Gamma_0 \times (0,T), \\ v(0) = v_0 \in H_0^2(\Gamma_0);\ v_t(0) &= v_1 \in H_{\gamma,0}^1(\Gamma_0). \end{aligned} \tag{5.3.23}$$

Free boundary conditions. The effect of rotational forces leads to a more complex form of both the equation and the boundary conditions. Indeed, in this case we have

$$\begin{aligned} v_{tt}(t) + \gamma \mathcal{A}_N v_{tt}(t) + \alpha \mathcal{A} v_t(t) + \mathcal{A} v(t) + \rho z_t(t)|_{\Gamma_0} &= \mathcal{B} u(t) \\ \text{on } \mathcal{D}(\mathcal{A}^{1/2})' \times (0,T), \\ v(0) = v_0 \in D(A^{1/2}), \quad v_t(0) = v_1 \in \mathcal{D}(A^{1/4}) &\text{ when } \gamma > 0, \end{aligned} \tag{5.3.24}$$

where $\mathcal{A}$ is given by (5.3.16). Thus, in this case the abstract stiffness operator $\mathcal{M}_\gamma$ coincides with $I + \gamma \mathcal{A}_N$ and $\mathcal{D}({\mathcal{M}_\gamma}^{1/2}) = \mathcal{D}({A_N}^{1/2}) \sim H^1(\Gamma_0)$, when $\gamma > 0$, and $\mathcal{D}({\mathcal{M}_\gamma}^{1/2}) = L_2(\Gamma_0)$, when $\gamma = 0$.

The PDE version of this equation becomes

$$\begin{aligned} v_{tt}(t) - \gamma \Delta v_{tt}(t) + \alpha \Delta^2 v_t(t) + \Delta^2 v(t) + \rho z_t(t)|_{\Gamma_0} &= \mathcal{B} u(t) \\ \text{on } \mathcal{D}(\mathcal{A}^{1/2})' \times (0,T), \\ (\Delta + B_1)(v + \alpha v_t) = 0; \quad \left(\frac{\partial}{\partial \nu}\Delta + B_2\right)(v + \alpha v_t) - \gamma \frac{\partial}{\partial \nu} v_{tt} &= 0. \\ \text{on } \partial\Gamma_0 \times (0,T), \\ v(0) = v_0 \in H^2(\Gamma_0),\ v_t(0) &= v_1 \in H_\gamma^1(\Gamma_0), \end{aligned} \tag{5.3.25}$$

where B_1, B_2 are given in (5.3.17).

Remark 5.3.3. *We note that in PDE models* (5.3.21) *and* (5.3.25) *one can consider a simpler form of the boundary conditions, namely,*

$$v = 0, \qquad \Delta v = 0 \text{ on } \partial\Gamma_0 \tag{5.3.26}$$

in the case of (5.3.21) *and*

$$(\Delta + B_1) v = 0, \quad \left(\frac{\partial}{\partial \nu}\Delta + B_2\right)(v + \alpha v_t) - \gamma \frac{\partial}{\partial \nu} v_{tt} = 0 \text{ on } \partial\Gamma_0 \tag{5.3.27}$$

in the case of (5.3.25). *In fact, it is easy to verify that these boundary conditions are invariant under the dynamics of the plate and, therefore, one can consider initial data complying with* (5.3.26) *or* (5.3.27), *respectively, which will then guarantee that the corresponding solutions satisfy the same boundary conditions (see also* [140, *p.* 209]).

When considering uncontrolled models $u = 0$ described above by (5.3.21), (5.3.23), and (5.3.25), the well-posedness of the corresponding plate dynamics (with $\rho = 0$) is well known and follows from the Lummer–Phillips theorem [175] (see also Corollary 2.3.3). If, in addition, we assume that $\alpha > 0$, then the corresponding semigroups are analytic and ($l > 0$ in (5.3.17)) exponentially stable [48] in the topology of $H_v = \mathcal{D}(\mathcal{A}^{1/2}) \times \mathcal{D}({\mathcal{M}_\gamma}^{1/2})$ with $\mathcal{A}, \mathcal{M}_\gamma$ appropriately defined (see above) for each set of the boundary conditions.

An important issue that needs to be addressed is the stability of the overall structural acoustic model, which consists of the wave equation (5.2.1) coupled with the plate equation (5.2.2). This topic is discussed later in section 5.4. Before doing this, we discuss models with a boundary damping affecting an edge of the plate.

5.3.3 Boundary feedback: Case $g \neq 0$ in (5.3.10) and related stability

If the structural damping parameter α is positive, it is shown in [48] that the plate equation (with $\rho = 0, u = 0$) is exponentially stable, and so in this case there is no need to introduce any other sources of damping to the plate model. However, when structural damping is nonexistent (or of nonuniform density), one needs to consider other means for stabilizing the system. An attractive form of stabilizers is boundary dampers. We now turn to the case when boundary damping is added to the model. This, of course, alters the boundary conditions and therefore has an effect on the definitions of the operators involved. We investigate each type of boundary condition separately, with the focus on models of relevance in applications. In particular, we focus on boundary feedbacks which provide uniform stability properties of the structurally undamped ($\alpha = 0$) plate and of the corresponding structural acoustic models. Necessarily then, the boundary feedback will need to be strong enough, that is, the feedback will have to be sufficiently unbounded, to guarantee uniform stability. In fact, it is well known [193, 194] that for hyperbolic-like plates relatively bounded—with respect to the state space feedbacks F (i.e., $R(\lambda, A)F \in \mathcal{L}(U \to H)$) which are finite rank—will not provide any change in the essential spectrum of the underlying operator, and thus such mechanisms cannot provide uniform stability of the system that is not stable to begin with.

In addition to our efforts to obtain stability of the PDE by means of boundary feedback control, we also want to consider the system as it is defined on the so-called finite energy space. In particular, we wish to take $H^2(\Gamma_0)$ to be the space in which the vertical displacements of the plate evolve (in time). The reason for this requirement is related to the use of smart materials and the particular form of the actuators used for controlling the vibrations of the plate, which are represented by derivatives of delta functions. In mathematical terms, this implies that Assumption 5.2.2 needs to be in force. A weaker topology for the state space will make this assumption often unrealistic in practice.

Summarizing the above discussion, we have two requirements for the boundary feedback: (i) it must be sufficiently strong (unbounded), and (ii) it needs to provide the desired stability properties on a finite energy space. On the other hand, as seen in Corollary 2.3.3, well-posedness considerations will force limitations on the degree of unboundedness of the boundary feedback (see Assumption 5.2.1 and also item 4 in Assumption 2.2.1). A catch-22 situation therefore arises: the need to boundary-stabilize the system may be at odds with the requirements necessary for well-posedness. The resolution may require an appropriate calibration of the spaces $\mathcal{H}, U(\partial\Gamma_0)$ in Assumption 5.2.1. Indeed, notice that the strength of the boundary feedback (i.e., its degree of unboundedness) depends directly on the structure of $\mathcal{G}^*$ operator, and this, in turn, depends on the topological choice of the spaces involved. Now the requirement we work on the finite energy space will dictate and fix the choice of $\mathcal{H}$ (which must coincide with $L_2(\Gamma_0)$). There is no flexibility here. In this way, $\mathcal{D}(\mathcal{A}^{1/2})$ (with $\mathcal{A}$ given in (5.2.2)) coincides with the $H^2(\Gamma_0)$ topology, as desired. However, there will be some freedom in specifying spaces $U(\partial\Gamma_0)$, where $U(\partial\Gamma_0)$ is determined by appropriate choices of $U_i(\partial\Gamma_0)$. However, in choosing $U(\partial\Gamma_0)$, we need to comply with the main Assumption 5.2.1 (which restricts the strength of the feedback) and to further determine that the resulting feedback is indeed stabilizing. Ideally, we would like to have stabilization uniform with respect to the rotational forces—represented by the parameter γ—which is to say that stabilization is effective for both the Euler and the Kirchhoff plate. The examples provided below illustrate how to play this game correctly. As we will see, the main issue is to select a correct form of the adjoint operators to $\mathcal{G}$, which, in turn, depends on the choice of the pivot spaces $U(\partial\Gamma_0)$ and spaces $\mathcal{H}$ of realization for the biharmonic operator Δ^2. Since we are interested in finite energy spaces, the correct space $\mathcal{H}$ should coincide with $L_2(\Gamma_0)$. And, in fact, this happens to be a natural choice for the cases of hinged and free boundary conditions. Instead, in the case of clamped boundary conditions with boundary feedback, the situation is much more complicated and the structure of $\mathcal{H}$ is more complex (it involves factor spaces with low regularity; see [121]). Since these spaces are of limited interest in the context of smart controls, we shall not pursue this analysis here and we shall limit our considerations to hinged and free boundary conditions. These are discussed below.

Hinged boundary conditions. In this case, we consider a feedback acting via moments only (this is sufficient for stabilization—see section 5.4). Thus, we have

$$\mathcal{G} = (0, \mathcal{G}_2),\ \mathcal{G}_2 : U_2(\partial\Gamma_0) \to L_2(\Gamma_0), \qquad g = (0, g_2),$$

where $\mathcal{G}_2, \mathcal{A}$ are given in (5.3.15) and (5.3.14).

As the choice for $U_2(\partial\Gamma_0)$ and $\mathcal{H}$ we simply take

$$\mathcal{H} = L_2(\Gamma_0),\quad U_2(\partial\Gamma_0) \equiv L_2(\partial\Gamma_0).$$

Since $\mathcal{G}_2 : L_2(\partial\Gamma_0) \to H^{5/2}(\Gamma_0) \cap H_0^1(\Gamma_0)$ is bounded by elliptic theory, and $H^{5/2}(\Gamma_0) \cap H_0^1(\Gamma_0) \subset \mathcal{D}(\mathcal{A}^{1/2})$ by [67], we obtain boundedness of the operator

$$\mathcal{A}^{1/2}\mathcal{G}_2 : L_2(\partial\Gamma_0) \to L_2(\Gamma_0) = \mathcal{H},$$

which verifies Assumption 5.2.1. By using Green's formula and the definition of $\mathcal{G}_2$

one easily computes the adjoint [128] $\mathcal{G}_2^* : L_2(\Gamma_0) \to L_2(\partial\Gamma_0)$:

$$\mathcal{G}_2^* v = \frac{\partial}{\partial\nu}\mathcal{A}^{-1} v|_{\partial\Gamma_0} \Rightarrow \mathcal{G}_2^*\mathcal{A}v = \frac{\partial}{\partial\nu} v|_{\partial\Gamma_0}, \qquad v \in D(\mathcal{A}^{1/2}) \sim H_0^1(\Gamma_0) \cap H^2(\Gamma_0).$$

The abstract version of the plate equation with boundary damping via moments takes the form for all $t > 0$,

$$\begin{aligned} v_{tt} + \gamma\mathcal{A}^{1/2} v_{tt} + \alpha\mathcal{A}v_t + \mathcal{A}v + \rho z_t|_{\Gamma_0} + g_2\mathcal{A}\mathcal{G}_2\mathcal{G}_2^*\mathcal{A}v_t &= \mathcal{B}u \\ \text{on } \mathcal{D}(\mathcal{A}^{1/2})' \qquad &\text{with the initial conditions} \\ v(0) = v_0 \in \mathcal{D}(\mathcal{A}^{1/2}) &\sim H^2(\Gamma_0) \cap H_0^1(\Gamma_0), \\ v_t(0) = v_1 \in \mathcal{D}({\mathcal{M}_\gamma}^{1/2}) &\sim H_0^1(\Gamma_0), \end{aligned} \tag{5.3.28}$$

where again the constant γ is positive and proportional to the square of the thickness of the plate. In this case then, the stiffness operator $\mathcal{M}_\gamma$ appearing in the abstract model (5.2.2) coincides with $I + \gamma\mathcal{A}^{1/2}$ and $\mathcal{D}({\mathcal{M}_\gamma}^{1/2}) = \mathcal{D}(\mathcal{A}^{1/4}) \sim H_0^1(\Gamma_0)$. When the constant $\gamma = 0$, then, of course, we have $\mathcal{D}({\mathcal{M}_\gamma}^{1/2}) = L_2(\Gamma_0)$ and $v_1 \in L_2(\Gamma_0)$.

With $\mathcal{A}$, $\mathcal{M}_\gamma$, and $\mathcal{G}$ as defined, we can invoke the well-posedness results stated in Corollary 2.3.3, to conclude that the plate model (5.3.28), with $\rho = 0, u = 0$, is well posed on $H_v = \mathcal{D}(\mathcal{A}^{1/2}) \times \mathcal{D}({\mathcal{M}_\gamma}^{1/2})$.

The PDE version of (5.3.28) is (see [128] where this model with $\rho = \alpha = 0, \gamma > 0$, was treated)

$$\begin{aligned} v_{tt} - \gamma\Delta v_{tt} + \Delta^2 v + \alpha\Delta^2 v_t + \rho z_t &= \mathcal{B}u \text{ in } \mathcal{D}(\mathcal{A}^{1/2})' \times (0,\infty), \\ v = 0, \ \Delta(v + \alpha v_t) &= -g_2\frac{\partial}{\partial\nu} v_t \text{ on } \partial\Gamma_0 \times (0,\infty), \\ v(0) = v_0 \in H^2(\Gamma_0) \cap H_0^1(\Gamma_0), \ \ &v_t(0) = v_1 \in H_{\gamma,0}^1(\Gamma_0). \end{aligned} \tag{5.3.29}$$

In line with Remark 5.3.3 we can replace the boundary conditions in (5.3.29) by $v = \Delta v = 0$ on $\partial\Gamma_0$.

Remark 5.3.4. *Going back to Assumption* 2.2.1, *we note that* $\mathcal{G}_2^*\mathcal{A}$ *is bounded and surjective,* $H^2(\Gamma_0) \to H^{1/2}(\partial\Gamma_0)$, *so the requirement in Assumption* 2.2.1 *is satisfied with*

$$H = L_2(\Gamma_0), \ U = L_2(\partial\Gamma_0), \ U_0 = H^{1/2}(\partial\Gamma_0),$$

where H, U, U_0 *are as in Assumption* 2.2.1. *This observation allows one to conclude, by virtue of Theorem* 2.3.1, *well-posedness of the plate model* (5.3.28) *with a nonlinear boundary feedback given by* $g(\frac{\partial}{\partial\nu}v_t)$, *where* g *is a monotone function subject to polynomial growth condition.*

Our next step of analysis is to determine whether the boundary feedback constructed above provides uniform stability for the uncontrolled plate model in (5.3.29) and the resulting structural acoustic model with $\alpha = 0$—that is, no structural damping—and all values of $\gamma \geq 0$. That such is the case follows from stability results in [81, 40, 41], which we recall next. We discuss immediately below the stability results valid for the plate model, while the stability issue for the entire structural acoustic model is addressed in section 5.4.4.

Theorem 5.3.1 (see [81, 128]). *Consider the plate equation* (5.3.29) *with* $\alpha = 0$, $\rho = 0$, $u = 0$, $g_2 > 0$. *The corresponding dynamics generates an exponentially stable semigroup on the space*

$$\mathcal{D}(\mathcal{A}^{1/2}) \times \mathcal{D}(\mathcal{M}_\gamma{}^{1/2}) \sim H^2(\Gamma_0) \cap H_0^1(\Gamma_0) \times H_\gamma^1(\Gamma_0).$$

Moreover, the rates of decay are uniform with respect to the parameter $0 \leq \gamma \leq M$.

Remark 5.3.5. *Theorem* 5.3.1 *was first proved in* [128] *for the case* $\gamma > 0$. *Uniformity of decay rates for all* $0 \leq \gamma < M$ *was established in* [81]. *Subsequently, the result of Theorem* 5.3.1 *was extended in* [88] *to the case of nonlinear boundary damping and nonlinear plate equations.*

Controllability and stabilization of plates via hinged boundary conditions can be found in [153, 102, 154] *and references therein. The results presented there require two boundary conditions as the stabilizers. To dispense with this restriction, microlocal estimates are employed in* [81].

Remark 5.3.6. *We emphasize that the boundary feedback inserted in* (5.3.29) *is effective for both the Euler–Bernoulli and the Kirchhoff models, although the feedback in* (5.3.29) *is not typically seen in the context of stabilizing the Euler equation* ($\gamma = 0$). *Indeed, it has a higher degree of unboundedness than a classical feedback used in Euler plate theory, i.e., the term* $\frac{\partial}{\partial\nu}\Delta^{-1}v_t$ [75, 104]. *However, the latter feedback, while effective for Euler–Bernoulli plate models and defined on the space of optimal regularity* [126], *where we have the boundedness and surjectivity of the control-to-state map, does not stabilize the Kirchhoff plate and leads to the state space that has topology too low* (H^1) *for the displacement component. This fact was a motivation behind the work of* [81], *where it was shown that a simple and local feedback* $\frac{\partial}{\partial\nu}v_t$ *provides robust stabilization for both the Euler and the Kirchhoff model. The papers* [40, 41] *extend this result from plates to the entire structural acoustic models.*

Free boundary conditions. We now consider the damping acting via moments or shears. Accordingly, we take $g = (g_1, g_2), \mathcal{G} = (\mathcal{G}_1, \mathcal{G}_2)$, given by (5.3.18) with a natural space $\mathcal{H} = L_2(\Gamma_0)$. The operator $\mathcal{A}$ is given by (5.3.16).

Since we know that the Euler–Bernoulli model can be stabilized by shear forces only (see section 4.4.3), our first choice is to take $g = (0, g_2)$ with the boundary space $U_2(\partial\Gamma_0) = L_2(\partial\Gamma_0)$. Computations of the adjoint $\mathcal{G}_2^* : L_2(\partial\Gamma_0) \to L_2(\Gamma_0)$ give

$$\mathcal{G}_2^* v = -\mathcal{A}v|_{\partial\Gamma_0} \Rightarrow \mathcal{G}_2^*\mathcal{A}v = -v|_{\partial\Gamma_0}, \qquad v \in \mathcal{D}(\mathcal{A}^{1/2}) \sim H^2(\Gamma_0).$$

It is immediate to verify that $\mathcal{G}_2^*\mathcal{A}^{1/2}$ is bounded (and also surjective) from $H^2(\Omega)$ onto $H^{3/2}(\partial\Gamma_0) \subset L_2(\partial\Gamma_0)$. Thus

$$\mathcal{A}^{1/2}\mathcal{G}_2 : L_2(\partial\Gamma_0) \to L_2(\Gamma_0)$$

is bounded, implying Assumption 5.2.1.

Note that Assumption 2.2.1 is satisfied with $U_0 = H^{3/2}(\partial\Gamma_0)$, $U = L_2(\partial\Gamma_0)$, which via Theorem 2.3.1 allows us to conclude the well-posedness for the model with nonlinear, monotone boundary conditions.

The effect of rotational forces leads to the following abstract formulation [34]:

$$
\begin{aligned}
v_{tt} + \gamma \mathcal{A}_N v_{tt} + \alpha \mathcal{A} v_t + \mathcal{A} v + \rho z_t|_{\Gamma_0} + g \mathcal{A} \mathcal{G}_2 \mathcal{G}_2^* \mathcal{A} v_t + b(x) v_t& \\
= \mathcal{B} u \text{ on } \mathcal{D}(\mathcal{A}^{1/2})' \times (0, \infty)&, \\
v(0) = v_0 \in H^2(\Gamma_0) = \mathcal{D}(\mathcal{A}^{1/2})&, \\
v_t(0) = v_1 \in H^1(\Gamma_0) = \mathcal{D}(\mathcal{A}^{1/4}), \qquad \gamma > 0&, \qquad (5.3.30)
\end{aligned}
$$

where $b(x) \geq 0$ a.e. in Γ_0.

Thus, in this case we have that the abstract stiffness operator $\mathcal{M}_\gamma$ coincides with $I + \gamma \mathcal{A}_N$ and $\mathcal{D}({\mathcal{M}_\gamma}^{1/2}) = D({A_N}^{1/2}) \sim H^1(\Gamma_0)$ when $\gamma > 0$ and when $\gamma = 0$, $\mathcal{D}({\mathcal{M}_\gamma}^{1/2}) = L_2(\Gamma_0), v_1 \in L_2(\Gamma_0)$.

Since Assumption 5.2.1 is satisfied, we are in a position to apply Corollary 2.3.3 to conclude the well-posedness of the dynamics described by (5.3.30) with $u = 0, \rho = 0$ on the state space $H^2(\Gamma_0) \times \mathcal{D}({\mathcal{M}_\gamma}^{1/2})$.

The PDE version of this equation becomes

$$
\begin{aligned}
v_{tt} - \gamma \Delta v_{tt} + \alpha \Delta^2 v_t + \Delta^2 v + \rho z_t|_{\Gamma_0} + b(x) v_t = \mathcal{B} u \text{ on } \mathcal{D}(\mathcal{A}^{1/2})' \times (0, \infty)& \\
(\Delta + B_1)(v + \alpha v_t) = 0&, \\
\left(\frac{\partial}{\partial \nu} \Delta + B_2 \right) (v + \alpha v_t) - \gamma \frac{\partial}{\partial \nu} v_{tt} = g v_t \text{ on } \partial \Gamma_0 \times (0, \infty)& \\
v(0) = v_0 \in H^2(\Gamma_0), \qquad v_t(0) = v_1 \in H^1_\gamma(\Gamma_0)&. \\
(5.3.31)&
\end{aligned}
$$

As a follow-up question one could ask whether the above uncontrolled and structurally undamped system (i.e., $\alpha = 0, u = 0$) is uniformly stable. If $\gamma = 0$, the answer is yes. In fact, the following result is a special case of a more general nonlinear result established in [109] (see also section 4.3).

Theorem 5.3.2 (see [109]). *Consider the plate model* (5.3.31) *with* $\alpha = 0, u = 0, \rho = 0, g > 0, \gamma = 0$, *and* $l > 0$ *in* (5.3.17). *The corresponding dynamics generate an exponentially stable semigroup on* $H^2(\Gamma_0) \times L_2(\Gamma_0)$.

Remark 5.3.7. *The stabilization problem for the Euler–Bernoulli plate model was considered in* [97]. *However, the result presented in* [97] *(and references therein) requires geometric conditions of the star-shaped type imposed on the boundary* $\partial \Gamma_0$. *Application of microlocal techniques in* [109] *allows us to dispense with these geometric constraints.*

Stability of the corresponding structural acoustic model is discussed in section 5.4 of this chapter.

Results presented above dealt with the case $\gamma = 0$. The situation is very different when $\gamma > 0$. In such case, it is no longer possible to stabilize plates with one boundary stabilizer only. In addition, the boundary feedback acting on the shears needs to be "more aggressive." For $\gamma > 0$, it is well known that a simple feedback $g v_t$ is too weak to provide uniform stabilization for the Kirchhoff model. Thus we need to look for more unbounded boundary feedbacks. The experience gained with plate equations in two dimensions tells us [99] that to stabilize the Kirchhoff plate, two stabilizers are needed—moments and torques.

We return to our algorithm but this time we consider the damping to be acting via moments and torques. We take $g = (g_1, g_2), \mathcal{G} = (\mathcal{G}_1, \mathcal{G}_2)$, given by (5.3.18). The natural space $\mathcal{H} = L_2(\Gamma_0)$ and boundary spaces are as follows:

$$U_1(\partial(\Gamma_0)) = L_2(\partial(\Gamma_0)), \quad U_2(\partial(\Gamma_0)) = H^{-1}(\partial(\Gamma_0)).$$

Note that the choice for a less regular $U_2(\partial(\Gamma_0))$ is obvious, in view of the fact that the previous choice of L_2 space led to feedbacks that were too regular.

Computations of the adjoints (see [140, pp. 305–309])

$$\mathcal{G}_1^* : L_2(\Gamma_0) \to L_2(\partial\Gamma_0), \ \mathcal{G}_2^* : L_2(\Gamma_0) \to H^{-1}(\partial\Gamma_0)$$

give

$$\mathcal{G}_1^* \mathcal{A} v = \frac{\partial}{\partial \nu} v|_{\partial\Gamma_0} : \ \ \mathcal{G}_2^* \mathcal{A} v = \left(\frac{\partial}{\partial \tau}\right)^2 v|_{\partial\Gamma_0}, \qquad v \in \mathcal{D}(\mathcal{A}^{1/2}) \sim H^2(\Gamma_0).$$

It is immediate to verify that $\mathcal{G}_1^*\mathcal{A}$ (resp., $\mathcal{G}_2^*\mathcal{A}$) are bounded and surjective from $H^2(\Gamma_0) = \mathcal{D}(\mathcal{A}^{1/2})$ onto $H^{1/2}(\partial\Gamma_0)$ (resp., $H^{-1/2}(\partial\Gamma_0)$). Thus Assumption 5.2.1 (and also Assumption 2.2.1) is satisfied with

$$U(\partial\Gamma_0) = L_2(\partial\Gamma_0) \times H^{-1}(\partial\Gamma_0), \ \mathcal{H} = L_2(\Gamma_0).$$

Moreover, Assumption 2.2.1 is satisfied with

$$U_0 = H^{1/2}(\partial\Gamma_0) \times H^{-1/2}(\partial\Gamma_0), \ U = L_2(\partial\Gamma_0) \times H^{-1}(\partial\Gamma_0).$$

The nonlinear theory presented in section 2.3 is now applicable for the model.

The effect of rotational forces leads to the following abstract formulation:

$$\begin{aligned} v_{tt} + \gamma \mathcal{A}_N v_{tt} + \alpha \mathcal{A} v_t + \mathcal{A} v + \rho z_t|_{\Gamma_0} + \mathcal{A} g \mathcal{G} \mathcal{G}^* \mathcal{A} v_t &= \mathcal{B} u \\ \text{on } \mathcal{D}(\mathcal{A}^{1/2})' &\times (0,\infty), \\ v(0) = v_0 \in H^2(\Gamma_0) &= \mathcal{D}(\mathcal{A}^{1/2}), \\ v_t(0) = v_1 \in H^1(\Gamma_0) = \mathcal{D}(\mathcal{A}^{1/4}), \qquad &\gamma > 0. \end{aligned} \tag{5.3.32}$$

The well-posedness for this model (with $u = 0, \rho = 0$) on finite energy space $H^2(\Gamma_0) \times H^1(\Gamma_0)$, $\gamma > 0$, and $H^2(\Gamma_0) \times L_2(\Gamma_0)$, $\gamma = 0$, follows from Corollary 2.3.3.

The PDE version of this equation is

$$\begin{aligned} v_{tt} - \gamma \Delta v_{tt} + \alpha \Delta^2 v_t + \Delta^2 v + \rho z_t|_{\Gamma_0} &= \mathcal{B} u \text{ on } \mathcal{D}(\mathcal{A}^{1/2})' \times (0,\infty), \\ (\Delta + B_1)(v + \alpha v_t) &= -g_1 \frac{\partial}{\partial \nu} v_t \text{ on } \partial\Gamma_0 \times (0,\infty), \\ \left(\frac{\partial}{\partial \nu}\Delta + B_2\right)(v + \alpha v_t) - \gamma \frac{\partial}{\partial \nu} v_{tt} &= -g_2 \left(\frac{\partial}{\partial \tau}\right)^2 v_t \text{ on } \partial\Gamma_0 \times (0,\infty), \\ v(0) = v_0 \in H^2(\Gamma_0), \qquad &v_t(0) = v_1 \in H^1_\gamma(\Gamma_0). \end{aligned} \tag{5.3.33}$$

We recognize now that the obtained boundary feedback is essentially that considered in [99] for the Kirchhoff plates. In fact, it was shown in [99] that under star-shaped conditions imposed on $\partial\Gamma_0$, this feedback stabilizes the plate equation with $\gamma > 0, \alpha = 0, \rho = 0, u = 0$.

By applying arguments and techniques in microlocal analysis, one can extend the result of [99] to general domains which do not need to satisfy the star-shaped condition. In fact, the following result is a specialization of a nonlinear result presented in [83, 84].

Theorem 5.3.3. *Consider the plate model* (5.3.33) *with* $\alpha = 0, u = 0, g_1, g_2 > 0, l > 0$. *Then, the corresponding dynamics are exponentially stable on* $H^2(\Gamma_0) \times H^1_\gamma(\Gamma_0)$, *where* $H^1_\gamma(\Gamma_0)$ *is a completion of* $C^\infty(\Gamma_0)$ *functions with respect to the norm given by* (4.4.73).

Remark 5.3.8. *We emphasize that for the case of free boundary conditions two stabilizers are needed to achieve simultaneous stabilization of the PDE* (5.3.33), *for either* $\gamma > 0$ *or* $\gamma = 0$. *This is unlike the case of hinged or clamped boundary conditions, where one single stabilizer provides uniform stability for either the Euler or the Kirchhoff model.*

The issue of stability for the entire structural acoustic model with boundary controls is addressed in section 5.4.4.

5.3.4 Shells: Curved-wall Γ_0

Since in many practical applications the walls of an airplane or helicopter are curved walls, a correct description of the wall is through shell equations defined on a curved boundary Γ_0; see Figures 1b and 2b. This is a bit complicated and it requires some notation. To simplify the exposition, we assume that there is no boundary damping present in the model. The shell model to be considered is a dynamic model for a thin, linear dynamic shell [29, 33, 50, 30, 31]. In what follows we assume that the shell is clamped at the edges. We begin by describing the notation to be used, where we follow [29]. Throughout this section, the summation convention is used with greek letters belonging to the set {1,2} and latin letters belonging to the set {1,2,3}. The middle surface S of the shell (which coincides in our case with Γ_0) is defined to be the image of a connected bounded open set $D \subset \mathcal{E}^2$ under the mapping $\mathbf{\Psi} : (\xi^1, \xi^2) \in \overline{D} \to \mathcal{E}^3$, where $\Psi \in [C^3(\overline{D})]^3$ and $\mathcal{E}^n$ is the n-dimensional Euclidean space. Then for any point on the surface of the shell, it is assumed that the two tangent vectors given by $\mathbf{a}_\alpha = \partial\mathbf{\Psi}/\partial\xi^\alpha$ are linearly independent. Moreover, these two vectors, along with the normal vector, $\mathbf{a}_3 = \frac{\mathbf{a}_1 \times \mathbf{a}_2}{|\mathbf{a}_1 \times \mathbf{a}_2|}$, define a covariant basis for a local reference frame on the surface of the shell. Hereafter, the notation $\mathbf{\Phi}_{,\alpha} = \partial\mathbf{\Phi}/\partial\xi^\alpha$ for any point $(\xi^1, \xi^2) \in \overline{D}$ is used.

The contravariant basis for the tangent plane at any point on the surface of the shell is given by the two vectors $\mathbf{a}^\beta$ defined by the relation $\mathbf{a}_\alpha \cdot \mathbf{a}^\beta = \delta^\beta_\alpha$. The contravariant and covariant vectors are connected by the well-known relationships

$$\mathbf{a}_\alpha = a_{\alpha\beta}\mathbf{a}^\beta, \quad \mathbf{a}^\alpha = a^{\alpha\beta}\mathbf{a}_\beta, \quad a_{\alpha\beta} = \mathbf{a}_\alpha \cdot \mathbf{a}_\beta = a_{\beta\alpha}, \quad \text{and} \quad a^{\alpha\beta} = \mathbf{a}^\alpha \cdot \mathbf{a}^\beta = a^{\beta\alpha},$$

where the matrix $(a_{\alpha\beta})$ represents the first fundamental form of the surface with its inverse given by the matrix $(a^{\alpha\beta})$.

The second fundamental form, denoted by $(b_{\alpha\beta})$ measures the normal curvatures of the middle surface of the shell. It is defined by $b_{\alpha\beta} = b_{\beta\alpha} = -\mathbf{a}_\alpha \cdot \mathbf{a}_{3,\beta} = \mathbf{a}_3 \cdot \mathbf{a}_{\alpha,\beta}$.

The contravariant basis of the second fundamental form is given by $b^\beta_\alpha \equiv a^{\beta\lambda} b_{\lambda\alpha}$.

The Christoffel symbols, $\Gamma^\alpha_{\beta\lambda} = \mathbf{a}^\alpha \cdot \mathbf{a}_{\beta,\lambda}$, give rise to the following covariant derivatives for the displacement vector of the middle surface $\mathbf{v}(\xi^1, \xi^2) = v_i \mathbf{a}^i$ in a fixed reference frame: $v_{\alpha|\beta} = v_{\alpha,\beta} - \Gamma^\lambda_{\alpha\beta} v_\lambda$ and $v_{3|\alpha\beta} = v_{3,\alpha\beta} - \Gamma^\lambda_{\alpha\beta} v_{3,\lambda}$.

For a more complete description of the geometrical considerations and defining characteristics of thin shells, see [29] or [33].

We represent the vector function for all three displacements by $\mathbf{v} = (v_1, v_2, v_3)$, while in-surface displacements are given by $\overrightarrow{v} = (v_1, v_2)$. The transverse displacement is denoted by v_3. In addition, the linear strain tensor is denoted by $\epsilon_{\alpha\beta}(\overrightarrow{v}) = \frac{1}{2}(v_{\alpha|\beta} + v_{\beta|\alpha})$. The middle surface strain tensor, $\gamma_{\alpha\beta}(\mathbf{v}) = \epsilon_{\alpha\beta}(\overrightarrow{v}) - b_{\alpha\beta} v_3$, and the change of curvature tensor,

$$\rho_{\alpha\beta}(\mathbf{v}) = v_{3|\alpha\beta} - b^\lambda_\alpha b_{\lambda\beta} v_3 + b^\lambda_{\alpha|\beta} v_\lambda + b^\lambda_\alpha v_{\lambda|\beta} + b^\lambda_\beta v_{\lambda|\alpha},$$

are defined in Koiter's development of shell theory [29]. These strain measures, along with an appropriate constitutive law, are used to derive corresponding stress measures. The exact form of the stress measures depends on the shell material. Thus, we assume the simplest possible scenario, which is based on the assumptions that the shell consists of elastic, homogeneous, and isotropic material and the strains are small everywhere [31]. Additionally, we assume that all nonzero stress components are imposed on surfaces that are parallel to the middle surface of the shell. Then the stress resultants and moments, respectively, take on the forms

$$n^{\alpha\beta}(\mathbf{v}) = eE^{\alpha\beta\lambda\mu}\gamma_{\lambda\mu}(\mathbf{v}) \quad \text{and} \quad m^{\alpha\beta}(\mathbf{v}) = \frac{e^3}{12} E^{\alpha\beta\lambda\mu}\rho_{\lambda\mu}(\mathbf{v}),$$

where e represents the thickness of the shell, and the tensor of elastic moduli is given by

$$E^{\alpha\beta\lambda\mu} = \frac{E}{2(1+\nu)}\left[a^{\alpha\lambda}a^{\beta\mu} + a^{\alpha\mu}a^{\beta\lambda} + \frac{2\nu}{1-\nu}a^{\alpha\beta}a^{\lambda\mu}\right].$$

Here, E is Young's modulus and ν is Poisson's ratio for the material. Note that the tensor of elastic moduli is positive [30, 29]. Additionally, we make use of the following notation:

$$(u, v) = \int_D uv\sqrt{a}\, d\xi_1 d\xi_2, \qquad |u|_{s,D} \equiv |u|_{H^s(D)}, \qquad (u, w)_{0,D} \equiv (u, w)_{L_2(D)},$$

where $a = det(a_{\alpha\beta}) \neq 0$. When $s = 0$, we may simply write $|u|_D \equiv |u|_{0,D}$. For simplicity of notation, we denote time derivatives by dots. Thus $u_t = \dot{u}, u_{tt} = \ddot{u}$.

To model the dynamics of a shell, we follow the arguments given in [32], where the dynamics are introduced with the help of a bilinear form given by

$$\begin{aligned} b(\mathbf{v}, \mathbf{u}) \equiv{}& \hat{\rho}e(a^{\alpha\beta}u_\alpha, v_\beta) + \hat{\rho}e(u_3, v_3) + \hat{\rho}\frac{e^3}{12}[((b^1_1 b^2_2 - b^2_1 b^1_2)a^{\alpha\beta}u_\alpha, v_\beta) \\ &+((b^1_1 b^2_2 - b^2_1 b^1_2)u_3, v_3) + (a^{\alpha\beta}(u_{3|\alpha} + b^\lambda_\alpha u_\lambda), v_{3|\beta} + b^\nu_\beta v_\nu) \\ &+(u_\alpha, v_{3|\beta} b^\nu_\nu) + (u_{3|\alpha}, v_\beta b^\nu_\nu) + 2(b^\lambda_\alpha u_\lambda, v_\beta b^\nu_\nu)], \end{aligned} \tag{5.3.34}$$

where the constant $\hat{\rho}$ is the density constant.

The model to be considered is given in the following variational form in which we look for solutions $\mathbf{v} = (v_1, v_2, v_3) = (\overrightarrow{v}, v_3) \in \left[H_0^1(D)\right]^2 \cap H_0^2(D)$ such that

$$\begin{aligned} b(\ddot{\mathbf{v}}, \Phi) &+ \left(eE^{\alpha\beta\lambda\mu}\gamma_{\alpha\beta}(\mathbf{v}), \gamma_{\lambda\mu}(\Phi)\right) + ed_1\left(E^{\alpha\beta\lambda\mu}\gamma\alpha\beta(\dot{\mathbf{v}}), \gamma_{\lambda\mu}(\Phi)\right) \\ &+ \left(\frac{e^3}{12}E^{\alpha\beta\lambda\mu}\rho_{\alpha\beta}(\mathbf{v}), \rho_{\lambda\mu}(\Phi)\right) + e^3 d_2\left(E^{\alpha\beta\lambda\mu}\rho_{\alpha\beta}(\dot{\mathbf{v}}), \rho_{\lambda\mu}(\Phi)\right) \\ &= \rho(\dot{z}, \phi_3) + (\mathcal{B}_1 u, \phi_\alpha) + (\mathcal{B}_2 u, \phi_3) \qquad (5.3.35)\end{aligned}$$

for all $\Phi = (\phi_1, \phi_2, \phi_3) = (\overrightarrow{\phi}, \phi_3) \in \left[H_0^1(D)\right]^2 \times H_0^2(D)$. The term proportional to the third powers of the thickness e represents rotational forces. In fact, by analogy with the plates, we introduce constant $\gamma \equiv \rho\frac{e^3}{12}$, which represents rotational inertia. Since this constant is small, it may be neglected in some models. The variable $\dot{z}$ in (5.3.35) represents back pressure generated by the interior acoustic field, while the control operators $\mathcal{B}_1$ and $\mathcal{B}_2$ in (5.3.35) model piezoceramic patches bonded to the wall, and they will be specified later. Mathematically speaking, the operators $\mathcal{B}_\alpha$ represent the first and second derivatives of the characteristic functions describing the region occupied by the patches bonded to the walls [72, 71]. As such, they are represented as a linear combination of delta functions ($\mathcal{B}_1$) and derivatives of delta functions ($\mathcal{B}_2$) which are supported on a closed curve in S. In the case of cylindrical shells, the precise structure of these terms has been determined [18]. The important feature of these operators is their regularity property (they are, of course, unbounded). In particular,

$$\mathcal{B}_1 u \in H^{-1/2-\epsilon}(D), \ \mathcal{B}_2 u \in H^{-3/2-\epsilon}(D) \text{ for } u \in [L_2(D_j)]_{j=1}^N, \qquad (5.3.36)$$

where $D_j \subset D$, $j = 1 \dots N$, are regions occupied by patches bonded to the wall. The value of ϵ can be taken arbitrary small.

The coefficients d_1 and d_2 in (5.3.35) are the damping coefficients and represent the presence of passive damping (Kelvin–Voight structural damping) in the structure of the shell. Without loss of generality we assume that $d_1 = d_2 \equiv \alpha \geq 0$.

What is important for our subsequent arguments is the fact that the shell equation can be represented in an abstract form as (5.2.2) with the appropriate choices for the operators $\mathcal{A}, \mathcal{M}_\gamma, \mathcal{C}, \mathcal{B}$.

Choice of the operators $\mathcal{A}$, $\mathcal{C}$, and $\mathcal{M}_\gamma$. Since the domains $\Gamma_0 = S$ and D are isomorphic, it suffices to define all the operators on D. Thus, with $\mathcal{H} \equiv [L_2(D)]^3$, we define the operator $\mathcal{A} : [L_2(D)]^3 \to [L_2(D)]^3$ by

$$(\mathcal{A}\mathbf{v}, \Phi)_{0,D} = \frac{e^3}{12}\left(E^{\alpha\beta\lambda\mu}\rho_{\alpha\beta}(\mathbf{v}), \rho_{\lambda\mu}(\Phi)\right) + e\left(E^{\alpha\beta\lambda\mu}\gamma_{\alpha\beta}(\mathbf{v}), \gamma_{\lambda\mu}(\Phi)\right) \qquad (5.3.37)$$

for all $\mathbf{v}, \Phi \in \left[H_0^1(D)\right]^2 \times H_0^2(D)$. We have

$$\mathcal{D}(\mathcal{A}) = \left\{v_\alpha \in H^2(D) \cap H_0^1(D); \ v_3 \in H^4(D) \cap H_0^2(D)\right\}.$$

Operator $\mathcal{A}$ is a self-adjoint and positive operator, where this latter condition follows from the strict positivity of the tensor $E^{\alpha\beta\lambda\mu}$. Moreover, by the shell version of Korn's inequality [29] we also have:

$$(\rho_{\alpha\beta}(\mathbf{v}), \rho_{\alpha\beta}(\mathbf{v})) + (\gamma_{\alpha\beta}(\mathbf{v}), \gamma_{\alpha\beta}(\mathbf{v})) \geq c(|v_3|_{2,D}^2 + |v_\alpha|_{1,D}^2). \qquad (5.3.38)$$

Thus, $c\left[|v_3|^2_{2,D} + |v_\alpha|^2_{1,D}\right] \le (\mathcal{A}\mathbf{v},\mathbf{v}) \le C\left[|v_3|^2_{2,D} + |v_\alpha|^2_{1,D}\right]$.
Moreover, it is known [67]

$$\mathcal{D}(\mathcal{A}^\theta) = \begin{cases} \left[H_0^{2\theta}(D)\right]^2 \times H_0^{4\theta}(D) & \text{for } \theta \le \frac{1}{2}, \\ \mathcal{D}(\mathcal{A}^{\frac{1}{2}}) \cap \left(\left[H^{2\theta}(D)\right]^2 \times H^{4\theta}(D)\right) & \text{for } \theta > \frac{1}{2}. \end{cases} \tag{5.3.39}$$

We rewrite our original dynamics as

$$\mathcal{M}_\gamma \ddot{\mathbf{v}} + \mathcal{A}\mathbf{v} + \alpha\mathcal{A}\dot{\mathbf{v}} + \mathcal{C}\dot{z} = \mathcal{B}u, \tag{5.3.40}$$

where the unbounded operators $\mathcal{M}_\gamma : [L_2(D)]^3 \to [L_2(D)]^3$ and $\mathcal{C} : L_2(\Omega) \to [L_2(D)]^3$ are given by

$$(\mathcal{M}_\gamma \mathbf{v}, \Phi)_{0,D} \equiv b(\mathbf{v}, \Phi),$$
$$(\mathcal{C}z, \Phi)_{0,D} \equiv -\left(z|_S, \phi_3\right),$$
$$(\mathcal{B}u, \Phi) \equiv (\mathcal{B}_1 u, \phi_\alpha) + (\mathcal{B}_2 u, \phi_3).$$

To describe the properties of the stiffness operator $\mathcal{M}_\gamma$ we find it convenient to introduce the spaces

$$H_\gamma(D) \equiv \begin{cases} H_0^1(D), & \gamma > 0, \\ L_2(D), & \gamma = 0. \end{cases}$$

Since the shell is assumed thin, the operator $\mathcal{M}_\gamma$ is positive, self-adjoint on $[L_2(D)]^3$. More precisely (see [32, Lemma 2.1]),

$$c[|u_\alpha|^2_{0,D} + |v_3|^2_{H_\gamma(D)}] \le (\mathcal{M}_\gamma \mathbf{v}, \mathbf{v})_{0,D} \le C[|u_\alpha|^2_{0,D} + |v_3|^2_{H_\gamma(D)}].$$

Hence

$$\mathcal{M}_\gamma^{-1} \in \mathcal{L}([L_2(D)]^2 \times [H_\gamma(D)]' \to [L_2(D)]^2 \times H_\gamma(D)),$$

where $[H_\gamma(D)]'$ denotes the dual of $H_\gamma(D)$ with respect to $L_2(D)$ topology. In what follows we topologize $[L_2(D)]^2 \times H_\gamma(D)$ by

$$|\mathbf{v}|_{[L_2(D)]^2 \times H_\gamma(D)} \equiv |\mathcal{M}_\gamma^{1/2}\mathbf{v}|_{0,D}.$$

It is well known that the semigroup corresponding to the structurally damped plates and beams equation (5.2.2) with $\mathcal{C} = 0$ generates an analytic semigroup, which is exponentially stable [48]. Thus, the corresponding acoustic structure (5.2.2) with $\alpha = d_1 = d_2 > 0$ represents coupling between hyperbolic and analytic semigroups. Moreover, it was shown in [117, 118] that the structural acoustic model described by (5.2.1) with $d_2, d_3, l > 0$ and the above shell equation with $\alpha > 0, u = 0$ is exponentially stable on $H^1(\Omega) \times L_2(\Omega) \times \mathcal{D}(\mathcal{A}^{1/2}) \times \mathcal{D}(\mathcal{M}_\gamma^{1/2})$. A precise statement for this result is given in section 5.4.

Choice of the control operator $\mathcal{B}$. We describe next the control operator $\mathcal{B}$. To accomplish this, let D_j^* denote a parametrization (in D) of a closed curve in S which delimits the area D_j where the piezoceramic patches are bonded to the surface of the shell. We define the operator $\mathcal{B} : [L_2(D_j)]_{j=1}^k \equiv \mathcal{U} \to \mathcal{D}(\mathcal{A})'$

$$(\mathcal{B}u, \Phi)_{0,D} = (\mathcal{B}_1 u, \phi_\alpha) + (\mathcal{B}_2 u, \phi_3)$$
$$= \sum_{j=1}^k \int_{D_j^*} u_j[a_j\phi_\alpha + b_j \nabla\phi_3 \cdot \nu] dD_j^* \text{ for } \Phi \in \mathcal{D}(\mathcal{A}^{1/2}), \tag{5.3.41}$$

where the coefficients a_j and b_j reflect the properties of the patches and the material and the vector ν denotes normal direction to the curve D_j^*. Control functions $u_i(t)$ denote the voltage applied to each patch. More precise expressions for these coefficients are given in [18] for the case of cylindrical shells and in [72, 71] for more general shells. However, for our purposes, only the general structure of the control operator is important. We note that the control operator is unbounded. However, the important property of the control operator is given by the relation

$$\mathcal{A}^{-r}(\mathcal{B}) \in \mathcal{L}(\mathcal{U}; [L_2(D)]^3) \quad \text{for } r > \frac{3}{8}. \tag{5.3.42}$$

Indeed, (5.3.42) follows from the definition given in (5.3.41), Sobolev's embeddings, and the characterization of fractional powers given by (5.3.39). To see this, it suffices to note that $H^{1/2+\epsilon}(D)$ functions, when restricted to D_j^*, are in $L_2(D_j^*)$. Similarly, restrictions to D_J^* of $H^{3/2+\epsilon}(D)$ functions are in $H^1(D_j^*)$, where ϵ can be taken arbitrary small.

5.4 Stability in Linear Structural Acoustic Models

In this section we discuss questions related to stability of the overall structural acoustic models, which consist of the wave equation (5.2.1) coupled with the plate equation (5.2.2). This particular issue is critical when studying infinite-horizon control problems. In what follows, we take $u = 0, f = 0$ in (5.2.1), (5.2.2).

In section 5.3 we have described several plate equations that model flexible walls of an acoustic chamber. The basic damping mechanism affecting these models is either structural damping ($\alpha > 0$) or mechanical boundary damping ($g > 0$). In both cases, we established conditions under which the plate alone is uniformly stable. The main task now is to show that this stability is transferred onto the entire structure. Although some of the results presented are special cases of the more general nonlinear theory presented earlier in Chapter 4, for the convenience of the reader we collect some of the major findings applicable to linear systems.

It is well known that to achieve this we must have some damping in the acoustic chamber. Damping on the plate alone will not suffice. Indeed, from [155, 7] we know that the structural acoustic model without any damping imposed on the wave component is only strongly stable and not uniformly stable. (Uniform stability can be achieved, as recently shown in [10], for a model with zero initial conditions on the wave component). For uniform stability of the entire structural acoustic model, the presence of some damping on the wave is necessary. In what follows we discuss the linear wave model (5.2.1) with damping (internal or boundary) introduced by the parameters $d_i, i = 1, 2, 3$. With the aim of minimizing the amount of damping acting on the wave component, we introduce certain geometric conditions satisfied by the domain Ω. These conditions are responsible for the nontrapping rays phenomenon.

Assumption 5.4.1 (see Figure 2). *Assume that Γ_1 is convex (that is, the level set function describing Γ_1 has a nonnegative Hessian in the neighborhood of Γ_1 in Ω) and satisfies the following star-shaped condition: There exists a point $x_0 \in R^n$ such that*

$$(x - x_0) \cdot \nu \le 0, \qquad x \in \Gamma_1,$$

where ν is an outward normal vector to Γ_1.

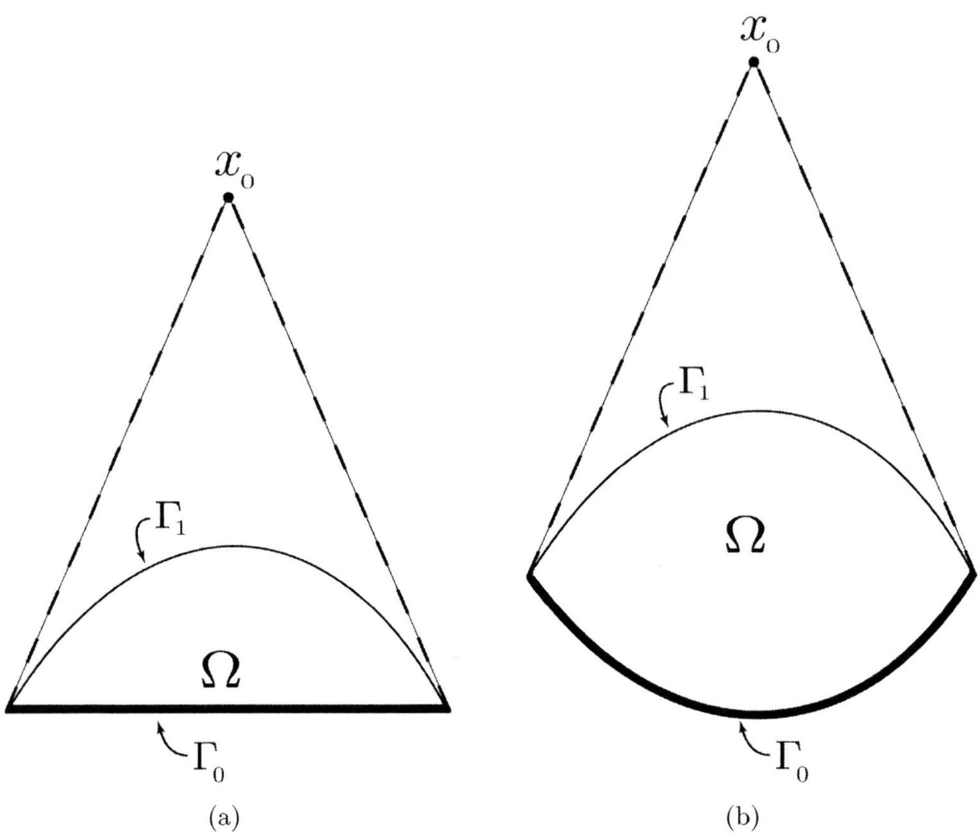

Figure 2: Cross sections of Ω: (a) $s = 1$, flat Γ_0. (b) $s = 3$, curved Γ_0.

Remark 5.4.1. *Assumption* 5.4.1 *implies* [112, 144] *an existence of the vector field* $\vec{h} = \nabla\phi \in C^2(\Omega)$ *such that*

$$\vec{h} \cdot \nu = 0 \text{ on } \Gamma_1,$$

$$Hessian\ (\phi) > 0 \text{ in } \Omega.$$

This particular vector field is needed for the construction of appropriate multipliers used in the proof of stabilization. Note that when the uncontrolled portion of the boundary satisfies zero Dirichlet conditions, the requirements for such a vector field are less stringent. Indeed, it suffices to take $h = x - x_0$ *such that* $(x - x_0) \cdot \nu \leq 0,\ x \in \Gamma_1$. *However, if the portion of the boundary is subject to Neumann data, we must have that* $h \cdot \nu = 0$ *on* Γ_1. *To achieve this, local convexity is sufficient (although it is known not to be necessary; see* [144].

5.4.1 Internal damping on the wall

Our first stability result deals with the situation when structural damping is active in the plate model. In what follows we use the following space describing compatibility conditions among the elements in the state space X:

$$X_0 = \Big\{ y = (z_1, z_2, v_1, v_2) \in X = H_z \times H_v :$$
$$\int_\Omega z_2 dx = \int_{\Gamma_0} [v_1 - d_3 z_1] ds - d_2 \int_{\Gamma_1} z_1 ds \Big\}. \qquad (5.4.43)$$

The spaces H_v correspond to plate dynamics; their specification will depend on the given boundary conditions. The role of compatibility conditions imposed in X_0 is to prevent steady states from the dynamics. In fact, it can be shown by an elementary compactness-uniqueness argument [42] that the underlying norm dictated by X_0 is a norm (and not a seminorm) on $H_1(\Omega) \times L_2(\Omega)$ for the wave component.

Theorem 5.4.1. *Consider the structural acoustic model described by* (5.2.1) *with* $f = 0$ *and the plate equation given by* (5.3.21) *or* (5.3.23) *or* (5.3.25) *with* $u = 0$ *and* $\alpha > 0$. *In addition, assume that* $d_1 + d_2 + d_3 > 0$, *and, in the case* $d_2 = d_1 = 0$, *we impose also geometric Assumption* 5.4.1. *Then, the dynamics described by the structural acoustic model described above generate an exponentially stable semigroup on the space* X_0 *given by* (5.4.43) *where the spaces* H_z *and* H_v *are the usual energy spaces associated with the wave and plate dynamics.* $H_z = H^1(\Omega) \times L_2(\Omega)$, $H_v = \mathcal{D}(\mathcal{A}^{1/2}) \times \mathcal{D}({\mathcal{M}_\gamma}^{1/2})$, *and* $\mathcal{A}, \mathcal{M}_\gamma$ *are specified, respectively, for each set of boundary conditions in* (5.3.21), (5.3.23), *or* (5.3.25). *In addition, the decay rates obtained for the respective semigroups are uniform with respect to* $0 \le \gamma \le M$.

The proof of this theorem follows by combining directly the arguments given in [2, 8] with those in [112, 114, 147], where various configurations of damping placement in the acoustic chamber were considered.

Remark 5.4.2. *We note that the result of Theorem* 5.4.1 *remains valid for any value of* $0 \le \theta \le 1/2$ *in* (5.2.2) [147].

We now turn to shell models. In fact, with structural damping active for the shell's equation, one can show that the resulting structural acoustic model is uniformly stable. The result in this direction is provided below.

Theorem 5.4.2 (see [117, 118]). *Consider the structural acoustic model described by* (5.2.1) *with* $f = 0$ *and the shell equation in* (5.3.35) *or, equivalently,* (5.3.40) *with* $\alpha > 0, u = 0$. *We assume that* $d_1 + d_2 + d_3 > 0$, *and in the case* $d_2 = d_1 = 0$ *we also impose geometric Assumption* 5.4.1; *see Figure* 2*b*. *If* $d_1 = d_3 = 0$, *we assume instead that Assumption* 5.4.1 *holds on* $\Gamma_D = D$. *Then, the resulting semigroup is exponentially stable on* X_0 *given by* (5.4.43) *with* $H_z = H^1(\Omega) \times L_2(\Omega)$ *and* $H_v = [H_0^1(D)]^2 \times H_0^2(D) \times [L_2(D)]^2 \times H_\gamma(D)$. *The corresponding decay rates are uniform with respect to* $0 \le \gamma \le M$.

To achieve exponential stability of structural acoustic models with internal damping, it is not necessary to have structural damping. Weaker forms of internal damping are sufficient for this purpose. In fact, a popular form of internal damping is viscous damping, which corresponds to the value of $\theta = 0$, when $\mathcal{M}_\gamma = I$ in (5.2.2).

In what follows we analyze the stability of structural acoustic models (5.2.2) with various values of $0 \le \theta \le 1/2$.

Theorem 5.4.3. *Consider the structural acoustic model described by* (5.2.1) *with* $f = 0$ *and the plate equation given by* (5.2.2) *with* $u = 0$, $\alpha > 0$, $0 \le \theta \le 1/2$. *We assume that*

- $d_1 + d_2 + d_3 > 0$ *and, in the case* $d_2 = d_1 = 0$, *we impose also geometric Assumption* 5.4.1,
- $|\mathcal{M}_\gamma{}^{1/2} v|_{[L_2(\Gamma_0)]^s} \le |\mathcal{A}^\theta v|_{[L_2(\Gamma_0)]^s}, \quad v \in \mathcal{D}(\mathcal{A}^{1/2})$.

Then, the dynamics described by the structural acoustic model described above generates an exponentially stable semigroup on the space X_0 *given by* (5.4.43) *with the spaces* H_z *and* H_v *the usual energy spaces associated with the wave and plate dynamics.* $H_z = H^1(\Omega) \times L_2(\Omega)$, $H_v = \mathcal{D}(\mathcal{A}^{1/2}) \times \mathcal{D}(\mathcal{M}_\gamma{}^{1/2})$. *The decay rates obtained are uniform with respect to* $0 \le \gamma \le M$.

The proof of this theorem follows from a standard argument based on equipartition of potential and kinetic plate energy combined with the wave stabilization method presented in [112].

Remark 5.4.3. *Note that the second (last) condition in Theorem* (5.4.3) *is redundant whenever* $\mathcal{M}_\gamma = I$. *Instead, when inertial moments are present in the model, the strength of the viscous damping (measured by* θ*) must be more substantial. For instance, in the case of hinged, clamped, or free plates with rotational forces accounted for, we must have* $1/4 \le \theta \le 1/2$. *A typical example of a hinged plate with viscous damping and inertial moments corresponds to taking* $\theta = 1/4$, $\mathcal{M}_\gamma = I + \gamma A_D, \mathcal{A} = A_D^2$, *hence* $\mathcal{A}^{1/2} = A_D$. *The resulting PDE model becomes*

$$
\begin{aligned}
v_{tt} - \gamma \Delta v_{tt} - \alpha \Delta v_t + \Delta^2 v + \rho z_t|_{\Gamma_0} &= \mathcal{B}u \ \text{on} \ \mathcal{D}(\mathcal{A}^{1/2})' \times (0,\infty), \\
v = 0, \quad \Delta v &= 0 \ \text{on} \ \partial\Gamma_0 \times (0,\infty), \\
v(0) = v_0 \in H^2(\Gamma_0) \cap H_0^1(\Gamma_0), \quad v_t(0) &= v_1 \in H_\gamma^1(\Gamma_0).
\end{aligned}
\tag{5.4.44}
$$

Remark 5.4.4. *The result of Theorem* 5.4.3 *can be extended to less rigid structures of viscous damping modeled by the operator* $D\mathbf{v}_t$ *where* $D^{1/2}$ *is comparable (in the sense of* [48]*) to* $\mathcal{M}_\gamma{}^{1/2}$.

5.4.2 Boundary damping on the wall

In the absence of structural or internal damping affecting the plate we are forced to consider other types of dissipation. In what follows we discuss the stability of structural acoustic models subject to boundary damping affecting the plate. The results depend on the type of boundary conditions imposed on the plate model.

Hinged plates. We consider the plate model given by (5.3.29) with $\alpha = 0, u = 0, g_2 > 0$. Our goal is the analysis of stability for the overall structural acoustic model. The result presented below describes the situation.

Theorem 5.4.4. *Consider the structural acoustic model consisting of an acoustic medium described by* (5.2.1) *and the wall described by* (5.3.29) *with* $\alpha = 0$, $u = 0$, $g_2 > 0$. *We assume that* $d_1 + d_2 + d_3 > 0$ *and, in the case* $d_2 = d_1 = 0$, *we impose the geometric Assumption* 5.4.1. *The corresponding dynamics generate an exponentially stable semigroup on the space* X_0 *given in* (5.4.43) *with* $H_z = H^1(\Omega) \times L_2(\Omega)$, $H_v = \mathcal{D}(\mathcal{A}^{1/2}) \times \mathcal{D}(\mathcal{M}_\gamma{}^{1/2}) \sim H^2(\Gamma_0) \cap H_0^1(\Gamma_0) \times H_\gamma^1(\Gamma_0)$. *Moreover, the decay rates of the resulting semigroup are uniform with respect to the parameter* $0 \le \gamma \le M$.

The proof of this theorem in the case when the damping is active on the full boundary (i.e., $d_2, d_3 > 0$) is given in [40, 41]. By combining the arguments of [112, 114] (wherein partial boundary damping is treated) with these in [41], one obtains the result stated in Theorem 5.4.4.

Free plates. We consider the plate model given by (5.3.31) with $\alpha = 0, u = 0, g > 0$, and $\gamma = 0$. The following result describes uniform stability for the corresponding structural acoustic model.

Theorem 5.4.5 (see [109]). *Consider the structural acoustic interaction given by the wave equation* (5.2.1) *with* $f = 0$ *and the plate equation* (5.3.31) *with* $\alpha = 0, u = 0, g > 0, \gamma = 0, l > 0$. *We assume that* $d_1 + d_2 + d_3 > 0$ *and, in the case* $d_2 = d_1 = 0$, *we impose the geometric Assumption* 5.4.1. *The corresponding dynamics generate an exponentially stable semigroup on* X_0 *The definition of* X_0 *is given in* (5.4.43) *with* $H_z = H^1(\Omega) \times L_2(\Omega)$ *and* $H_v = H^2(\Gamma_0) \times L_2(\Gamma_0)$.

Remark 5.4.5. *In the special case when* $d_3 = d_2 > 0$, *the result above is included in Theorem* 4.3.1. *As we saw there, the proof of this (and Theorem* 5.4.5*) relies heavily on microlocal arguments and, in particular, on recent results in* [189] *pertaining to microlocal decomposition of boundary operators for higher-order PDEs. This technique, applied to free boundary conditions, is key in allowing stabilization of the plate with one control only, without assuming any geometric restrictions on the shape of the plate.*

If rotational forces are present in the model, stabilization via free boundary conditions requires, as for the uncoupled plates, two stabilizers: moments and torques. The following result pertains to the case $\gamma > 0$.

Theorem 5.4.6. *Consider the structural acoustic interaction given by the wave equation* (5.2.1) *with* $f = 0$ *and the plate equation* (5.3.33) *with* $\alpha = 0, u = 0, g_1, g_2 > 0, l > 0$. *We assume that* $d_1 + d_2 + d_3 > 0$ *and, in the case* $d_2 = d_1 = 0$, *we impose the geometric Assumption* 5.4.1. *The corresponding dynamics generates an exponentially stable semigroup on* X_0. *The definition of* X_0 *is given in* (5.4.43) *with* $H_z = H^1(\Omega) \times L_2(\Omega)$ *and* $H_v = H^2(\Gamma_0) \times H^1(\Gamma_0)$.

The proof of Theorem 5.4.6 follows by combining the argument used in the proof of Theorem 5.4.5 with the one developed in the context of boundary stabilization of Kirchhoff plates with free boundary condition [84].

Remark 5.4.6. *One should be able to prove that the decay rates obtained in Theorem* 5.4.6 *are uniform with respect to* $0 \le \gamma \le M$ *when taking the form of the boundary feedback in* (5.3.33) *with positive constants* g, g_1, g_2

$$(\Delta + B_1)v = -g_1\sqrt{\gamma}\frac{\partial}{\partial\nu}v_t,$$

$$\left(\frac{\partial}{\partial\nu}\Delta + B_2\right)v - \gamma\frac{\partial}{\partial\nu}v_{tt} = gv_t - g_2\sqrt{\gamma}\left(\frac{\partial}{\partial\tau}\right)^2 v_t. \tag{5.4.45}$$

In fact, this independence of the rates with respect to γ *with the feedback given in* (5.4.45) *can be proved by following* [99], *where the corresponding result is proved for the plates under an additional assumption* $(x - x_0) \cdot \nu > 0$ *on* $\partial\Gamma_0$. *To dispense* *with this geometric restriction, one needs to evoke microlocal analysis arguments. The difficulty in the free case is that the control of the constants depending on γ is necessary at the microlocal level. This point is technical and has not been carried out yet.*

Remark 5.4.7. *One may also consider cases when boundary damping on the plate is active only on a portion of the plate boundary. In such cases, suitable geometric conditions on the nondissipative portion of the boundary are needed.*

5.5 Comments and Open Problems

- The structural acoustic model in (5.2.2) does not cover thermal plates. However, this can be easily remedied by introducing another equation describing the effects of the temperature. Since this added generalization will make the notation more complicated, we opted not to include thermal plates at the abstract level. However, the arguments of this section can be easily adjusted to cover the case of thermal plates as well; see [148].
- As mentioned in section 5.3, two problems that deserve complete analysis are (i) uniform stabilization of the structural acoustic model with the boundary damping in the clamped boundary conditions, and (ii) stabilization, uniform with respect to the parameter γ, of the Kirchhoff version of the structural acoustic model with the damping in the free boundary conditions via moments and shears. Particular emphasis should be paid to the uniformity of the decay rates with respect to the parameter γ and with the feedback given in (5.4.45).

  Regarding the stabilization with boundary damping in clamped boundary conditions, the results existing for uncoupled plates [168, 23] provide a construction of a boundary damping which yields exponential decay rates in topology of optimal regularity [126], which, however, is much below the finite energy level. Restudy of the problem with a proper focus on obtaining stability estimates at the finite energy level appears to be an outstanding task.
- It would be interesting to consider boundary stabilization of the same models as considered in this chapter with an added internal damping (viscous or structural). The goal should be to show that the decay rates of the energy (obtained via boundary feedback) are uniform with respect to the amount (very small) of this internal damping. (Note that in Theorems 5.4.4, 5.4.5 and 5.4.6, the parameter responsible for structural damping α equals zero.) The addition of viscous damping should not present any technical difficulties. This is because viscous damping is modeled by a bounded operator and as such is robust with respect to the multipliers calculations. However, the addition of structural damping is a more delicate issue. The techniques used in proving uniform boundary stability may not be robust enough to support the presence in the model of uncontrolled structural damping. Positive results have been obtained so far for the cases of clamped or hinged boundary conditions in thermoelastic plates and boundary damping imposed on the wave component [147].
- A question that deserves attention is the issue of mixed boundary conditions imposed on the walls of an acoustic chamber. One would like to consider

a combination of Dirichlet and Neumann boundary conditions prescribed on appropriate portions of Γ. The difficulty arises because of potential singularities developed in elliptic solutions. To cope with the issue, one should use new results in [69], wherein geometrical conditions imposed on Ω prevent development of such singularities. For instance, a combination of Neumann data prescribed on a flat wall Γ_0 with Dirichlet boundary conditions given on Γ_1 yield sufficiently regular solutions if the angle at the intersection of Γ_1 and Γ_0 is less than π (typically one has $\pi/2$). In the case of regular domains, singular solutions do develop. However, even in this case there is a way to justify multipliers computations, if the right geometric star-shaped conditions are imposed on Ω [69]. Detailed study of this issue in the context of structural acoustic problems remains an interesting and worthwhile endeavor.

Chapter 6

Feedback Noise Control in Structural Acoustic Models: Finite Horizon Problems

6.1 Orientation

In this chapter we study an optimal control problem, defined over a finite time horizon, for two classes of interactive systems that exemplify either hyperbolic-parabolic coupling or hyperbolic-hyperbolic coupling. The prototypes for these models are structure acoustic interactions introduced in Chapter 5. We focus on the two canonical cases: structurally damped walls, that is, the parameter α in (5.2.2) is strictly positive, and undamped walls, with the parameter $\alpha = 0$.

The main goal here is to provide a meaningful optimal synthesis of the optimal control problem, in terms of quantities such as the Riccati operator and the decoupling function $r(t)$, where the first is defined via solutions of an appropriate differential Riccati equation (DRE), and the latter is defined in terms of the forced nonautonomous dynamics governed by the adjoint to the Riccati feedback generator.

These concepts are well known in the context of finite dimensional theory and have also been generalized within the framework of C_0 semigroups, albeit with bounded control operators [14, 152]. Thus, if the control operator B, appearing in the description of controlled dynamics (5.2.9), were bounded from the control space to the state space, there would be no need here to study the problem (at least from the mathematical point of view). It would suffice to borrow the existing results and translate them in terms of the quantities describing the C_0 semigroup in (5.2.9). However, this is not the case. Indeed, the control operator arising in structural acoustic models are very badly unbounded (at least within the context of models proposed in [16, 72] and references therein), and therefore the standard control theory does not apply.

On the other hand, the last 15 years or so have witnessed a vigorous development of control theory in the context of unbounded control operators, motivated mostly by boundary and point control problems for PDEs. (We mean unbounded here in the sense that the range of the control operator B is not in the basic state space.) Of course, one could simply (and probably naively) bypass this difficulty by redefining the state space to be large enough to contain the range of B and then attempt to treat the entire control problem on a larger space. In doing this, however, the observation operator very often becomes unbounded and so we are merely exchanging one problem for another (unless the quantities observed are very smooth, which is seldom the case). In addition, the change of topology of the underlying state space changes entirely the equation's stabilizability and controllability properties, which are critical for the infinite horizon case. Moreover, and most important, even if one could successfully account for the unboundedness of B by enlarging the state space, the corresponding Riccati equation would be defined only on an artificial space. Such an abstract solution to (5.2.9) would be rather useless as it would give no theoretical or physical insight into the original applied problem, nor could the corresponding control quantities be measured or numerically constructed in any meaningful way.

This was the main and prime motivation for introducing the theory of optimal control with unbounded controls, with a view to setting this theory on proper mathematical footing. A comprehensive treatment of the topic can be found in recent books [129, 140, 28, 66, 21]. The principal lesson to be gleaned from these works is that the generalization from bounded control theory to unbounded control constitutes a technical quantum leap; it is a much more difficult and involved task than the previous generalization of finite-dimensional theory to C_0 semigroup theory but with bounded control operators. This is not meant to downplay the earlier and necessary generalization, which is technical in its own right. We point out that in the study of unboundedly controlled PDEs, one is unavoidably confronted with new phenomena which have no analogues in previous control theories. To make true headway in this endeavor, one must incorporate a vast array of PDE tools, including the microlocal analysis and pseudodifferential operator theory needed to obtain sharp PDE inequalities. Therefore, for the case of unbounded control operators, the corresponding theory is not simply a rehash of what has been done in the past; where given A to be either a matrix or generator of a semigroup, one could, without much thought, freely construct such terms as $e^{At}B$. The difference between B bounded and unbounded are eminent and obvious. We point out the most glaring ones. For the unbounded case, we may no longer have the DRE defined in any reasonable sense. In addition, the gain operator may not be defined on the entire domain, and thus one needs to make sure when and where it is defined at all. Indeed, for hyperbolic and exactly controllable systems, the gain operator cannot be bounded unless the control operator B itself is. No pathologies of this type can occur in the case of bounded control operators. Moreover, properties of the control quantities, including the gain, are absolutely critical when considering numerical-implementations algorithm design. It is an expected fact that properties and pathologies of the controller and characterizing Riccati equation, if available, depend heavily and critically on the type of the dynamics considered as well as on the degree of unboundedness of B. For example, the case of parabolic dynamics (where terms qualitatively are more regular) presents a situation drastically different from that corresponding to hyperbolic dynamics (wherein the effects of propagation of singularities must be fully accounted for to accurately determine the behavior of the solution [129, 28, 140]).

In reference to our canonical example of interactive systems, we have already noted that structural acoustic interactions display a mixed behavior of parabolicity and hyperbolicity. Thus, the previously developed theories, which are keenly tuned to the features of each respective type of dynamics (hyperbolic versus parabolic [140]), can no longer be applied to resolve the structural acoustic problem. Moreover, there is no way to predict how the component characteristics (hyperbolic or parabolic) will manifest themselves in the entire structure, and so for interactive systems the following questions loom: Do we have a Riccati equation or not? What is the meaning of the gain operator? Even in the case of hyperbolic-hyperbolic coupling, where one might surmise that the interactive system has many features of hyperbolicity, the coupling mechanism between the interacting components brings forth new phenomena that will not be present in a typical hyperbolic configuration.

To obtain the best possible results, we treat two distinct classes of structural acoustic problem: the respective configurations of hyperbolic-parabolic and hyperbolic-hyperbolic dynamics.

The control problem to be considered is formulated in section 6.2. Section 6.3 provides the statements of the main results. Sections 6.4–6.6 are devoted to their proofs. In particular, section 6.4 describes the results at the general abstract level. Section 6.5 deals with a more specialized type of dynamics exhibiting controlled singular behavior of the operator $e^{At}B$. The results of section 6.5 are then applied, in section 6.6, in the context of structure acoustic interactions.

6.2 Optimal Control Problem

To introduce the control problem it is convenient to define the state space $H \equiv H_z \times H_v$, where H_z, H_v are given by (5.2.6). Note that $H = X$ in the notation of Chapter 5. We introduce the state variable $\mathbf{y} \equiv [z, z_t, \mathbf{v}, \mathbf{v}_t]$.

Let R be a bounded operator from $H \to Z$, where Z is another Hilbert space (space of observations).

The following control problem is studied in the context of the models introduced in Chapter 5.

Control problem. Minimize the functional

$$J(u, \mathbf{y}) \equiv \int_0^T [|R\mathbf{y}|_Z^2 + |u|_U^2]\, dt \tag{6.2.1}$$

for all $u \in L_2(0, T; U)$ and $\mathbf{y} \in L_2(0, T; H)$ which satisfy (5.2.9) or equivalently the coupled equations (5.2.1), (5.2.2). From (5.2.7) and (5.2.9), this can be written as

$$\mathbf{y}_t - A\mathbf{y} = Bu + F, \qquad \mathbf{y}(0) \in H, \tag{6.2.2}$$

where the operator $A : H \to H$ was shown to generate a C_0 semigroup on H, and $B : U \to [D(A^*)]'$ represents the unbounded control operator. The forcing term $F \in L_1(0, T; H)$ describes the effect of the deterministic noise. In what follows we identify $\mathcal{U} = U$.

The time T is assumed in this chapter to be finite. In this case, we refer to the problem as the finite horizon control problem.

The main goal of this chapter is to provide feedback representation of the optimal control which, as is well known, involves a solution of an appropriate Riccati equation.

At this point, we note that if the control operator B were bounded from U to H, then the problem of the optimal synthesis would be a straightforward consequence of the existing optimal control theory established for C_0 semigroups with bounded control actions (see [14, 28, 159]). But the system (6.2.2) represents a PDE under the influence of an unbounded control operator. In general, for B unbounded, the issues of solvability of Riccati equations and the meaning of the feedback *gain operator*, B^*P—where unbounded B^* is adjoint of B and P is the Riccati operator P—are more delicate than in the bounded case. The analytical difficulties stem from precisely the unboundedness of the control operator B into the state space; in applications of interest, B is very badly unbounded. Therefore the gain operator may not even be properly defined on a dense subset of the state space (see counterexamples for hyperbolic systems given in [198] and references therein). If A in (6.2.2) were the generator of analytic semigroups, one could appeal to a rich theory that not only provides the meaning to the gain operator, but moreover shows that the gain operator B^*P is in fact bounded (see [140, 28, 129]). The latter conclusion is due to the regularizing effect of analyticity, which is subsequently inherited by the optimal solution; thus $B^*P(t)$ is more regular (in fact analytic in time t) than optimization predicts. Unfortunately, although only some components of the structure (6.2.2) may be analytic, the entire system is not analytic.

Thus, the novelty and interest of the present problem lies in the fact that the control operator is intrinsically unbounded and the overall dynamics is not analytic.

It might be expected, however, that the analyticity of some components of (6.2.2) will be partially propagated onto the entire structure, ultimately benefiting the regularity of the optimal control. Our goal in the analysis is to show that this transfer of regularity, from the (wall) component to the entire structure, is precisely what occurs in the case of structurally damped models. Conversely, for the case of undamped structures, there will be less regularity associated with the structural component, and, consequently, the gain operator will have a diminished regularity (see [43, 44]).

To put our control problem into a proper mathematical framework, the following questions need to be answered:

- Is the map *control* $\Rightarrow$ *state* well defined with the values in the state space H? In particular, does the control operator B satisfy the inequality

$$\int_0^T |B^* e^{A^*t} x|_U dt \leq C|x|_H^2, \qquad x \in D(A^*)? \tag{6.2.3}$$

- Is the gain operator B^*P densely defined on the state space? What is $D(B^*P)$?
- What is the precise statement of well-posedness of solutions to Riccati equations and of the optimal synthesis?

The remainder of this chapter provides answers to these questions.

6.3 Formulation of the Results

Our main task is to show that the dynamics described by (6.2.2) or, equivalently, (5.2.9) along with the functional cost given in (6.2.1) admits an optimal synthesis in the form of an appropriate Riccati feedback operator. The results obtained depend quantitatively on the type of the coupling. Thus we distinguish two cases: hyperbolic-parabolic coupling and hyperbolic-hyperbolic coupling. In the first case the results are applicable to control operators $\mathcal{B}$ subject to Assumption 5.2.2. In the latter case, instead, we need to impose additional restrictions on the control operators, which, however, turn compatible with the dynamics of the structural acoustic models [140, pp. 824–901].

6.3.1 Hyperbolic-parabolic coupling

We recall that the wall model under consideration is given by (5.2.2) with $\alpha > 0, \theta = 1/2$. The control operator $\mathcal{B}$ satisfies Assumption 5.2.2 and the additional Assumption 6.3.1 stated below.

Assumption 6.3.1. *We assume that the parameter r in Assumption 5.2.2 satisfies one of the following conditions:*

(i) *Either $r \leq 5/12$, if Ω is a smooth domain, $r \leq 7/16$ if Ω is a parallelepiped,*
(ii) *or $r < 1/2$, and if vector $\mathbf{v} \in \mathcal{D}({\mathcal{M}_\gamma}^{1/2}) \to v \in H^{1/3}(\Gamma_0)$.*

(We recall that v is the last component of vector $\mathbf{v}$.)

Remark 6.3.1. *We note that the above assumptions are always satisfied for PDE examples introduced in Chapter 5. Indeed, the restriction on the values of r are always satisfied by control operators appearing in the context of smart materials where $\mathcal{B}$ is represented by derivatives of delta functions. This follows from $\mathbf{v} \in \mathcal{D}(\mathcal{A}^{3/8+\epsilon}) \to v \in H^{3/2+\epsilon}(\Gamma_0)$ for all $\epsilon > 0$. Thus, in this case any $3/8 < r < 1/2$ satisfies Assumption 5.2.2. Regarding the restrictions on $\mathcal{D}((\mathcal{M}_\gamma)^{1/2})$, in the case when rotational forces are accounted for, Assumption 6.3.1(ii) is always satisfied, since the last component of $\mathcal{D}({\mathcal{M}_\gamma}^{1/2})$ is always in $H^1(\Gamma_0) \subset H^{1/3}(\Gamma_0)$. Thus, the conditions imposed by Assumption 6.3.1 do not restrict our framework introduced in Chapter 5. These technical conditions are needed, however, for a rigorous proof of Theorem 6.3.1.*

Theorem 6.3.1 ($\alpha > 0$). *Consider the control problem governed by the dynamics described in (5.2.9), with $\alpha > 0, \theta = 1/2$, and the functional cost given in (6.2.1). The control operator $\mathcal{B}$ is subject to Assumption 5.2.2 and Assumption 6.3.1. Moreover, we assume that $F \in L_2(0,T;H)$. Then, for any initial condition $\mathbf{y}_0 \in H$, there exists a unique optimal pair $(u^0, \mathbf{y}^0) \in L_2(0,T;U\times H)$ with the following properties:*

(i) *Regularity of the optimal pair.*

$$u^0 \in C([0,T];U), \qquad \mathbf{y}^0 \in C([0,T];H).$$

(ii) *Regularity of the gains and optimal synthesis. There exist a self-adjoint positive operator $P(t) \in \mathcal{L}(H)$ with the property*

$$B^*P(.) \in \mathcal{L}(H \to C([0,T];U))$$

and an element $r \in C([0,T];H)$ *(depending on* F*) with the property*

$$B^* r \in C([0,T];U)$$

such that

$$u^0(t) = -B^* P(t)\mathbf{y}^0(t) - B^* r(t), \qquad t \geq 0.$$

(iii) *Feedback evolution. The operator* $A_{P(t)} \equiv A - BB^* P(t) : \mathcal{D}(A_{P(t)} \subset H \to D(A^*)'$ *generates a strongly continuous evolution on* H.

(iv) *Riccati equation. The operator* $P(t)$ *is a unique (within the class of self-adjoint positive operators subject to the regularity in part* (ii)*) solution of the following operator* DRE*:*

$$\begin{aligned}(P_t(t)x,y)_H = (A^* P(t)x,y)_H + (P(t)Ax,y)_H + (R^* Rx,y)_H \\ - (B^* P(t)x, B^* P(t)y)_U \text{ for } x,y \in D(A). \qquad (6.3.4)\end{aligned}$$

(v) *Equation for* r*. With* $A_P(t)$ *defined in part* (iii)*, the element* $r(t)$ *satisfies the differential equation*

$$\begin{aligned}r_t(t) = -A_P^*(t)r(t) - P(t)F \quad \text{on } [D(A)]' \\ \text{with the terminal condition } r(T) = 0. \qquad (6.3.5)\end{aligned}$$

Remark 6.3.2. *We note that the results of Theorem* 6.3.1*, parts* (i) *and* (ii)*, provide more regularity properties of the optimal solution than optimization alone predicts. Indeed, the synthesis is defined pointwise in time with the gains represented by bounded operators. This is not a usual property of solutions to control problems with unbounded control actions (see* [129]*). On the other hand, these regularity results are needed to give sense to the Riccati equation* (6.3.4).

Remark 6.3.3. *In the case* $f = 0$ *Theorem* 6.3.1 *was proved in* [4].

Remark 6.3.4. *The result of Theorem* 6.3.1 *applies to all models in section* 5.3 *with* $\alpha > 0, \theta = 1/2$.

Proof of Theorem 6.3.1 is given in sections 6.4–6.6.

6.3.2 Hyperbolic-hyperbolic coupling: General case

We next turn to a much less regular case, when the coupling on the interface is hyperbolic-hyperbolic, which corresponds to the case $\alpha = 0$.

Theorem 6.3.2 ($\alpha = 0$). *Consider the control problem governed by the dynamics described in* (5.2.9)*, with* $\alpha = 0$*, along with the functional cost given in* (6.2.1)*. The control operator* $\mathcal{B}$ *is subject to Assumption* 5.2.2 *and in addition we assume the following.*

Assumption 6.3.2.

$$\int_0^T |B^* e^{At}\mathbf{y}|_U^2 dt \leq C|\mathbf{y}|_H^2, \qquad \mathbf{y} \in D(A^*).$$

We also assume that $F \in L_2(0,T;H)$*. Then, for any initial condition* $\mathbf{y}_0 \in H$*, there exists a unique optimal pair* $(u^0, \mathbf{y}^0) \in L_2(0,T;U \times H)$ *with the following properties:*

(i) *Regularity of the optimal pair.*

$$u^0 \in L_2(0,T;U); \ \mathbf{y}^0 \in C([0,T];H).$$

(ii) *Optimal synthesis. There exist a self-adjoint positive operator* $P(t) \in \mathcal{L}(H)$ *and an element* $r \in C([0,T];H)$ *such that* $P(t)\mathbf{y}^0(t) + r(t) \in D(B^*)$ *and*

$$u^0(t) = -B^*[P(t)\mathbf{y}^0(t) + r(t)] \ for\ t \geq 0\ a.e.$$

Proof of Theorem 6.3.2 is given in section 6.4.

Remark 6.3.5. *In the hyperbolic-hyperbolic case, Theorem* 6.3.2 *does not provide any information on the well-posedness of the gain operator* $B^*P(t)$*. This is in line with general hyperbolic theory* [129]. *However, if there is some additional (minimal) smoothness of the observation, then the gain operator* $B^*P(t)$ *is shown to be densely defined and to satisfy an appropriate Riccati equation. We do not pursue this direction here and refer the reader to the literature* [123, 140].

Note that the formulation of Theorem 6.3.2 requires Assumption 6.3.2. In general, such assumption may not hold, under the sole Assumption 5.2.2. However, in a special but important case when the control operator is given by the derivatives of the delta functions, and the structural dynamics (5.2.2) are given by the Kirchhoff equation, Assumption 6.3.2 holds true [44]. This particular situation is exploited below.

6.3.3 Hyperbolic-hyperbolic coupling: Special case of the Kirchhoff plate with point control

In this subsection we focus on a rather special structural acoustic model that serves as a prototype for applications of Theorem 6.3.2. The main feature of this model is that the control operator satisfies Assumption 6.3.2.

Following [44, 141] and [140, p. 885], we consider a two- or three-dimensional acoustic chamber, where the flat elastic wall Γ_0 is modeled as a Kirchhoff equation. This is a hyperbolic undamped equation, which accounts for rotational forces (see $\gamma > 0$ in (6.3.7) below). It is arguably a more accurate model than the Euler–Bernoulli model. As we have already noted, the basic structure of acoustic flow models has been known for a long time (see, e.g., [166, Example, p. 263], [24]). Perhaps the key contribution in modeling of smart material technology, as supplied, e.g., by [16, 15, 17, 72], is the presence of an even number of δ'-distributions (dipoles) supported at different points of the moving wall Γ_0 when $\dim \Gamma_0 = 1$. Mathematically, it suffices to incorporate only one such δ'; see (6.3.7).

PDE model with flat Kirchhoff wall Γ_0. Let Ω be a two-dimensional domain (the chamber). The case of a three-dimensional chamber is similar and considered in detail in [44, 197]. We consider explicitly two cases: (i) either Ω is a two-dimensional rectangle with three consecutive hard walls comprising the boundary Γ_1 and one vibrating wall comprising the boundary Γ_0 fixed at its extremes, where $\Gamma_0 \cup \Gamma_1 = \partial\Omega$; (ii) or else Ω is a general two-dimensional bounded domain, where the smooth boundary Γ is divided into two parts Γ_0 and Γ_1, $\Gamma = \Gamma_0 \cup \Gamma_1$, with Γ_1

acting as the hard wall and Γ_0 the flat portion acting as the moving wall fixed at its extremes.

If $z(t,x)$ denotes the acoustic wave (unwanted noise) in the chamber, and $v(t,x), x \in \Gamma_0$, denotes the displacement of Γ_0, then the relevant system of PDE equations describing the given problem is

$$\begin{aligned} z_{tt} &= \Delta z + f \text{ in } Q \equiv \Omega \times (0,T), \\ \frac{\partial}{\partial \nu} z &= 0 \text{ on } \Sigma_1 \equiv \Gamma_1 \times (0,T), \\ \frac{\partial}{\partial \nu} z &= v_t \text{ on } \Sigma_0 \equiv \Gamma_0 \times (0,T), \end{aligned} \tag{6.3.6}$$

$$\begin{aligned} v_{tt} - \gamma \Delta v_{tt} + \Delta^2 v + z_t = \delta'(x_0) u(t) \text{ in } \Sigma_0, \\ \text{either clamped boundary condition } v = \frac{\partial}{\partial \nu} v = 0 \text{ on } \partial\Gamma_0 \times (0,T), \\ \text{or else hinged boundary condition } v = \Delta v = 0 \text{ on } \partial\Gamma_0 \times (0,T). \end{aligned} \tag{6.3.7}$$

Equations (6.3.6) and (6.3.7) are associated with the initial conditions:

$$z(0) = z_0, \quad z_t(0) = z_1, \quad v(0) = v_0, \quad v_t(0) = v_1. \tag{6.3.8}$$

Here ν is the unit outward normal vector at $x \in \Gamma_1$ and $\gamma > 0$ is a constant. In (6.3.7), x_0 is a chosen point on Γ_0, $u(t)$ is the scalar control function, and $\delta'(x_0)$ denotes the derivative of the Dirac measure at x_0, which models the action of the piezoceramic patch. Because of $\gamma > 0$, and since there is no structural damping (represented in abstract equation (5.2.2) by the term $\mathcal{A}v_t$), this model differs significantly from the structural component in section 6.3.1. In our present situation, there is no analyticity of the associated uncoupled equation for v; it is in fact hyperbolic with a finite speed of propagation.

We report from [44, 197, 141] and [140, p. 805], the following results, which further specialize the abstract semigroup setup presented in section 5.2. Let $\mathbf{y}(t) = [z(t), z_t(t), v(t), v_t(t)]$. Then the coupled PDE system (6.3.6), (6.3.7) can be written abstractly as the equation

$$\mathbf{y}_t = A\mathbf{y} + Bu + F \in [D(A^*)]', \quad \mathbf{y}(0) = \mathbf{y}_0 = [z_0, z_1, v_0, v_1] \in H, \tag{6.3.9}$$

where H is the Hilbert space

$$H = H^1(\Omega) \times L_2(\Omega) \times H^2(\Gamma_0) \times H^1(\Gamma_0). \tag{6.3.10}$$

Indeed, the last two components in (6.3.10) are refined as $H_0^2(\Gamma_0) \times H_0^1(\Gamma_0)$ for the clamped boundary condition and $[H^2(\Gamma_0) \cap H_0^1(\Gamma_0)] \times H_0^1(\Gamma_0)$ for the hinged boundary condition.

(a) More specifically, in the case of the hinged boundary condition we topologize H as

$$H = D({A_N}^{1/2}) \times L_2(\Omega) \times D(\mathcal{A}^{1/2}) \times D({\mathcal{M}_\gamma}^{1/2}) \equiv H_z \times H_v.$$

The operators $A : H \supset D(A) \to H$ in (6.3.9) are given by

$$A \equiv \begin{pmatrix} 0 & I & 0 & 0 \\ -A_N & 0 & 0 & (A_N + I)N(\cdot|_{\Gamma_0}) \\ 0 & 0 & 0 & I \\ 0 & -{\mathcal{M}_\gamma}^{-1}N^*(A_N + I) & -{\mathcal{M}_\gamma}^{-1}\mathcal{A} & 0 \end{pmatrix} \tag{6.3.11}$$

with its natural domain [44]. The operators $B : U \to [D(A^*)]'$, $B^* : D(A^*) \to U$ in (6.3.9) are given by

$$Bu = \begin{pmatrix} 0 \\ 0 \\ 0 \\ {\mathcal{M}_\gamma}^{-1}\delta'(x_0)u \end{pmatrix}, \quad B^* \begin{pmatrix} y_1 \\ y_2 \\ y_3 \\ y_4 \end{pmatrix} = -\frac{d}{dx} y_4|_{x=x_0}. \qquad (6.3.12)$$

Finally, the forcing term has the representation $F = [0, f, 0, 0]^T$. Throughout, we take $U = R$, and we used the operators $\mathcal{A}, A_N, N$, defined in section 4.3.1 with $\mathcal{A}$ given by (5.3.12) or else (5.3.14) and

$$\mathcal{M}_\gamma \equiv I + \gamma \mathcal{A}^{1/2}.$$

We note that the operator ${\mathcal{M}_\gamma}^{-1}\mathcal{A}$ is positive, self-adjoint on the space $D({\mathcal{M}_\gamma}^{1/2})$ topologized by the inner product:

$$(x, y)_{D({\mathcal{M}_\gamma}^{1/2})} = ((I + \gamma\mathcal{A}^{1/2})x, y)_{L_2(\Gamma_0)}, \qquad x, y \in D(\mathcal{A}^{1/4}),$$

and moreover that

$$D(\mathcal{A}^{1/2}) = H^2(\Gamma_0) \cap H_0^1(\Gamma_0); \quad D(\mathcal{A}^{1/4}) = H_0^1(\Gamma_0).$$

When H is topologized by the equivalent norm given by (6.3.10), then the operator A is skew adjoint, $A^* = -A$, and thus generates a unitary group e^{At} on H, $t > 0$ (conservative homogeneous problem):

$$e^{A^*t} = e^{-At}; \quad ||e^{A^*t}||_{\mathcal{L}(X)} = ||e^{-At}||_{\mathcal{L}(H)} \equiv 1,$$

$$Re(Ax, x)_H = Re(A^*x, x)_H \equiv 0, \qquad x \in D(A) = D(A^*).$$

(b) The conclusion is similar for the case of clamped or free boundary conditions.

As a consequence, the solution of the abstract version (5.2.9), of the coupled PDE problem (6.3.6), (6.3.7), is then

$$\mathbf{y}(t) = e^{At}\mathbf{y}_0 + (Lu)(t) + (KF)(t),$$

$$(Lu)(t) = \int_0^t e^{A(t-s)}Bu(s)ds, \ (KF)(t) = \int_0^t e^{A(t-s)}F(s)ds. \qquad (6.3.13)$$

Since the operator B is not bounded from $U = R$ to H but instead satisfies the regularity $B : U \to [D(A^*)]'$, it is necessary to specify the regularity of the operator L in (6.3.13). This is given next. The main result of the present section is the following regularity theorem for (6.3.13).

Theorem 6.3.3 (see [44, 197]). *With reference to the abstract model* (6.3.9) *of the coupled PDE system* (6.3.6), (6.3.7), *and the space H given by* (6.3.10), *we have that for each $0 < T < \infty$, the operator L in* (6.3.13) *satisfies the property*

$$L : continuous\ L_2(0, T; U) \to C([0, T]; H). \qquad (6.3.14)$$

In PDE terms, the meaning of (6.3.14) *is as follows. Set $z_0 = z_1 = v_0 = v_1 = 0$ in* (6.3.6), (6.3.7); *then the corresponding solution satisfies*

$$u \in L_2(0, T) \to [z(t), z_t(t), v(t), v_t(t)] \in C([0, T]; H).$$

Equivalently, by duality, then the following abstract trace regularity holds true: given $T > 0$, *there exist* $C_T > 0$ *such that*

$$\int_0^T |B^* e^{A^* t} \mathbf{y}|_U^2 dt \le C_T |\mathbf{y}|_H^2, \qquad y \in H. \tag{6.3.15}$$

In PDE terms the meaning of (6.3.15) *is as follows. Let* $u = 0$ *in the original equation; then the corresponding homogeneous problem satisfies the estimate*

$$\int_0^T |v_{tx}(t, x_0; \mathbf{y}_0)|^2 dt \le C_T |\mathbf{y}_0|_H^2, \tag{6.3.16}$$

$y_0 = [z_0, z_1, v_0, v_1]$, *where* $v_{tx}(t, x_0; \mathbf{y}_0)$ *is the second partial derivative of the solution* v *in* t *and* x, *evaluated at the point* $x = x_0 \in \Gamma_0$ *(point observation), and due to the initial condition* y_0 *and to* $u = 0$.

Remark 6.3.6. *The regularity result stated in Theorem* 6.3.3 *is also valid for the wave and plate equations* (6.3.6) *and* (6.3.7) *subject to boundary damping. This is to say that one may consider absorbing boundary conditions on* Γ_1

$$\frac{\partial}{\partial \nu} z + d z_t = v_t, \qquad d \ge 0,$$

and on $\partial \Gamma_0$

$$v = 0, \qquad \Delta v = -k \frac{\partial}{\partial \nu} v_t, \qquad k \ge 0.$$

In fact, when the damping is active on Γ_0 *the arguments are even simpler* [140, *Chapter* 2].

Remark 6.3.7. *Theorem* 6.3.3 *is sharp. The regularity* $[v, v_t] \in C([0, T], H^2(\Gamma_0) \times H^1(\Gamma_0))$ *of the Kirchhoff component (the one subject directly to the control action* u*) of the coupled problem is exactly the same as the regularity for the uncoupled problem (see* [196, 195]*). Such regularity is* $1/2+\epsilon$ *higher in Sobolev space units, with respect to the space variable, over the regularity that is obtained by the variation of parameter formula of the corresponding semigroup, based simply on the membership property that*

$$\delta'(x_0) \in [H^{\frac{3}{2}+\epsilon}(\Gamma_0)]' \subset [D(\mathcal{A}^{\frac{3}{8}+\frac{1}{4}\epsilon})]'$$

or $\mathcal{A}^{-3/8-\frac{1}{4}\epsilon} \delta'(x_0) \in L_2(\Gamma_0)$, *where* $\dim \Gamma_0 = 1$ *and where* $\mathcal{A}$ *is the biharmonic operator defined in the hinged case. A similar loss of* $1/2 + \epsilon$ *would occur, if one used directly and analogously the abstract formula* (6.3.13), *with* e^{At} *the strongly continuous semigroup with* B *given by* (6.3.12). *Similarly,* (6.3.16) *does not follow from energy methods: it requires a combination of sharp regularity results for the uncoupled Kirchhoff part (Theorem* 3.1 *in* [196]*) and for the uncoupled wave part (Theorem* 3.2 *from* [130]; *see also* [127]*).*

A direct combination of the results in Theorems 6.3.3 and 6.3.2 gives the final result for the structural acoustic problem with $\alpha = 0$ and the physically relevant choice of $B = \delta'$.

Corollary 6.3.1. *With reference to the model given in* (6.3.6) *and* (6.3.7), *the conclusions of Theorem* 6.3.2 *apply.*

Comments about the proof of Theorem 6.3.3. Two proofs of Theorem 6.3.3 may be given: a proof of estimate (6.3.16) [197] (dual problem) and a direct proof of the original regularity for the operator L [44]. The latter appears more streamlined than the former. Both proofs, however, rely critically on the sharp regularity results of the two basic dynamical components of the noise reduction model: [196] for the Kirchhoff equation with point control and [127, 130] for related results for the wave equation with Neumann control. More specifically, the two main ingredients of the proofs of [44] are the following sharp regularity results.

1. Let $\phi(t,x)$ denote the solution of the following mixed problem for the Kirchhoff equation (which is problem (6.3.7) without coupling):

$$\begin{aligned} \phi_{tt} - \gamma\Delta\phi_{tt} + \Delta^2\phi = \delta'(x_0)u(t) \text{ in } \Sigma_0, \\ \text{either clamped boundary condition } \phi = \frac{\partial}{\partial\nu}\phi = 0 \text{ on } \partial\Gamma_0 \times (0,T), \\ \text{or else hinged boundary condition } \phi = \Delta\phi = 0 \text{ on } \partial\Gamma_0 \times (0,T), \\ \phi(t=0) = \phi_0,\ \phi_t(t=0) = \phi_1 \text{ in } \Gamma_0. \end{aligned} \tag{6.3.17}$$

For this problem we have the following optimal regularity result.

Lemma 6.3.1 (see [196]). *Recall that* $\dim \Gamma_0 = 1$. *Consider the* ϕ *problem in* (6.3.17) *with, say, hinged boundary condition and thus with*

$$\{\phi_0,\phi_1\} \in [H^2(\Gamma_0)\cap H_0^1(\Gamma_0)] \times H_0^1(\Gamma_0), \qquad u \in L_2(0,T). \tag{6.3.18}$$

Then, continuously,

$$\{\phi,\phi_t,\phi_{tt}\} \in C([0,T];[H^2(\Gamma_0)\cap H_0^1(\Gamma_0)] \times H_0^1(\Gamma_0) \times L_2(0,T,L_2(\Gamma_0)). \tag{6.3.19}$$

For the clamped boundary condition, the first component space is $H_0^2(\Gamma_0)$.

The proof of Lemma 6.3.1 is given in [196], along with the observation that the space regularity with $\delta'(\cdot)$ is one Sobolev unit less than the space regularity result with $\delta(\cdot)$ given in [196]. We also note that the dual version of this result valid for the wave equation only was shown in [162] and [153]. The regularity result related to the two-dimensional version of Lemma 6.3.1 is given in [201].

2. Let $\psi(t,x)$ denote the solution to the wave equation (which is the problem (6.3.6) without coupling and without the forcing term f):

$$\begin{aligned} \psi_{tt} = c^2\Delta\psi \text{ in } Q \equiv \Omega \times (0,T), \\ \psi = 0 \text{ on } \Sigma_1 \equiv \Gamma_1 \times (0,T), \\ \frac{\partial}{\partial\nu}\psi = g \text{ on } \Sigma_0 \equiv \Gamma_0 \times (0,T), \\ \psi(t=0) = 0,\ \psi_t(t=0) = 0 \text{ in } \Omega. \end{aligned} \tag{6.3.20}$$

A sharp regularity theory for problem (6.3.20) is given in [130, 127], from which we quote below: henceforth, α is a constant taking up the following values (where $\epsilon > 0$ is arbitrary):

$$\begin{aligned} \alpha = 3/5 - \epsilon \quad &\text{for general smooth domains } \Omega \in R^n, \quad n \geq 2, \\ \alpha = 3/4 \quad &\text{for a parallelepiped of } R^n, \quad n \geq 2, \\ \alpha = 2/3 \quad &\text{for a sphere of } R^n, \quad n \geq 2. \end{aligned} \tag{6.3.21}$$

Lemma 6.3.2 (see [130, 127, 135]). *With reference to problem* (6.3.20) *(where actually* $\dim \Omega = 2$ *), we have*

(i) *Interior regularity. Let* $g \in H^1(0,T; L_2(\Gamma_0)) \cap C([0,T]; H^{\alpha-1/2}(\Gamma_0)), g(t=0) = 0$. *Then, continuously,*

$$\{\psi, \psi_t, \psi_{tt}\} \in C([0,T]; H^{\alpha+1}(\Omega) \times H^{\alpha}(\Omega) \times H^{\alpha-1}(\Omega)). \tag{6.3.22}$$

(ii) *Boundary regularity. Let* $g \in H^1(\Sigma_0); g(t=0) = 0$. *Then, continuously,*

$$\psi|_\Gamma \in H^{2\alpha}(\Sigma). \tag{6.3.23}$$

The derivation of the sharp (optimal) regularity results Lemmas 6.3.1, 6.3.2 for the hyperbolic mixed PDE problems (6.3.17), (6.3.20), besides representing the first fundamental step in the analysis of coupled dynamics (6.3.6), (6.3.7), have key implications in the study of associated optimal control problems. In fact, these results are the main building blocks behind the proof of Theorem 6.3.3.

The remaining sections of this chapter are devoted to the proofs of the main results formulated in Theorems 6.3.1 and 6.3.2.

6.4 Abstract Optimal Control Problem: General Theory

In this section we consider the general optimal control problem that is governed by a strongly continuous semigroup with unbounded control operators. Our goal is to recall some of the first results pertaining to solvability of this control problem. These results will be a starting point for an analysis of structural acoustic control problem and the proofs of Theorems 6.3.1 and 6.3.2.

6.4.1 Formulation of the abstract control problem

Let H, U, Z be given Hilbert spaces and let the following operators be given:

- A is a generator of a strongly continuous semigroup on H with domain $D(A) \subset H \subset D(A^*)'$.
- The operator $B : U \to D(A^*)'$ is bounded. Moreover, B satisfies the following condition:

$$|R(\lambda, A)Bu|_H \leq C_\alpha |u|_U, \qquad \lambda = \alpha + i\beta, \quad \beta \in R, \quad \lambda \in \rho(A). \tag{6.4.24}$$

- The operator $R : H \to Z$ is bounded.
- $F \in L_1(0,T; H)$ is a given element.

With these quantities we consider the following *dynamics*:

$$\mathbf{y}_t = A\mathbf{y} + Bu + F \text{ on } D(A^*)', \qquad \mathbf{y}(0) = \mathbf{y}_0 \in H. \tag{6.4.25}$$

Associated with (6.4.25) is the functional cost

$$J(u, \mathbf{y}) \equiv \int_0^T [|Ry|_Z^2 + |u|_U^2]\, dt. \tag{6.4.26}$$

Optimal control problem. *Minimize the functional* $J(u, \mathbf{y})$ *for all* $u \in L_2(0,T;U)$ *and* $\mathbf{y} \in L_2(0,T;H)$ *which satisfy* (6.4.25).

By the standard optimization argument [19, 14] one deduces the existence and uniqueness of optimal solution $u^0 \in L_2(0,T;U), y^0 \in L_2((0,T);H)$ to the optimal control problem. Our main aim is to derive the optimal synthesis for the control problem along with a characterization of optimal control via an appropriate Riccati equation.

6.4.2 Characterization of the optimal control

For the reader's convenience we quote several results and formulas applicable to the optimal control problem stated above. These are taken from [140, 163].

To do this we introduce the so-called solution operator, often also referred to as the control-to-state map:

$$(Lu)(t) \equiv \int_0^t e^{A(t-s)} Bu(s)ds. \tag{6.4.27}$$

Condition (6.4.24) is equivalent to the statement that the control-to-state operator L is topologically bounded L_2 in time. This is to say, for all $T > 0$

$$L \in \mathcal{L}(L_2(0,T;U) \to L_2(0,T;H)). \tag{6.4.28}$$

We also use the L^* adjoint of L given by

$$(L^*w)(t) = B^* \int_t^T e^{A^*(s-t)} w(s)ds,$$

and by duality, we obtain that

$$L^* \in \mathcal{L}(L_2(0,T;H) \to L_2(0,T;U)). \tag{6.4.29}$$

The effect of the deterministic noise is represented by the element

$$\mathcal{F}(t) \equiv \int_0^t e^{A(t-s)} F(s)ds,$$

and from standard semigroup theory [175] with $F \in L_1(0,T;H)$,

$$\mathcal{F} \in C([0,T];H).$$

By using the above notation one obtains ([156 and 134, section 6.2.3]) the explicit formulas for the optimal control. These are collected in the lemma below.

Lemma 6.4.1. *With reference to the control problem stated in* (6.2.1), *and arbitrary initial condition* $\mathbf{y}_0 \in H$ *and* $F \in L_1(0,T;H)$, *there exists a unique optimal pair denoted by* $(u^0, \mathbf{y}^0)$ *with the following properties:*

(i) $u^0 = -L^*R^*R\mathbf{y}^0 \in L_2(0,T;U)$,
(ii) $u^0 = -[I + L^*R^*RL]^{-1}L^*R^*R[e^{A(\cdot)}\mathbf{y}_0 + \mathcal{F}] \in L_2(0,T;U)$,
(iii) $\mathbf{y}^0 = [I + LL^*R^*R]^{-1}[e^{A(\cdot)}\mathbf{y}_0 + \mathcal{F}] \in L_2(0,T;H)$,
(iv) $u^0(t) = -B^*[P(t)\mathbf{y}^0(t) + r(t)]$ *a.e. in* $t \in (0,T)$,

where $P(t), r(t)$ *are givenby formulas* (6.4.33), (6.4.35) *below.*

Proof. The first three statements follow directly from optimality conditions (see, for instance, [140]). To provide an expression for optimal synthesis it is convenient to introduce the evolution operator $\Phi(t,s)$ defined by

$$\Phi(.,s)x \equiv [I + L_s L^* R^* R]^{-1} e^{A(.-s)} x \in L_2(s,T;H) \text{ for } x \in H, \tag{6.4.30}$$

where

$$(L_s u)(t) \equiv \int_s^t e^{A(t-\tau)} Bu(\tau) d\tau.$$

The following properties of the operators introduced above are used frequently (see [66, 176, 140]):

$$[I + L^* R^* R L_s]^{-1} \in \mathcal{L}(L_2(s,T;U)). \tag{6.4.31}$$

Since $[I + L_s L^* R^* R]^{-1} = -L_s[I + L^* R^* R L_s]^{-1} L^* R^* R + I$, a fortiori we have

$$[I + L_s L^* R^* R]^{-1} \in \mathcal{L}(L_2(s,T;H)) \text{ uniformly in } s \in [0,T]. \tag{6.4.32}$$

Thus from the representation (6.4.30)

$$\Phi(.,s)x \in L_2((s,T);H) \text{ for } x \in H \text{ uniformly in } s.$$

Based on the formulas in Lemma 6.4.1, we recognize that $\Phi(t,s)x$ coincides with the optimal trajectory corresponding to $F = 0$ and originating at the time s with the initial datum x.

The evolution operator allows us to define the Riccati operator $P(t)$ given by the formula [14, 129, 66, 140]

$$P(t) \equiv \int_t^T e^{A^*(s-t)} R^* R \Phi(s,t) ds. \tag{6.4.33}$$

Clearly, $P(\cdot) \in \mathcal{L}(H, C([0,T],H))$. Moreover, it is standard to show that $P(t)$ is self-adjoint and positive on H [66]. We define next the adjoint state

$$p(t) \equiv \int_t^T e^{A^*(s-t)} R^* R \mathbf{y}^0(s) ds \in C([0,T];H).$$

We have

$$u^0 = -B^* p \in L_2(0,T;H). \tag{6.4.34}$$

Indeed, convex optimization and the Liusternik–Lagrange multiplier theorem give the characterization [140, 129]

$$u^0 = -[I + L^* R^* R L]^{-1} L^* R^* R(e^{A(\cdot)} \mathbf{y}_0 + \mathcal{F}),$$

which is equivalent to

$$u^0 = -L^* R^* R(e^{A(\cdot)} \mathbf{y}_0 + L u^0 + \mathcal{F}) = -L^* R^* R \mathbf{y}^0,$$

which, in turn, gives $B^* p = L^* R^* R y^0$. The definition of the adjoint state p and regularity of L^* given in (6.4.29) complete the proof of the assertion (6.4.34).

Finally, we define the variable

$$r(t) \equiv p(t) - P(t)\mathbf{y}^0(t) \in L_2(0,T;H). \tag{6.4.35}$$

By (6.4.34) we have

$$P(t)\mathbf{y}^0(t) + r(t) \in D(B^*) \text{ a.e. in } t$$

with

$$B^*[P(t)\mathbf{y}^0(t) + r(t)] = -u^0(t). \qquad \square$$

6.4.3 Additional properties under the hyperbolic regularity assumption

To provide more information about a solution to the optimal control problem, the following hyperbolic regularity condition (introduced first in [122] in the context of the wave equation) imposed on the operator B is used.

Assumption 6.4.1.

$$\int_0^T |B^* e^{A^* t} \mathbf{y}|_U^2 dt \le C_T |\mathbf{y}|_H^2, \qquad \mathbf{y} \in D(A^*), \tag{6.4.36}$$

where $(Bu, v)_H = (u, B^* v)_U, \; u \in U, v \in D(B^*) \supset D(A^*)$.

Remark 6.4.1. *Clearly condition* (6.4.36) *implies* (6.4.24), *but the converse is not true. A simple counterexample can be provided by a one-dimensional heat equation with Dirichlet boundary control* [140]. *However, in the case of time reversible dynamics (e.g., hyperbolic problems), both conditions are equivalent; see* [125].

The above condition (6.4.36) is equivalent to the statement that the solution operator L is continuous in time [123, 129, 140]. This is to say

$$L_s \in \mathcal{L}(L_2(s,T;U)) \to C([s,T];H) \text{ for all } s \ge 0. \tag{6.4.37}$$

This property implies, in particular,

$$\mathbf{y}^0 \in C([0,T];H),$$

which is an important property of the optimal trajectory.

Moreover, condition (6.4.36) can be used to show the following additional regularity of control operators:

$$[I + L_s L^* R^* R]^{-1} \in \mathcal{L}(C(s,T;H)) \cap \mathcal{L}(L_2(s,T;H)) \quad \text{uniformly in } s \in [0,T]. \tag{6.4.38}$$

Thus, an important consequence of the hyperbolic regularity condition (6.4.36) is that the operator $\Phi(t,s)$ is continuous in time, i.e.,

$$\Phi(.,s) \in \mathcal{L}(H \to C([s,T];H)) \text{ for each } 0 \le s \le T \text{ uniformly in } s. \tag{6.4.39}$$

It can also be easily verified that $\Phi(t,s)$ possesses the evolution property [66], i.e.,

$$\Phi(t,s) = \Phi(t,\tau)\Phi(\tau,s), \quad 0 \le s \le \tau \le t.$$

Continuity of optimal trajectory implies also continuity of the variable r. That is to say, under condition (6.4.36) we have

$$r(t) \equiv p(t) - P(t)\mathbf{y}^0(t) \in C([0,T];H). \tag{6.4.40}$$

The following relations between the Riccati operator, adjoint variable p, and r are known.

Lemma 6.4.2 (see [163, Lemma 2.2.1]). *Under the Assumption* 6.4.1, *we have for any* $F \in L_1(0,T;U)$ *and initial condition* $\mathbf{y}_0 \in H$

(i)

$$p_t = -A^* p - R^* R \mathbf{y}^0 \ in \ [D(A^*)]', \qquad p(T) = 0. \tag{6.4.41}$$

(ii) $r(t)$ *does not depend on the initial condition* $\mathbf{y}_0$.

(iii)

$$P(t)\mathbf{y}^0(t) + r(t) \in D(B^*) \ a.e. \ in \ t,$$
$$u^0(t) = -B^* p(t) = -B^*[P(t)\mathbf{y}^0(t) + r(t)] \in L_2(0,T;U).$$

(iv) $\mathbf{y}^0 \in C([0,T];H), \quad r \in C([0,T];H)$.

(v) $\Phi(\cdot,s) \in \mathcal{L}(H;C([s,T];H)), \quad \Phi(t,\cdot) \in \mathcal{L}(H;C([0,t];H)), \qquad 0 \le s,t \le T.$

continuously with respect to both variables s *and* t.

A few comments are in order. Lemma 6.4.2 provides the desired optimal synthesis of the optimal control in terms of the optimal state and the contribution due to the disturbance. Thus, in some sense, one can say that the feedback control problem has been solved by virtue of Lemma 6.4.2. However, there are three major drawbacks of this theory:

(i) There is no characterization of the Riccati operator via the Riccati equation (which would then provide a tool for finding $P(t)$ independent of the optimal solution, in contrast to the formula defining $P(t)$, which depends on the optimal state). Here, we note, that the main technical difficulty in justifying the derivation of the Riccati equation is due to the fact that the nonlinear term in Riccati equation $P(t)BB^*P(t)$ has an unbounded operator B^*.

(ii) In the formula describing the optimal synthesis (see (iii) in the lemma), the unbounded operator B^* is outside the square bracket and, in general, cannot be moved inside. That is to say that the gain operators $B^*P(t)$ and $B^*r(t)$ may not be defined at all. In fact, there are examples of control problems where the gain operator B^*P (for the case $T = \infty$) is not well defined even on the domain of any power of the generator A [204]. This, of course, undermines the usefulness of feedback control theory, where one would like to compute independently the feedback due to the initial datum and the feedback due to the disturbance f.

(iii) While we have constructed the evolution operator corresponding to the unperturbed problem (without the disturbance), it is not clear at all that this evolution has a generator. In fact, in the case of the time-dependent problem, it is well known that the existence of the continuous evolution does not imply the existence of the generator (there are counterexamples even in the finite-dimensional cases).

The two main underlying technical issues in establishing the Riccati equation are the differentiability of the evolution operator with respect to the second variable, and the differentiability of the Riccati operator. We show that the required regularity properties for $\Phi, P(t)$ are valid under stronger regularity conditions imposed on the operator B. This is the goal of sections 6.4.4 and 6.5.

6.4.4 DRE, feedback generator, and regularity of the gains B^*P, B^*r

In this section we make the following hypothesis ([169] and [134, Chapter 9]):

$$\int_0^T |R^* Re^{At} Bu|_H dt \le C_T |u|_U. \tag{6.4.42}$$

We note that this estimate may still hold for some classes of unbounded observation operators R. For the sake of simplicity of exposition, we do not consider such cases and restrict ourselves to the case when R is bounded.

It was shown in [176, 140] that subject to the validity of conditions (6.4.36) and (6.4.42) the following result holds true.

Lemma 6.4.3 (see [176]). *Assume that the estimates* (6.4.36) *and* (6.4.42) *hold true. In addition to all conclusions in Lemma* 6.4.2 *the following properties hold:*

(i) *The gain operator is bounded, that is to say,*

$B^*P(t) \in \mathcal{L}(H \to U)$ *for all* $0 \le t \le T$ *and* $B^*P(.)x \in C([0,T];U)$ *for* $x \in H$.

(ii) *The operator* $A_P(t) \equiv A - BB^*P(t)$ *defined from* H *into* $[D(A^*)]'$ *is a generator of a strongly continuous evolution* $\Phi(t,s)$, *where* $\Phi(t,s)$ *is defined by the formula* (6.4.30). *This means, in particular, that for any* $x \in H$, $y(t,s) \equiv \Phi(t,s)x \in C([s,T];H)$ *satisfies the equation*

$$\begin{aligned}(y_t(t,s), \phi)_H &= ([I - A^{-1}BB^*P(t)]y(t,s), A^*\phi)_H \\ &= 0 \text{ for all } \phi \in D(A^*), \qquad y(s,s) = x.\end{aligned} \tag{6.4.43}$$

(iii) *The DRE*

$$\begin{aligned}(P_t(t)x, y)_H + (A^*P(t)x, y)_H + (P(t)Ax, y)_H + (R^*Rx, y)_H \\ = (B^*P(t)x, B^*P(t)y,)_U \text{ for } x, y \in D(A)\end{aligned} \tag{6.4.44}$$

admits a unique solution in the class of self-adjoint positive operators $P(t) \in \mathcal{L}(H)$ *and such that the regularity requirement in part* (i) *holds true.*

(iv) *The operator* $P(t)$ *defined by the formula* (6.4.33) *is a unique solution of Riccati equation* (6.4.44) *in the class described by* (i).

The result of the lemma is critical. It provides a meaning for the gain operator $B^*P(t)$ (which turns out to be bounded) and yields the existence and uniqueness of solutions to the classical Riccati equation with unbounded coefficients. We note that this type of result is typical in the case of analytic semigroups. What is somewhat unexpected here is the fact that the same is true (see Theorem 6.3.1) for the structural acoustic problem, which is not analytic.

Remark 6.4.2. *The result of Lemma* 6.4.3 *has been known for some time to apply to systems of the so-called Pritchard–Salomon class (see* [54] *and references therein), where a stronger requirement imposed on* R, *namely, the earlier theory, requires that*

$$\int_0^T |Re^{At}Bu|_Z^2 dt \le C_T|u|_U. \tag{6.4.45}$$

Thus, systems of this particular class require almost twice as much in smoothing effect (compensation) from the observation R *than required by* (6.4.42).

6.5 Riccati Equations Subject to the Singular Estimate for $e^{At}B$

In this section we consider the same abstract problem as introduced in section 6.3.1. However, we study the problem under an additional singular type estimate imposed on the $|e^{At}B|_{\mathcal{L}(U;H)}$. While this type of estimate is typical for analytic semigroups, it is far less common in the hyperbolic-like problems. Our motivation for considering this case comes from structural acoustic models where the validity of this estimate is shown by microlocal analysis methods. This is despite the fact that the underlying generator for the structural acoustic model is not of the analytic type.

6.5.1 Formulation of the results

A standing assumption in this section is the following.

Assumption 6.5.1. *There exists a constant $0 \le r_0 < 1$ such that for some $T > 0$*

$$|e^{At}B|_{\mathcal{L}(U;H)} \le \frac{C_T}{t^{r_0}} \quad \text{for} \quad 0 < t \le T. \tag{6.5.46}$$

Remark 6.5.1. *Note that the estimate in Assumption* 6.5.1 *is reminiscent of the estimates usually satisfied by analytic semigroups* [129, 28, 140]. *However, there are semigroups that are not analytic yet the above estimate still holds. An example of interest to us is a structural acoustic problem with hyperbolic-parabolic coupling.*

We note that Assumption 6.5.1 automatically implies the validity of the hypothesis in (6.4.42) (with R bounded) (but not (6.4.45)). Therefore, we obtain the following corollary.

Corollary 6.5.1. *Assume that Assumptions* 6.4.1 *and* 6.5.1 *hold true. Then all the conclusions of Lemma* 6.4.3 *remain valid.*

Remark 6.5.2. *If $r_0 < 1/2$ in Assumption* 6.5.1, *then automatically condition* (6.4.36) *holds and all the conclusions of Lemma* 6.4.3 *remain valid.*

In view of the corollary above, to provide a meaningful and regular optimal synthesis, it suffices to analyze the contribution of $B^*r(t)$. The main result in this direction is given below.

Lemma 6.5.1. *Under the validity of Assumptions* 6.4.1 *and* 6.5.1, *in addition to all conclusions of Lemma* 6.4.3 *we have the following.*

(i) *For any $F \in L_1(0,T;H)$, $\mathbf{y}_0 \in H$, we have*

$$u^0 \in C([0,T];U), \quad \mathbf{y}^0 \in C([0,T];H), \quad B^*r \in C([0,T];U).$$

(ii) *$r \in C([0,T];H)$ satisfies the following equation:*

$$r_t(t) = -[A^* - P(t)BB^*]r(t) - P(t)F(t) \text{ on } [D(A)]', \qquad r(T) = 0.$$

Equivalently, $r(t) = \int_t^T \Phi^*(\tau,t)P(\tau)F(\tau)d\tau$.

By combining the results of Lemma 6.5.1 and Lemma 6.4.2 we obtain the following.

Corollary 6.5.2. *Assume that Assumptions* 6.4.1 *and* 6.5.1 *are valid. Then, there exists a unique optimal pair* $u^0 \in C([0,T];U), \mathbf{y}^0 \in C([0,T];H)$ *such that*

$$u^0(t) = -B^*P(t)\mathbf{y}^0(t) - B^*r(t), \qquad t \geq 0,$$

and $P(t), r(t)$ *satisfy all the properties listed in Lemmas* 6.4.3 *and* 6.5.1.

While Assumption 6.5.1 implies (6.4.42), it does not generally imply Assumption 6.4.1 (unless $r_0 < 1/2$; see Remark 6.5.2). Thus it makes sense to ask whether the singular estimate in Assumption 6.5.1 alone is sufficient to imply all the conclusions in Corollary 6.5.2. The answer to this is positive, although the affirming details are challenging. Recall that one of the main advantages of assuming (6.4.36) was that this condition implied almost for free the continuity of the evolution $\Phi(t,s)$. Yet it turns out that the continuity of $\Phi(t,s)$ holds under the sole Assumption 6.5.1; however, the proof of this fact is more involved. Of course in the absence of Assumption 6.4.1, one must go through the process of reproving Lemma 6.4.3, under the sole Assumption 6.5.1. This can be done best by taking advantage of singular integral theory (as in the analytic case [140]). Ultimately, the result of Corollary 6.5.2 can be shown to hold under Assumption 6.5.1 alone with the condition (6.4.36) being unnecessary. However, to simplify the exposition and to connect this theory with results existing already in the literature, we first provide a complete proof in a more restrictive framework, i.e., we first proceed under condition (6.4.36). At the end of the section we outline the main steps of the proof for a more general case, without assuming (6.4.36). This result is stated as thus.

Theorem 6.5.1. *Assume that Assumption* 6.5.1 *is valid. Then, there exists a unique optimal pair* $u^0 \in L_2(0,T;U), y^0 \in C([0,T];H)$ *such that*

$$u^0(t) = -B^*P(t)y^0(t) - B^*r(t), \qquad t \geq 0,$$

and $P(t), r(t)$ *satisfy all the properties listed in Lemmas* 6.4.3 *and* 6.5.1.

6.5.2 Proof of Lemma 6.5.1

We begin by introducing the following singular spaces, which are a cornerstone of the analysis in the case of analytic semigroups [140, 28, 132]. Let X be a Banach space:

$$\begin{aligned} Z_r(s,T;X) \equiv \{f \in C((s,T];X) \text{ such that } &|f|_{Z_r(s,T;X)} \\ &= \sup_{z\in(s,T]} |f(z)|_X |z-s|^r < \infty\}. \end{aligned}$$

It is known that the space $Z_r(s,T;X)$ is a Banach space. The following proposition collects various singular estimates known in the literature [65, 129, 140, 28].

Proposition 6.5.1. *Let the operator* B *satisfy Assumption* 6.5.1 *with some* r, $0 \leq r < 1$. *The following operators are bounded uniformly with respect to the parameter* $0 \leq s \leq T$, *but the bounds generally depend on* r, T*:*

- $L_s : C([s,T];U) \to C([s,T];H)$,
- $L_s^* : C([s,T];H) \to C([s,T];U)$,
- $[I + L^*R^*RL_s]^{-1} \in \mathcal{L}(C([s,T];U))$,
- $[I + L_sL^*R^*R]^{-1} \in \mathcal{L}(C([s,T];H))$,
- $L_s : Z_r(s,T;U) \to Z_{2r-1}(s,T;H) \subset Z_r(s,T;H)$,
- $L_s^* : Z_r(s,T;H) \to Z_{2r-1}(s,T;U) \subset Z_r(s,T;U)$,
- $[I + L^*R^*RL_s]^{-1} \in \mathcal{L}(Z_r(s,T;U))$,
- $[I + L_sL^*R^*R]^{-1} \in \mathcal{L}(Z_r(s,T;H))$.

Proof. The first two properties are from Proposition 4.4.1 in [140] (see also Theorem 3.3 in [132]). The third property follows from Theorem 4.4.4 in [140] (see also Theorem 3.4 in [132]). The last property follows simply by using algebraic decomposition

$$[I + L_sL^*R^*R]^{-1} = -L_s[I + L^*R^*RL_s]^{-1}L^*R^*R + I$$

together with the first three properties in Proposition 6.5.1. □

Proper Proof of Lemma 6.5.1. By using explicit representation of optimal control u^0 and optimal trajectory $\mathbf{y}^0$ given by Lemma 6.4.1 together with regularity properties listed in Proposition 6.5.1 we obtain

$$u^0 \in C([0,T];U), \quad \mathbf{y}^0 \in C([0,T];H). \tag{6.5.47}$$

We note that the regularity of $B^*P(t)$, given in part (i) of Lemma 6.4.3, together with the fact that $B^*p \in L_2(0,T;U)$ (see (6.4.34)) yields immediately via (6.4.35) that

$$B^*r \in L_2(0,T;U). \tag{6.5.48}$$

But actually the feedback representation of optimal control given in Lemma 6.4.1 together with (6.5.47) yield a stronger conclusion $B^*r \in C([0,T];U)$, which gives the statement in part (i) of Lemma 6.5.1.

To prove part (ii) of Lemma 6.5.1, it suffices to derive the differential equation satisfied by r. To this end we need to establish more regularity properties, in particular those of the evolution $\Phi(t,s)$. For this purpose, we need to study the properties of operators L_s, L_s^* as acting on singular spaces $Z_{r_0}(s,T;H)$. This is done below.

Step 1. We establish a singular estimate for the norm of the operator $\Phi(t,s)B$ which will imply that the operator B is singular with respect to $A_P(t) \equiv A - BB^*P(t)$ with the same order of singularity as $e^{At}B$. Consequently, $\Phi(t,s)B$ is shown to be well defined for all $t > s$ as a bounded operator $U \to H$.

Proposition 6.5.2. *Let $T > 0$ be arbitrary. Assume Assumption* 6.4.1 *and* 6.5.1 *with a constant $r_0 < 1$. Then*

$$|\Phi(t,s)B|_{\mathcal{L}(U;H)} = |B^*\Phi^*(t,s)x|_{\mathcal{L}(H;U)} \le \frac{C_T}{(t-s)^{r_0}}, \qquad r_0 < 1, \; s < t < T, \tag{6.5.49}$$

and the constant C_T does not depend on s. If, in addition, $r_0 < 1/2$, then

$$\int_s^T |B^*\Phi^*(t,s)x|_U^2 dt \le C_T|x|_H^2. \tag{6.5.50}$$

Proof. To prove Proposition 6.5.2 we consider the following equation defined originally on $[\mathcal{D}(A^*)]'$:

$$y_t(t,s) - (A - BB^*P(t))\, y(t,s) = 0 \text{ for } y(s,s) = Bu, \qquad u \in U. \tag{6.5.51}$$

Our goal is to show that there exists a unique solution to (6.5.51) that satisfies

$$|y(t,s)|_H \leq \frac{C_T}{(t-s)^{r_0}} |u|_U, \qquad t > s. \tag{6.5.52}$$

Since, a fortiori, $y(t,s) = \Phi(t,s)Bu$, by duality (see [125]) (6.5.49) is equivalent to (6.5.52). To prove (6.5.52), we will be solving the integral equation

$$y(t,s) = \int_s^t e^{A(t-\tau)} BB^*P(\tau)y(\tau)d\tau + e^{A(t-s)}Bu = L_s(B^*P(\cdot)y(\cdot))(t) + e^{A(t-s)}B \tag{6.5.53}$$

on $[\mathcal{D}(A^*)]'$, which is a weak formulation of (6.5.51).

To solve this it suffices to show that for every $u \in U$ the map $\mathcal{T} : Z_{r_0}(s,T;H) \to Z_{r_0}(s,T;H)$ defined by

$$\mathcal{T}y(\cdot,s)(t) \equiv -L_s(B^*P(\cdot)y(\cdot,s))(t) + e^{A(t-s)}Bu$$

is a contraction. From Assumption 6.5.1 we have

$$|e^{At}Bu|_H \leq \frac{C_T}{t^{r_0}} |u|_U \text{ for } 0 < t \leq T. \tag{6.5.54}$$

This estimate and Corollary 6.5.1 imply that $B^*P(t) \in \mathcal{L}(H;U)$, uniformly in $t \in (s,T)$. Consequently, we are in a position to apply the contraction mapping principle with respect to the space $Z_{r_0}(s,T;H)$ (see [28]). Indeed, by Proposition 6.5.1, Lemma 6.4.3, (6.5.53), and (6.5.54) we have the following pointwise estimate for all $t > s$:

$$\begin{aligned} |y(t,s)|_H &\leq |L_s(B^*Py(\cdot,s)|_H + \frac{C_T}{(t-s)^{r_0}}|u|_U \\ &\leq C_T \frac{1}{|t-s|^{2r_0-1}} |L_s|_{\mathcal{L}(Z_{r_0}(s,T;H);C(s,T;H))} \\ &\times |B^*P(\cdot)|_{\mathcal{L}(H;L_\infty(0,T;H))} |y(\cdot,s)|_{Z_{r_0}(s,T;H)} \\ &\qquad + \frac{C_T}{(t-s)^{r_0}} |u|_U, \qquad t > s. \end{aligned} \tag{6.5.55}$$

(Note that by Lemma 6.4.3 and the Banach–Steinhaus theorem, $\sup_{0 \leq t \leq T} |B^*P(t)|_{\mathcal{L}(H)} \leq \infty$.) Then (6.5.55) implies that

$$|y(\cdot,s)|_{Z_{r_0}(s,T;H)} \leq C_T |t-s|^{1-r_0} |y(\cdot,s)|_{Z_{r_0}(s,T;H)} + C_T|u|_U \tag{6.5.56}$$

and shows that the map $\mathcal{T}$ maps $Z_{r_0}(H)$ into itself. Since $1 - r_0 > 0$, one easily shows that by taking first sufficiently small T, the map $\mathcal{T}$ is a contraction. This contraction property can be propagated in finitely many steps to an arbitrary large interval $(0,T)$.

The argument above yields an existence and uniqueness of solutions to (6.5.53) residing in the space $Z_r(H)$. Moreover, from the definition of the spaces $Z_{r_0}(s,T;H)$ we obtain the estimate

$$|y(t,s)|_H \leq \frac{C_T}{|t-s|^{r_0}}|u|_U, \qquad s<t\leq T, \tag{6.5.57}$$

which in turn proves (6.5.52) and hence the assertion in (6.5.49).

The inequality in (6.5.50) is a simple consequence of (6.5.49) and the fact that $r_0 < 1/2$. □

Step 2. We prove next that the variable r satisfies the desired differential equation. We note here that the main difficulty (due to the unboundedness of B) is a justification of the performed calculations, which otherwise would be only formal.

We start with the characterization of $r(t)$ given in Lemma 6.4.1:

$$r(t) = p(t) - P(t)\mathbf{y}^0(t). \tag{6.5.58}$$

Since the variable r does not depend on the initial condition, we can replace $\mathbf{y}^0(t)$ by $y^*(t)$, where the variable y^* corresponds to the optimal trajectory with zero initial data at time $t=0$, i.e., $\mathbf{y}(0)=0$ in (6.4.25). Thus, $r(t)$ can be written as (cf. (6.5.58))

$$r(t) = \int_t^T e^{A^*(s-t)}R^*Ry^*(s)ds - \left(\int_t^T e^{A^*(s-t)}R^*R\Phi(s,t)ds\right)y^*(t), \tag{6.5.59}$$

where we used the evolution property of $\Phi(t,s)$ and the definitions of $p(t), P(t)$.

Step 3. We show the following proposition.

Proposition 6.5.3. *Assume Assumptions* 6.4.1 *and* 6.5.1. *Then the following differentiability property of the evolution operator is valid:*

(i) $\frac{d}{dt}[\Phi(\cdot,t)y^*(t)] = \Phi(\cdot,t)[-BB^*r(t)+F(t)]$ *in* $Z_{r_0}(t,T;H)$,
(ii) *if* $r_0 < 1/2$, *then* $\frac{d}{dt}[\Phi(.,t)y^*(t)] \in L_2(t,T;H)$.

Proof. We go back to the explicit representation of the optimal trajectory and evolution operator. Denoting $X(s,t) \equiv \Phi(s,t)y^*(t)$ from Lemma 6.4.1 we obtain

$$X(s,t) = ([I+L_tL^*R^*R]^{-1}e^{A(.-t)}y^*(t))(s).$$

Hence

$$([I+L_tL^*R^*R]X(.,t))(s) = e^{A(s-t)}y^*(t). \tag{6.5.60}$$

We note that in $\mathcal{D}(A^*)'$

$$\begin{aligned}\frac{d}{dt}[e^{A(s-t)}y^*(t)] &= -Ae^{A(s-t)}y^*(t) + e^{A(s-t)}Ay^*(t)\\ &\quad - e^{A(s-t)}BB^*P(t)y^*(t) - e^{A(s-t)}BB^*r(t) + e^{A(s-t)}F(t)\end{aligned} \tag{6.5.61}$$

after using (6.5.59). All the quantities on the right-hand side of (6.5.61) are well defined on $[D(A^*)]'$ and for $s>t$. Indeed, the latter follows from the regularity of

$B^*P(t)$, which is pointwise defined (Lemma 6.4.3), and from the $L_2(0,T;U)$ regularity of B^*r (see (6.5.48)) and singular estimate for $e^{At}B$ assumed in Assumption 6.5.1. Hence,

$$\frac{d}{dt}[e^{A(s-t)}y^*(t)] = -e^{A(s-t)}[BB^*P(t)y^*(t) + BB^*r(t) - F(t)], \tag{6.5.62}$$

where the above relation can be extended, by density and Assumption 6.5.1, to hold on the entire H for $s > t$ and almost everywhere in time t. Moreover, by direct computations

$$\begin{aligned}\left(\left[\frac{d}{dt}L_t\right] L_t^* R^* RX(.,t)\right)(s) &= -e^{A(s-t)}B(L_t^* R^* RX(.,t))(t) \\ &= -e^{A(s-t)}BB^* \int_t^T e^{A^*(z-t)}R^* RX(z,t)dz \\ &= -e^{A(s-t)}BB^* \int_0^T e^{A^*(z-t)}R^* R\Phi(z,t)dz y^*(t) \\ &= -e^{A(s-t)}BB^*P(t)y^*(t),\end{aligned} \tag{6.5.63}$$

where, by virtue of Assumption 6.5.1 and regularity of B^*P, the last term is well defined in $Z_{r_0}(t,T:H)$.

Now we are in a position to differentiate in time the operator equation (6.5.60) (using implicitly the fact that $\frac{d}{dt}L_t^* = 0$). This leads to

$$([I + L_t L_t^* R^* R]X_t(.,t))(s) = -\left(\left[\frac{d}{dt}L_t\right] L_t^* R^* RX(.,t)\right)(s) + \frac{d}{dt}[e^{A(s-t)}y^*(t)]. \tag{6.5.64}$$

Invoking (6.5.62) and (6.5.63) yields

$$-\left(\left[\frac{d}{dt}L_t\right] L_t^* R^* RX(.,t)\right)(s) + \frac{d}{dt}[e^{A(s-t)}y^*(t)] = -e^{A(s-t)}[BB^*r(t) - F(t)], \tag{6.5.65}$$

and from (6.5.64)–(6.5.65),

$$([I + L_t L_t^* R^* R]X_t(.,t))(s) = -e^{A(s-t)}[BB^*r(t) - F(t)].$$

Since $e^{A(s-t)}BB^*r(t) \in Z_{r_0}(t,T,H)$, for almost all t, on account of Proposition 6.5.1 along with Assumption 6.5.1 we can apply the inverse of $[I + L_t L^* R^* R]$ to both sides of the equation above to have

$$X_t(.,t) = [I + L_t L^* R^* R]^{-1} e^{A(.-t)}[-BB^*r(t) + F(t)] \in Z_{r_0}(t,T;H) \quad \text{for a.e. } 0 < t < T.$$

But this is equivalent (recall the characterization of Φ given by (6.4.30)) to

$$X_t(s,t) = \Phi(s,t)[-BB^*r(t) + F(t)],$$

where the above expression is well defined for $s > t$. (Here we use the regularity of $\Phi(s,t)B$ given by Proposition 6.5.2.) Recalling the definition of $X(t,s)$ completes the proof of part (i) in the proposition. Part (ii) follows directly from part (i) and the fact that $r_0 < 1/2$. □

Step 4. *Proper proof of part* (ii), *Lemma* 6.5.1. We are now in a position to differentiate the formula (6.5.59) for r (in $[D(A^*)]'$, which gives

$$\begin{aligned}
r_t(t) &= -R^*Ry^*(t) - A^* \int_t^T e^{A^*(s-t)} R^*Ry^*(s)ds + R^*R\Phi(t,t)y^*(t) \\
&+ A^* \int_t^T e^{A^*(s-t)} R^*R\Phi(s,t)y^*(t)ds - \int_t^T e^{A^*(s-t)} R^*R\frac{d}{dt}[\Phi(s,t)y^*(t)]ds \\
&= -A^*r(t) - \int_t^T e^{A^*(s-t)} R^*R\frac{d}{dt}[\Phi(s,t)y^*(t)]ds \\
&\qquad \text{using the result of Proposition 6.5.3} \\
&= -A^*r(t) - \left(\int_t^T e^{A^*(s-t)} R^*R\Phi(s,t)ds\right)[-BB^*r(t) + F(t)] \\
&= -A^*r(t) - P(t)[-BB^*r(t) + F(t)],
\end{aligned} \tag{6.5.66}$$

which, due to the boundedness of $P(t)B, \Phi(s,t)B, B^*r$ (see Lemma 6.4.3 and the estimates (6.5.48), (6.5.49)) is the desired equation satisfied on $C([0,T]; D(A)']$. The proof of Lemma 6.5.1 is thus completed. □

Remark 6.5.3. *The regularity of the gain function $r(t)$ can also be studied directly from the properties of the differential equation satisfied by r. Since this argument is very convenient and will be used later, we provide the details.*

By the result (ii) *in Lemma* 6.5.1 *and the equation for the adjoint evolution operator*

$$\frac{d}{dt}\Phi^*(s,t) = -A_P^*(t)\Phi^*(s,t)$$

we have the following representation:

$$r(t) = \int_t^T \Phi^*(\tau,t)P(\tau)F(\tau)d\tau. \tag{6.5.67}$$

The estimate (6.5.49) *in Step* 1 *now gives*

$$\begin{aligned}
|B^*r(t)|_U &= |B^* \int_t^T \Phi^*(\tau,t)P(\tau)F(\tau)d\tau|_U \\
&\leq C_T \int_t^T \frac{1}{(\tau-t)^{r_0}} |P(\tau)|_{\mathcal{L}(H)}|F(\tau)|_H d\tau \\
&\leq C_T|F|_{L_\infty(0,T;H)}.
\end{aligned} \tag{6.5.68}$$

As we see, the representation formula (6.5.67) *yields a simple proof of regularity of B^*r, but the result obtained this way is not optimal. It requires that $F \in L_\infty(0,T;H)$ rather than $F \in L_1(0,T;H)$.*

6.5.3 Proof of Theorem 6.5.1

We recall that the statement of Theorem 6.5.1 is the same as that of Corollary 6.5.1, except that in the former the hyperbolic regularity condition (6.4.36) is assumed. As mentioned, the main issues to be settled now are the proof of continuity (in time) of the evolution operator $\Phi(t, s)$ and the regularity of the gain $B^*P(t)$, which under condition (6.4.36) follows simply from Lemma 6.4.3. In the present scenario, with only Assumption 6.5.1 in force, the results of Lemma 6.4.3 need to be reproved.

Proof of Theorem 6.5.1. We recall that the hyperbolic regularity condition (6.4.36) allowed us to use the abstract result in Lemma 6.4.3, which in turn provides critical boundedness of the gains $B^*P(t)$. Without assuming hyperbolic regularity, we need to establish this property by a different argument. In what follows we present the key arguments leading to the aforementioned regularity of the gains. A starting point is Lemma 6.4.1, whose validity requires only condition (6.4.24). As mentioned earlier, condition (6.4.24) is implied by Assumption 6.5.1. In fact, this follows from L_2 boundedness of Hilbert transforms. By using the explicit representation of optimal control u^0 and optimal trajectory y^0 given by Lemma 6.4.1 and combining these with the regularity properties of Proposition 6.5.1 one obtains the following.

Lemma 6.5.2. *For every* $\mathbf{y}_0 \in H$ *and* $F \in L_1(0, T; H)$

$$u^0 \in C([0, T]; U), \quad \mathbf{y}^0 \in C([0, T]; H).$$

The actual proof of Lemma 6.5.2 is rather technical, but it follows along steps (using the bootstrap argument) that are identical to the analytic case studied in the literature [140].

Our next step is to analyze the regularity of the gain $B^*P(t)$.

Lemma 6.5.3. *Under Assumption* 6.5.1 *we have*

$$B^*P \in \mathcal{L}(H \to C([0, T]; U)). \tag{6.5.69}$$

Proof. We recall the control-to-state operator $L_\tau : L_2(\tau, T; U) \to L_2(\tau, T; H)$ defined by

$$L_\tau u(t) \equiv \int_\tau^t e^{A(t-s)} Bu(s)ds.$$

Due to the singular estimate in Assumption 6.5.1, it is immediate to show that L_τ is a bounded operator on L_p spaces. This is to say,

$$L_\tau \in \mathcal{L}(L_p(\tau, T; U) \to L_p(\tau, T; H)), \qquad 1 \le p \le \infty,$$
$$L_\tau \in \mathcal{L}(C([\tau, T]; U) \to C([\tau, T]; H)). \tag{6.5.70}$$

By exploiting a more singular estimate in Assumption 6.5.1 together with Young's inequality we are able to show [132, 140] more regularity for L_τ. These are given below.

$$L_\tau \in \mathcal{L}(L_2(\tau, T; U) \to C([\tau, T]; H)) \text{ for } r_0 < 1/2,$$
$$L_\tau \in \mathcal{L}(L_p(\tau, T; U) \to C([\tau, T]; H)) \text{ for } p > 1,\ r_0 + \frac{1}{p} < 1,$$
$$L_\tau \in \mathcal{L}(L_p(\tau, T; U) \to L_{\bar{p}}(\tau, T; H)),$$
$$\text{for } r_0 \ge 1/2,\ p > 1,\ r_0 + \frac{1}{p} \ge 1,\ \bar{p} = \frac{p}{(r_0 - 1)p + 1} > p. \tag{6.5.71}$$

It is important to note that in the last inequality we always have $\bar{p} - p > 0$, a fact that reflects a certain smoothing mechanism on L_p spaces. The above property allows us to show that the evolution corresponding to the unforced problem (with $F = 0$) is continuous in time. To see this, let $\Phi(t, \tau)$ denote, as before, the evolution operator corresponding to the optimization problem with $F = 0$. We already know from Lemma 6.4.1 that $\Phi(t, \tau)$ admits the following representation:

$$\Phi(t, \tau) = [I + L_\tau L_\tau^* R^* R]^{-1} e^{A(\cdot - \tau)}, \tag{6.5.72}$$

where we always have that

$$[I + L_\tau L_\tau^* R^* R]^{-1} e^{A(\cdot - \tau)} \in \mathcal{L}(H; L_2(\tau, T; H))$$

uniformly in τ. By using properties in (6.5.71) together with the bootstrap argument akin to what is done in [132], one can show that

$$\Phi(\cdot, \tau) : H \to C([\tau, T]; H) \text{ is bounded for all } 0 \le \tau \le T. \tag{6.5.73}$$

(This is how Lemma 6.5.2 is derived.) The Riccati operator $P(t) : H \to H$, which is bounded, positive, and self-adjoint, admits the following explicit representation (see (6.4.33)) in terms of the evolution $\Phi(t, \tau)$:

$$P(t) = \int_t^T e^{A^*(s-t)} R^* R \Phi(s, t) ds.$$

This representation together with the continuity properties of the evolution $\Phi(t, \tau)$ and the singular estimate in Assumption 6.5.1 allow us to conclude the assertion in Lemma 6.5.3. Indeed, we evaluate

$$|B^* P(t) x|_U \le \int_t^T |B^* e^{A^*(s-t)}|_{\mathcal{L}(H \to U)} |R^* R \Phi(s, t) x|_H ds$$

$$\le C \int_t^T \frac{1}{(s-t)^{r_0}} |\Phi(s, t) x|_H ds \le C_T \int_t^T \frac{1}{(s-t)^{r_0}} ds |x|_H \le C_T |x|_H, \tag{6.5.74}$$

where we have used the singular estimate in Assumption 6.5.1 and the fact that $r_0 < 1$. □

Lemma 6.5.4. *Under the Assumption* 6.5.1 *with* $A_P(t) \equiv A - BB^* P(t)$, *we have that* $A_P(t)$ *generates a strongly continuous evolution operator* $\Phi(\tau, t)$ *on* H.

Proof. To prove the lemma it suffices to establish unique solvability in $C([\tau, T]; H)$ of the following integral equation:

$$y(t) = e^{A(t-\tau)} x - \int_\tau^t e^{A(t-s)} BB^* P(s) y(s) ds$$

with a given initial condition $x \in H$. This can be accomplished by applying the contraction principle in the same manner as in [132, 28, 140]. To see this, it is enough to note the estimate

$$\int_\tau^t |e^{A(t-s)} BB^* P(s) y(s)|_H ds \le C \int_\tau^t \frac{1}{(t-s)^{r_0}} |B^* P(s)|_{\mathcal{L}(H \to U)} |y(s)|_H ds$$

$$\le C \int_\tau^t \frac{1}{(t-s)^{r_0}} |y(s)|_H ds \le C (t-\tau)^{1-r_0} |y|_{C([\tau,T],H)}. \tag{6.5.75}$$

The above estimate allows us to apply the contraction mapping principle on some small interval (τ, T_0). This local result can be extended to a global one (with $T > 0$ arbitrary) in finitely many steps. □

The next key point is to establish a singular estimate for the evolution $\Phi(t, \tau)$.

Lemma 6.5.5. *For $\Phi(t,\tau)$ as defined by (6.5.72) we have for fixed $0 \le \tau < T$*

$$|\Phi(t,\tau)B|_{\mathcal{L}(H;U)} \le C_T \frac{1}{(t-\tau)^{r_0}}, \qquad \tau < t \le T. \tag{6.5.76}$$

Proof. We estimate $y(t) \equiv \Phi(t,\tau)Bu$, which solves the following integral equation a fortiori (by virtue of the evolution property for Φ):

$$y(t) = e^{A(t-\tau)}Bu - \int_\tau^t e^{A(t-s)}BB^*P(s)y(s)ds. \tag{6.5.77}$$

To establish the solvability of this equation (with a given $u \in U$) we use the singular spaces Z_r, as introduced in Proposition 6.5.1:

$$Z_r(\tau, T; H) \equiv \left\{ f \in C((\tau, T], H);\ \sup_{\tau < t \le T} |f(t)(t-\tau)^r|_H < \infty \right\}.$$

Applying a fixed-point argument in combination with Assumption 6.5.1 and the result of Lemma 6.5.3, an argument very much in line with that used in the proof of Proposition 6.5.2 (see also [132]), one proves the existence of a unique solution $y(t) \in Z_{r_0}(\tau, T; H)$ satisfying (6.5.77) and the estimate

$$|y(t)|_H \le \frac{C_T}{(t-\tau)^{r_0}} |u|_U \ \text{ for arbitrary } u \in U.$$

This, in turn, implies the estimate stated in the lemma. □

The estimate in Lemma 6.5.5 together with the inequality in Lemma 6.5.3 are critical in verifying the differentiability properties of $\Phi(t, s)$ with respect to the second argument s. The argument here is similar to that in Proposition 6.5.3 (see also [132, 140]) and leads to the following equation:

$$\frac{d}{ds}\Phi(t,s)x = -\Phi(t,s)[A - BB^*P(s)]x, \qquad x \in D(A). \tag{6.5.78}$$

This representation for $\frac{d}{ds}\Phi(t,s)$ on $D(A)$ allows a rigorous justification of the version of the DRE, given in Lemma 6.4.3. To see this we differentiate the formula

$$P(t) = \int_t^T e^{A^*(s-t)} R^* R \Phi(s,t) ds$$

on $D(A)$ with the values in $D(A^*)'$. We obtain for all $x, y \in D(A)$

$$\begin{aligned} (P_t(t)x, y)_H &= -(A^*P(t)x, y)_H - (R^*Rx, y)_H \\ &- \left(\int_t^T e^{A^*(s-t)} R^* R \Phi(s,t)[A - BB^*P(t)]x, y \right)_H, \end{aligned} \tag{6.5.79}$$

where the key observation is that by Lemma 6.5.5

$$|\Phi(s,t)Bu|_H \le C_T \frac{1}{(t-s)^{r_0}}$$

and consequently $\int_t^T e^{A^*(s-t)} R^* R \Phi(s,t) Bu ds$ is a well-defined Bochner integral for $u \in U$. Thus we obtain

$$(P_t(t)x, y)_H = -(A^* P(t)x, y)_H - (R^* R x, y)_H \\ - (P(t)Ax, y)_H + (B^* P(t)x, B^* P(t)y)_U. \tag{6.5.80}$$

This completes the derivation of DRE and justifies all the statements made in Lemma 6.4.3 under the sole Assumption 6.5.1.

The same arguments as in Step 4 of Proposition 6.5.3 now apply to derive the equation for r stated in Lemma 6.5.1 (no need for Assumption 6.4.1).

Our final step is to establish the regularity of function r. Indeed, since $A_P(t)$ generates strongly continuous evolution, the equation

$$r_t = -A_P^*(t)r - P(t)F : \; r(T) = 0$$

admits a unique solution $r \in C([0,T]; H)$ with any $F \in L_1(0,T;H)$. We need to establish stronger regularity properties for $r(t)$.

Lemma 6.5.6. *For any* $F \in L_1(0,T;H)$, $\mathbf{y}_0 \in H$, *we have*

$$B^* r \in C([0,T]; U). \tag{6.5.81}$$

Proof. We already know that $u^0 \in C([0,T]; U)$. From optimality conditions furnished by Lemma 6.4.1 we obtain

$$B^* r = -u^0 - B^* P(t) y^0(t).$$

Lemma 6.5.2 and the regularity of $B^* P(t)$ imply now the final conclusion of Lemma 6.5.6. □

The final statement of Theorem 6.5.1 follows by applying regularity properties established in Lemmas 6.5.3 and 6.5.6 together with Lemma 6.4.2. □

6.6 Back to Structural Acoustic Problems: Proofs of Theorems 6.3.1 and 6.3.2

To prove the results of these theorems, we apply abstract results given in Lemma 6.4.2, Lemma 6.5.1, and Theorem 6.5.1. To this end we need to verify all the hypotheses imposed on abstract operators.

Theorem 6.3.2 follows directly from Lemma 6.4.2. (Proof of this theorem for the case when $f = 0$ is given in [140].) Thus, we concentrate on the proof of Theorem 6.3.1.

Proof of Theorem 6.3.1. Here we deal with the case of structurally damped wall, that is, the system (5.2.9) models a hyperbolic-parabolic coupling. In this case ($\alpha > 0$), we show that Assumptions 6.4.1 and 6.5.1 are both satisfied for this specific problem. In consequence, the conclusion of Theorem 6.3.1 follows directly from the application of Corollary 6.5.2 (and also Theorem 6.5.1).

Remark 6.6.1. *We note that in view of Theorem* 6.5.1 *there is no real need for verification of condition* (6.4.36) *whenever Assumption* 6.5.1 *is in force.*

For simplicity of exposition we treat the model (5.2.9) without the boundary damping. In fact, this scenario represents the most challenging case, as the addition of the damping (particularly on Γ_0) simplifies some of the arguments [70]. Thus, in what follows we assume that the damping parameters are equal to zero, i.e., $d_i = g_i = 0$ for all i.

We are working within the framework of dynamics described by (5.2.9), $\alpha > 0$, with the operators A given by (5.2.7) and operator $\mathcal{B}$ subject to Assumption 5.2.2 and Assumption 6.3.1. For the convenience of the reader we recall the notation:

$$A_z : H_z \to H_z; \quad A_v : H_v \to H_v; \quad H = H_z \times H_v$$

$$A_z \equiv \begin{pmatrix} 0 & -I \\ A_N & 0 \end{pmatrix}, \quad A_v \equiv \begin{pmatrix} 0 & -I \\ \mathcal{M}_\gamma{}^{-1}\mathcal{A} & \alpha\mathcal{M}_\gamma{}^{-1}\mathcal{A} \end{pmatrix},$$

$$-A \equiv \begin{pmatrix} A_z & -c^2\rho^{-1}C^* \\ C & A_v \end{pmatrix}, \tag{6.6.82}$$

$$C^* \begin{pmatrix} \mathbf{v} \\ \mathbf{v}_t \end{pmatrix} = \begin{pmatrix} 0 \\ c^{-2}\rho(A_N + I)Nv_t \end{pmatrix}, \tag{6.6.83}$$

$$Bu = \begin{pmatrix} 0 \\ 0 \\ 0 \\ \mathcal{M}_\gamma{}^{-1}\mathcal{B}u \end{pmatrix}, \quad F = \begin{pmatrix} 0 \\ f \\ 0 \\ 0 \end{pmatrix}. \tag{6.6.84}$$

With the above notation, the semigroup solution $e^{A^*t}\mathbf{y}$ can be described in the following manner:

$$\mathbf{y} = \begin{bmatrix} \vec{z} \\ \vec{\mathbf{v}} \end{bmatrix} = e^{A^*t} \begin{bmatrix} \vec{z_0} \\ \vec{\mathbf{v}_0} \end{bmatrix}, \quad x \equiv \begin{bmatrix} \vec{z_0} \\ \vec{\mathbf{v}_0} \end{bmatrix},$$

where $\vec{z} = \begin{pmatrix} z \\ z_t \end{pmatrix}$, $\vec{\mathbf{v}} = \begin{pmatrix} \mathbf{v} \\ \mathbf{v}_t \end{pmatrix}$, and $z, \mathbf{v}$ satisfy the following second-order equation:

$$\begin{aligned} z_{tt} + A_N z - (A_N + I)Nv_t &= 0, \\ \mathcal{M}_\gamma \mathbf{v}_{tt} + \mathcal{A}\mathbf{v} + \alpha\mathcal{A}\mathbf{v}_t + c^2\rho\mathcal{C}z_t|_{\Gamma_0} &= 0 \\ [z(0), z_t(0), \mathbf{v}(0), \mathbf{v}_t(0)] &= [\vec{z_0}, \vec{\mathbf{v}_0}]. \end{aligned} \tag{6.6.85}$$

Our first step is to quantify the unboundedness of the control operator B. In fact, the first result deals with the hyperbolic regularity of the control operator B-condition (6.4.36).

6.6.1 Verification of Assumption 6.4.1

Validity of Assumption 6.4.1 was verified in [140, pp. 906–908], when $B = \delta'$. Below we follow the more general argument given in [70].

Proposition 6.6.1. *Let $\alpha > 0$. Subject to Assumption 5.2.2, the control operator B is admissible, i.e., for any $T > 0$ the following inequality is satisfied:*

$$\int_0^T |B^* e^{A^* t} x|_U^2 dt \leq C_T |x|_H^2, \qquad x \in D(A^*). \tag{6.6.86}$$

Proof. *Step* 1. Without loss of generality we take $c = \rho = 1$. Multiplying the first equation in (6.6.85) by z_t, integrating the result over $\Omega \times (0,T)$ (these computations are first performed for smooth initial conditions and then extended by the density to all space), and taking advantage of (5.2.4) yields

$$|z_t(T)|_{0,\Omega}^2 + |A_N{}^{1/2} z(T)|_{0,\Omega}^2 - 2\int_0^T (v_t, z_t)_{\Gamma_0}\, dt \\ = |z_t(0)|_{0,\Omega}^2 + |A_N{}^{1/2} z(0)|_{0,\Omega}^2. \tag{6.6.87}$$

Multiplying the second equation in (6.6.85) by $\mathbf{v}_t$ and integrating the result over $\Gamma_0 \times (0,T)$ and accounting for the relation in (6.6.83) and (5.2.4) yields

$$|\mathcal{M}^{1/2}\mathbf{v}_t(T)|_{0,\Gamma_0}^2 + |\mathcal{A}^{1/2}\mathbf{v}(T)|_{0,\Gamma_0}^2 + 2\int_0^T |\mathcal{A}^{1/2}\mathbf{v}_t|_{0,\Gamma_0}^2 dt \\ = -2\int_0^T (z_t, v_t)_{\Gamma_0}\, dt + |\mathcal{M}^{1/2}\mathbf{v}_t(0)|_{0,\Gamma_0}^2 + |\mathcal{A}^{1/2}\mathbf{v}(0)|_{0,\Gamma_0}^2. \tag{6.6.88}$$

From (6.6.87), (6.6.88) and noting the cancellation of two boundary terms, we obtain

$$\int_0^T |\mathcal{A}^{1/2}\mathbf{v}_t|_{0,\Gamma_0}^2 dt \leq C(|z_t(0)|_{0,\Omega}^2 + |A_N{}^{1/2} z(0)|_{0,\Omega}^2 \\ + |\mathcal{M}^{1/2}\mathbf{v}_t(0)|_{0,\Gamma_0}^2 + |\mathcal{A}^{1/2}\mathbf{v}(0)|_{0,\Gamma_0}^2) \leq C \left\| \begin{bmatrix} \vec{z_0} \\ \vec{\mathbf{v}_0} \end{bmatrix} \right\|_H^2. \tag{6.6.89}$$

Step 2. We compute explicitly the adjoint operator B^*. Indeed,

$$\left(u, B^* \begin{pmatrix} z \\ z_t \\ \mathbf{v} \\ \mathbf{v}_t \end{pmatrix} \right)_U = \left(Bu, \begin{pmatrix} z \\ z_t \\ \mathbf{v} \\ \mathbf{v}_t \end{pmatrix} \right)_H$$

$$= \left(\mathcal{M}^{-1}\mathcal{B}u, \mathbf{v}_t\right)_{D(\mathcal{M}_\gamma{}^{1/2})} = (\mathcal{B}u, \mathbf{v}_t)_{\Gamma_0} = \left(\mathcal{A}^{-r}\mathcal{B}u, \mathcal{A}^r \mathbf{v}_t\right)_{0,\Gamma_0}.$$

Hence, by using Assumption 5.2.2 we obtain

$$\left| B^* \begin{pmatrix} z \\ z_t \\ \mathbf{v} \\ \mathbf{v}_t \end{pmatrix} \right|_U \leq C|\mathcal{A}^r \mathbf{v}_t|_{0,\Gamma_0}. \tag{6.6.90}$$

Collecting (6.6.89) and (6.6.90) gives

$$\int_0^T |B^* e^{A^* t} x|_U^2 dt \le C \int_0^T |\mathcal{A}^r \mathbf{v}_t|_{0,\Gamma_0}^2 \le C \int_0^T |\mathcal{A}^{1/2} \mathbf{v}_t|_{0,\Gamma_0}^2, \tag{6.6.91}$$

where we have used the fact that $r < 1/2$. This proves the inequality in Proposition 6.6.1. □

Remark 6.6.2. *We recall that the regularity inequality in Proposition* 6.6.1, *via duality, is equivalent (see* [125]*) to the continuity of the control map $u \to \mathbf{y}$ from $L_2(0, T, U) \to C([0, T]; H)$. This result substantially improves the results of* [17], *where the continuity of this map was shown to hold in the space of distributions only, i.e., on $D(A^*)'$.*

6.6.2 Verification of Assumption 6.5.1

We recall that the wall model under consideration is given by (5.2.2) with $\alpha > 0$, $\theta = 1/2$. The control operator $\mathcal{B}$ is subject to Assumption 5.2.2 and the additional Assumption 6.3.1.

Proposition 6.6.2. *Under Assumptions* 5.2.2 *and* 6.3.1 *with $\theta = 1/2$ the following estimate is valid for all $T > 0$:*

$$|e^{At} B|_{\mathcal{L}(U;H)} \le \frac{C_T}{t^r} \text{ for } 0 < t \le T, \tag{6.6.92}$$

where r is as given in Assumption 5.2.2.

Remark 6.6.3. *We note that the estimate of the type as given in* (6.6.92) *is typical within the framework of analytic semigroups (see* [175, 14]*). However, in our case, the semigroup e^{At} is not analytic.*

Corollary 6.5.2 in combination with Proposition 6.6.2 and Proposition 6.6.1 yield the result of Theorem 6.3.1. Thus, it remains to prove Proposition 6.6.2. This is accomplished below, where we follow closely the arguments presented in [4].

Proof of Proposition 6.6.2. To begin with, we state two results used critically in the proof. The first result deals with sharp regularity properties of traces of solutions to the wave equation. The second result is related to interpolation theory and a characterization of fractional powers for structurally damped operators.

The following regularity result for the wave equation, which is also of independent interest in the context of PDEs, will be used in the sequel.

Lemma 6.6.1 (see [3, 130, 127, 188]). *Let z be a solution to the wave equation:*

$$\begin{aligned} z_{tt} &= c^2 \Delta z \text{ in } \Omega \times (0, T), \\ \frac{\partial}{\partial \nu} z &= g \in L_2(0, T; H^\beta(\Gamma)) \text{ on } \Gamma \times (0, T), \\ z(0) &= z_t(0) = 0 \text{ in } \Omega. \end{aligned} \tag{6.6.93}$$

We assume that $\beta = 1/4$ in the case of parallelepiped and $\beta = 1/3$ in the case of general smooth domains. Then we have

$$[z, z_t, z_t|_\Gamma] \in C(0, T; H^1(\Omega)) \times C(0, T; L_2(\Omega)) \times L_2(0, T; H^{-\beta}(\Gamma)). \tag{6.6.94}$$

The original proof of Lemma 6.6.1 with $\beta = 2/5$, for the smooth domain, is given in [127, 130, 135]. A sharper version of this result, with $\beta = 1/3$, was proved later in [188]. The corresponding result for rectangular domains is given in [3]. Note that this regularity with $\beta = 1/2$ was proved earlier in [167]. However, for our arguments it will be essential that $\beta < 1/2$; hence the regularity result of [167] is not sufficient.

Remark 6.6.4. *We note that the result of the Lemma* 6.6.1 *improves the classical* [151, 167] *regularity results on several accounts. Indeed, standard results* [151] *give only* $H^{1/2+\beta}(\Omega)$ *regularity provided*

$$g \in H^{\beta,\beta}(\Gamma \times (0,T)).$$

Thus, much less spatial regularity is available assuming more (time) regularity of the boundary data. Regarding the regularity of the traces of the velocity, classical PDE theory [151] *does not provide any information about* $z_t|_\Gamma$ *for solutions with the low regularity (as considered above). The paper* [167] *provides* $H^{-1/2}(\Gamma \times (0,T))$ *regularity for* $z_t|_\Gamma$ *with* $(z, z_t) \in H^1(\Omega) \times L_2(\Omega)$. *However, this is a much weaker result (by at least a* 1/6 *derivative) than the one stated in Lemma* 6.6.1.

Another important result, critical for the proof of Proposition 6.6.2, is the following characterization of fractional powers for the elastic operator:

$$-A_0 \equiv \begin{bmatrix} 0 & I \\ -\mathring{A} & -\alpha\mathring{A} \end{bmatrix} : H_0 \equiv D(\mathring{A}^{1/2}) \times \mathcal{H} \to H_0, \qquad \alpha > 0,$$

where $\mathring{A} : \mathcal{H} \to \mathcal{H}$ is any self-adjoint, positive defined operator on a Hilbert space $\mathcal{H}$.

Lemma 6.6.2 (see [49]). *With reference to the operator* A_0 *defined above, we have the following:*

1. *for any* $0 \le \theta \le 1/2$, *the characterization for the domains of* A_0^θ
$$D(A_0^\theta) = D(\mathring{A}^{1/2}) \times D(\mathring{A}^\theta),$$
2. *the map*
$$e^{-A_0 t} \in \mathcal{L}(H_0 \to L_2(0,T; [D(\mathring{A}^{1/2})]^2).$$

Remark 6.6.5. *We note that the formula for the fractional powers of the operator* A_0 *breaks down for* $\theta > 1/2$. *See* [49].

The main idea behind the proof of Proposition 6.6.2 is to engage into a suitable fixed-point argument that will establish a priori regularity for the velocity of the wall component. To accomplish this, we apply the trace regularity result in Lemma 6.6.1 (in particular, the regularity of the trace of the velocity z_t) and the properties of the elastic operator A_0 given in Lemma 6.6.2. The details of this argument are given below.

The most demanding case is when the stiffness operator $\mathcal{M}_\gamma$ is bounded and $\mathcal{D}(\mathcal{M}_\gamma)$ does not satisfy any additional regularity requirements. In fact, this is not surprising because the additional regularity of $\mathcal{D}(\mathcal{M}_\gamma)$ in $L_2(\Gamma_0)$ (exemplified by the presence of rotational inertia in the model) implies the additional regularity of the velocity component of the plate or shell equation.

For this reason, we first prove Proposition 6.6.2 assuming that condition (i) in Assumption 6.3.1 is satisfied. The case when rotational inertia are active (condition (ii) in Assumption 6.3.1) is treated later. It turns out that the arguments in the latter case are simpler.

Proof of Proposition 6.6.2 *under part* (i) *of Assumption* 6.3.1.

Step 1: Let β be the parameter specified in Lemma 6.6.1. We introduce the integral operator

$$K : L_2(0,T; D(A_v^{\frac{\beta}{4}})) \to L_2(0,T; D(A_v^{\frac{\beta}{4}})$$

given by

$$K(\vec{\mathbf{v}})(t) \equiv \int_0^t e^{-A_v(t-s)} C \int_0^s e^{-A_z(s-\tau)} C^* \vec{\mathbf{v}}(s) ds.$$

Recall that $-A_v$ generates an analytic semigroup and its fractional powers are well defined [175]. Our claim is as follows.

Lemma 6.6.3. *The following operators are bounded:*

1. $K : L_2(0,T; D(A_v^{\frac{\beta}{4}})) \to C([0,T]; D(A_v^{1/2-\frac{\beta}{4}-\epsilon})$ *for all* $\epsilon > 0$,
2. $[I-K]^{-1} : L_2(0,T; D(A_v^{\frac{\beta}{4}})) \to L_2(0,T; D(A_v^{\frac{\beta}{4}})$.

Proof of Lemma 6.6.3. We define

$$T(\vec{\mathbf{v}})(t) = C \int_0^t e^{-A_z(t-s)} C^* \vec{\mathbf{v}}(s) ds.$$

Then, a fortiori,

$$T(\vec{\mathbf{v}})(t) = \begin{pmatrix} 0^s \\ \rho c^{-2} \mathcal{M}_\gamma{}^{-1}(z_t(t)|_{\Gamma_0}) \end{pmatrix},$$

where $z(t)$ is a solution to the wave equation:

$$\begin{aligned} z_{tt} &= c^2 z \text{ in } \Omega \times (0,T), \\ \frac{\partial}{\partial \nu} z &= 0 \text{ on } \Gamma_1 \times (0,T), \\ \frac{\partial}{\partial \nu} z &= v_t \text{ on } \Gamma_0 \times (0,T), \\ z(t=0) &= z_t(t=0) = 0 \text{ in } \Omega. \end{aligned} \tag{6.6.95}$$

Applying Lemma 6.6.1 to this PDE we obtain

$$|z_t|_{\Gamma_0}|_{L_2(0,T;H^{-\beta}(\Gamma_0))} \le C|v_t|_{L_2(0,T;H^\beta(\Gamma_0))} \le C|v_t|_{L_2(0,T;\mathcal{D}(\mathcal{A}^{\frac{\beta}{4}}))}. \tag{6.6.96}$$

In what follows we use the identification of domains of fractional powers of the operator $A_v = A_0$ given by Lemma 6.6.2 with $\mathcal{H} \equiv \mathcal{D}(\mathcal{M}_\gamma{}^{1/2})$, $\mathring{A} \equiv \mathcal{M}_\gamma{}^{-1}\mathcal{A}$, $H_0 = H_v = \mathcal{D}(\mathcal{A}^{1/2}) \times \mathcal{D}(\mathcal{M}_\gamma{}^{1/2})$, where we note that $\mathcal{D}(\mathring{A}^{1/2}) = \mathcal{D}(\mathcal{A}^{1/2})$.

By applying Lemma 6.6.2 one obtains the following characterization of fractional powers of A_v:

$$\mathcal{D}(A_v^\theta) = \mathcal{D}(\mathcal{A}^{1/2}) \times [\mathcal{D}(\mathcal{A}^{1/2}), \mathcal{D}(\mathcal{M}_\gamma{}^{1/2})]_{2\theta} \text{ for } 0 \le \theta \le 1/2 \tag{6.6.97}$$

(note that we always have $\mathcal{D}(\mathcal{A}^{1/2}),\subseteq \mathcal{D}({\mathcal{M}_\gamma}^{1/2})$). To see this, we have from Lemma 6.6.2 that

$$\mathcal{D}(A_v^\theta) = \mathcal{D}(\mathcal{A}^{1/2}) \times \mathcal{D}(\mathring{A}^\theta), \qquad 0 \le \theta \le 1/2.$$

Hence, in particular

$$\mathcal{D}(A_v^{1/2}) = \mathcal{D}(\mathcal{A}^{1/2}) \times \mathcal{D}(\mathcal{A}^{1/2}).$$

Moreover,

$$\mathcal{D}(A_v^0) = \mathcal{D}(\mathcal{A}^{1/2}) \times \mathcal{D}({\mathcal{M}_\gamma}^{1/2}).$$

Interpolating between these two equalities and recalling that $\mathring{A}$ is self-adjoint on $\mathcal{D}({\mathcal{M}_\gamma}^{1/2})$ yields (6.6.97) (see also Definition 2.1 in [151]).

From (6.6.97) we obtain for $0 \le \theta \le 1/2$

$$\mathcal{D}(A_v^\theta) \subset \mathcal{D}(\mathcal{A}^{1/2}) \times [\mathcal{D}(\mathcal{A}^{1/2}), [L_2(\Gamma_0)]^s]_{2\theta} = \mathcal{D}(\mathcal{A}^{1/2}) \times \mathcal{D}(\mathcal{A}^\theta). \tag{6.6.98}$$

Applying the containment (6.6.98) to (6.6.96) yields

$$|z_t|_{\Gamma_0}|_{L_2(0,T;H^{-\beta}(\Gamma_0))} \le C|\vec{\mathbf{v}}|_{L_2(0,T;\mathcal{D}(A_v^{\frac{\beta}{4}})}. \tag{6.6.99}$$

Let $\psi = (\psi_1, \psi_2) \in \mathcal{D}(A_v^{\frac{\beta}{4}})$. We compute

$$\begin{aligned} |(T\vec{\mathbf{v}}, \psi)_{H_v}| &= \rho c^{-2}|({\mathcal{M}_\gamma}^{-1} z_t|_{\Gamma_0}, \psi_2)_{\mathcal{D}({\mathcal{M}_\gamma}^{1/2})}| \\ = \rho c^{-2}|(z_t|_{\Gamma_0}, \psi_2)_{L_2(\Gamma_0)}| &= \rho c^{-2}|(\mathcal{A}^{-\frac{\beta}{4}} z_t|_{\Gamma_0}, \mathcal{A}^{\frac{\beta}{4}}\psi_2)_{L_2(\Gamma_0)}| \\ \le C|z_t|_{\Gamma_0}|_{H^{-\beta}(\Gamma_0)}|\mathcal{A}^{\frac{\beta}{4}}\psi_2|_{L_2(\Gamma_0)} &\le C|z_t|_{\Gamma_0}|_{H^{-\beta}(\Gamma_0)}|A_v^{\frac{\beta}{4}}\psi|_{H_v}, \end{aligned} \tag{6.6.100}$$

where we have used the inclusion in (6.6.98). Combining (6.6.100) with (6.6.99) gives

$$|T(\vec{\mathbf{v}})|_{L_2(0,T;\mathcal{D}(A_v^{\frac{\beta}{4}})')} \le C|\vec{\mathbf{v}}|_{L_2(0,T;\mathcal{D}(A_v^{\frac{\beta}{4}}))}. \tag{6.6.101}$$

Since $-A_v$ is a generator of an analytic semigroup on H_v, the following estimate is standard [28]:

$$\left|\int_0^t e^{-A_v(t-s)} f(s)ds\right|_{C([0,T];\mathcal{D}(A_v^{1/2-\theta-\epsilon}))} \le C|f|_{L_2(0,T;\mathcal{D}(A_v^\theta)')}, \tag{6.6.102}$$

where $\epsilon > 0$.

Using this and the construction

$$K(\vec{\mathbf{v}})(t) = \int_0^t e^{-A_v(t-s)}(T\vec{\mathbf{v}})(s)ds,$$

(6.6.101) and (6.6.102) yield

$$\begin{aligned} |K(\vec{\mathbf{v}})|_{C([0,T];\mathcal{D}(A_v^{1/2-\frac{\beta}{4}-\epsilon}))} &\le C|T(\vec{\mathbf{v}})|_{L_2(0,T;\mathcal{D}(A_v^{\frac{\beta}{4}})')} \\ &\le C|\vec{\mathbf{v}}|_{L_2(0,T;\mathcal{D}(A_v^{\frac{\beta}{4}}))}, \end{aligned} \tag{6.6.103}$$

which proves the desired assertion in the first part of Lemma 6.6.3.

The second part of the lemma follows by observing that $1/2 - \frac{1}{4}\beta - \epsilon > \frac{1}{4}\beta$ for ϵ sufficiently small, which in turn implies

$$|K(\vec{\mathbf{v}})|_{C([0,T];D(A_v^{\frac{\beta}{4}}))} \leq C|\vec{\mathbf{v}}|_{L_2(0,T;D(A_v^{\frac{\beta}{4}}))}. \tag{6.6.104}$$

Hence

$$|K(\vec{\mathbf{v}})|_{L_2(0,T;D(A_v^{\frac{\beta}{4}}))} \leq C_T|T(\vec{\mathbf{v}})|_{L_2(0,T;D(A_v^{\frac{\beta}{4}}))} \leq C_T|\vec{\mathbf{v}}|_{L_2(0,T;D(A_v^{\frac{\beta}{4}}))}, \tag{6.6.105}$$

where $C_T \to 0$ when $T \to 0$. This estimate allows the bounded invertibility of $I-K$, first on a small interval $(0, T_0)$; repeated use of this estimate eventually provides the inverse for arbitrary $T > 0$, after finitely many steps. □

Step 2. From the regularity assumed in Assumption 5.2.2 it follows that

$$|\mathcal{B}u|_{\mathcal{D}(\mathcal{A}^{*r})'} = |\mathcal{A}^{-r}\mathcal{B}u|_U \leq C|u|_U.$$

We shall prove

$$\left| \begin{array}{c} 0 \\ \mathcal{M}_\gamma{}^{-1}\mathcal{B}u \end{array} \right|_{\mathcal{D}(A_v^r)'} \leq C|u|_U. \tag{6.6.106}$$

To see this it suffices to take $\psi = (\psi_1, \psi_2) \in D(A_v^r)$ and compute with fixed u

$$\begin{aligned} \left(\begin{array}{c} 0 \\ \mathcal{M}_\gamma{}^{-1}\mathcal{B}u \end{array}, \psi \right)_{H_v} &= (\mathcal{M}_\gamma{}^{-1}\mathcal{B}u, \psi_2)_{\mathcal{D}(\mathcal{M}_\gamma{}^{1/2})} \\ &= (\mathcal{B}u, \psi_2)_{[L_2(\Gamma_0)]^s} = (\mathcal{A}^{-r}\mathcal{B}u, \mathcal{A}^r\psi_2)_{[L_2(\Gamma_0)]^s} \\ &\leq C|\mathcal{A}^r\psi_2|_{[L_2(\Gamma_0)]^s}|u|_U \leq C|A_v^r\psi|_{H_v}|u|_U, \end{aligned} \tag{6.6.107}$$

where in the last step we used (6.6.98). This estimate implies (6.6.106). In turn, (6.6.106), analyticity of $e^{-A_v t}$, and Lemma 6.6.2 yield for fixed $u \in U$

$$\begin{aligned} \int_0^T \left| e^{-A_v(t)} \left(\begin{array}{c} 0 \\ \mathcal{M}_\gamma{}^{-1}\mathcal{B}u \end{array} \right) \right|^2_{\mathcal{D}(A_v^{1/2-r})} dt &\leq C|u|_U^2, \\ \left| e^{-A_v(t)} \left(\begin{array}{c} 0 \\ \mathcal{M}_\gamma{}^{-1}\mathcal{B}u \end{array} \right) \right|_{H_v} &\leq \frac{C}{t^r}|u|_U \text{ for } 0 < t \leq T. \end{aligned} \tag{6.6.108}$$

Step 3. Denote

$$\left(\begin{array}{c} \vec{z} \\ \vec{\mathbf{v}} \end{array} \right)(t) = e^{At}Bu. \tag{6.6.109}$$

Then, it is easy to verify that $\vec{z} = (z, z_t)$ satisfies the wave equation (6.6.95) and

$$\vec{\mathbf{v}}(t) = e^{-A_v(t)} \left(\begin{array}{c} 0 \\ \mathcal{M}_\gamma{}^{-1}\mathcal{B}u \end{array} \right) + K(\vec{\mathbf{v}})(t).$$

Hence

$$[I-K]\vec{\mathbf{v}}(t) = e^{-A_v(t)} \left(\begin{array}{c} 0 \\ \mathcal{M}_\gamma{}^{-1}\mathcal{B}u \end{array} \right).$$

Applying the second result in Lemma 6.6.3 and (6.6.108) yields

$$\int_0^T |\vec{\mathbf{v}}(t)|^2_{D(A_v^{\frac{\beta}{4}})}$$

$$\leq C|[I-K]^{-1}|^2_{\mathcal{L}(L_2(0,T;D(A_v^{\frac{\beta}{4}}))}\left|e^{-A_v(t)}\begin{pmatrix}0\\ \mathcal{M}_\gamma{}^{-1}\mathcal{B}u\end{pmatrix}\right|^2_{L_2(0,T;D(A_v^{\frac{\beta}{4}}))}$$

$$\leq C\int_0^T \left|e^{-A_v(t)}\begin{pmatrix}0\\ \mathcal{M}_\gamma{}^{-1}\mathcal{B}u\end{pmatrix}\right|^2_{\mathcal{D}(A_v^{1/2-r})} dt \leq C|u|^2_U, \quad (6.6.110)$$

where in the last step we used the fact that $1/2 - r \geq \beta/4$ implied by $r \leq 5/12$, when $\beta = 1/3$ and $r \leq 7/16$, when $\beta = 1/4$. We have thus obtained the a priori regularity for the variable $\vec{\mathbf{v}}$,

$$\vec{\mathbf{v}} \in L_2(0,T;D(A_v^{\frac{\beta}{4}})), \quad |\vec{\mathbf{v}}|_{L_2(0,T;D(A_v^{\frac{\beta}{4}}))} \leq C|u|_U. \quad (6.6.111)$$

Step 4. With the regularity (6.6.111) for the structural component we go back to Lemma 6.6.1, which allows us to read off the regularity for the acoustic component $\vec{z}$ in (6.6.109):

$$|\vec{z}|_{C([0,T];H_z)} \leq C|u|_U. \quad (6.6.112)$$

Moreover, by using Lemma 6.6.3 and recalling $1/2 - \frac{1}{4}\beta > 0$ we also obtain

$$K\vec{\mathbf{v}} \in C([0,T];\mathcal{D}(A_v^{1/2-\frac{1}{4}\beta-\epsilon})) \subset C([0,T];H_v),$$

$$|K\vec{\mathbf{v}}|_{C([0,T];H_v)} \leq C|u|_U.$$

Finally, since for all $0 \leq t \leq T$

$$\vec{\mathbf{v}}(t) = K(\vec{\mathbf{v}})(t) + e^{-A_v(t)}\begin{pmatrix}0\\ \mathcal{M}_\gamma{}^{-1}\mathcal{B}u\end{pmatrix}, \quad (6.6.113)$$

we also obtain by (6.6.108)

$$|\vec{\mathbf{v}}(t)|_{H_v} \leq C|\vec{\mathbf{v}}|_{L_2(0,T;D(A_v^{\frac{\beta}{4}}))} + C\frac{1}{t^r}|u|_U \leq C\left[1+\frac{1}{t^r}\right]|u|_U. \quad (6.6.114)$$

Combining the inequalities in (6.6.112) and (6.6.114) yields the desired conclusion in Proposition 6.6.2 under the first part of Assumption 6.3.1.

The case when rotational forces are included in the model is treated next. In this case we have more regularity corresponding to the velocity component v_t. This allows us to obtain the result without additional restrictions on the parameter r (except for our standing assumption $r < 1/2$). The proof in this case follows essentially the same steps as before with appropriate modifications reflecting different topologies for the underlying spaces.

Proof of Proposition 6.6.2 *under condition* (ii) *of Assumption* 6.3.1.
Step 1(a).

Lemma 6.6.4. *The following operators are bounded:*

1. $K : L_2(0,T;H_v) \to C([0,T];D(A_v^{1/2-\epsilon})$ *for all* $\epsilon > 0$,
2. $[I-K]^{-1} \in \mathcal{L}(L_2(0,T;H_v))$.

Proof of Lemma 6.6.4. With

$$T(\vec{\mathbf{v}})(t) = C\int_0^t e^{-A_z(t-s)}C^*\vec{\mathbf{v}}(s)ds$$

we have

$$T(\vec{\mathbf{v}})(t) = \begin{pmatrix} 0^s \\ \rho c^{-2}\mathcal{M}_\gamma{}^{-1}(z_t|_{\Gamma_0}) \end{pmatrix},$$

where $z(t)$ is a solution to the wave equation (6.6.95). By Lemma 6.6.1

$$|z_t|_{\Gamma_0}|_{L_2(0,T;H^{-\beta}(\Gamma_0))} \le C|v_t|_{L_2(0,T;H^\beta(\Gamma_0))}. \tag{6.6.115}$$

By Assumption 6.3.1 and (6.6.115)

$$\begin{aligned}|[T(\vec{\mathbf{v}})]_2|_{L_2(0,T;\mathcal{D}(\mathcal{M}_\gamma{}^{1/2})} &\le C|\mathcal{M}_\gamma{}^{-1/2}\mathcal{C}(z_t|_{\Gamma_0})|_{L_2(0,T;L_2(\Gamma_0))} \\ &\le C|(z_t|_{\Gamma_0})|_{L_2(0,T;H^{-\beta}(\Gamma_0))} \le C|v_t|_{L_2(0,T;H^\beta(\Gamma_0))}.\end{aligned} \tag{6.6.116}$$

Combining this estimate with condition (ii) of Assumption 6.3.1 yields

$$|T(\vec{\mathbf{v}})|_{L_2(0,T;H_v)} \le C|v_t|_{L_2(0,T;H^\beta(\Gamma_0))} \le C|\vec{\mathbf{v}}|_{L_2(0,T;H_v)}. \tag{6.6.117}$$

Since $-A_v$ is a generator of an analytic semigroup on H_v, for arbitrary $f \in L_2(0,T;H_v)$ we have

$$\left|\int_0^t e^{-A_v(t-s)}f(s)ds\right|_{C([0,T];D(A_v^{1/2-\epsilon}))} \le C|f|_{L_2(0,T;H_v)}, \tag{6.6.118}$$

where $\epsilon > 0$.

(6.6.117), (6.6.118) now give for arbitrary $\mathbf{v} \in L_2(0,T;H_v)$

$$|K(\vec{\mathbf{v}})|_{C([0,T];D(A_v^{1/2-\epsilon}))} \le C|T(\vec{\mathbf{v}})|_{L_2(0,T;H_v)} \le C|\vec{\mathbf{v}}|_{L_2(0,T;H_v)}, \tag{6.6.119}$$

which proves the desired assertion in the first part of the lemma.

From this estimate moreover, we also have easily

$$|K(\vec{\mathbf{v}})|_{L_2(0,T;H_v)} \le C_T|\vec{\mathbf{v}}|_{L_2(0,T;H_v)}, \tag{6.6.120}$$

where $C_T \to 0$ when $T \to 0$. This allows one (as before) to obtain the asserted invertibility of $I - K$, first on a small interval $(0, T_0)$, and then for an arbitrary $T > 0$ in finitely many steps. This finishes the proof of Lemma 6.6.4. □

Step 2(a). By the same arguments as in Step 2 (see (6.6.106), (6.6.108)) we obtain for all $u \in U$

$$\begin{aligned}&\left|\begin{matrix} 0 \\ \mathcal{M}_\gamma{}^{-1}\mathcal{B}u \end{matrix}\right|_{\mathcal{D}(A_v^r)'} \le C|u|_U, \\ &\int_0^T \left|e^{-A_v(t)}\begin{pmatrix} 0 \\ \mathcal{M}_\gamma{}^{-1}\mathcal{B}u \end{pmatrix}\right|^2_{\mathcal{D}(A_v^{1/2-r})} dt \le C|u|_U^2, \\ &\left|e^{-A_v(t)}\begin{pmatrix} 0 \\ \mathcal{M}_\gamma{}^{-1}\mathcal{B}u \end{pmatrix}\right|_{H_v} \le \frac{C}{t^r}|u|_U; 0 < t \le T.\end{aligned} \tag{6.6.121}$$

Step 3(a). Denote

$$\begin{pmatrix} \vec{z} \\ \vec{\mathbf{v}} \end{pmatrix}(t) = e^{At}Bu. \tag{6.6.122}$$

Then, it is easy to verify that $\vec{z} = (z, z_t)$ satisfies the wave equation (6.6.95) and

$$\vec{\mathbf{v}}(t) = e^{-A_v(t)} \begin{pmatrix} 0 \\ \mathcal{M}_\gamma{}^{-1}\mathcal{B}u \end{pmatrix} + K(\vec{\mathbf{v}})(t).$$

Hence

$$[I - K]\vec{\mathbf{v}}(t) = e^{-A_v(t)} \begin{pmatrix} 0 \\ \mathcal{M}_\gamma{}^{-1}\mathcal{B}u \end{pmatrix}.$$

Applying the result of Lemma 6.6.4 and (6.6.121) and noting that $1/2 - r > 0$ yields

$$\int_0^T |\vec{\mathbf{v}}(t)|^2_{H_v} \le C|[I-K]^{-1}|^2_{\mathcal{L}(L_2(0,T;H_v))} \left| e^{-A_v(t)} \begin{pmatrix} 0 \\ \mathcal{M}_\gamma{}^{-1}\mathcal{B}u \end{pmatrix} \right|^2_{L_2(0,T;H_v)}$$
$$\le C|[I-K]^{-1}|^2_{\mathcal{L}(L_2(0,T;H_v))} \left| e^{-A_v(t)} \begin{pmatrix} 0 \\ \mathcal{M}_\gamma{}^{-1}\mathcal{B}u \end{pmatrix} \right|^2_{L_2(0,T;\mathcal{D}(A_v^{1/2-r})} \le C|u|^2_U. \tag{6.6.123}$$

Thus we have obtained the a priori regularity for the variable $\vec{\mathbf{v}}$

$$\vec{\mathbf{v}} \in L_2(0,T;H_v), \quad |\vec{\mathbf{v}}|_{L_2(0,T;H_v)} \le C|u|_U.$$

Step 4(a). With the above regularity, and recalling again that $\mathbf{v} \in \mathcal{D}(\mathcal{M}_\gamma{}^{1/2}) \to v \in H^{1/3}(\Gamma_0)$ we go back to Lemma 6.6.1, which allows us to read off the regularity for the acoustic component $\vec{z}$ in (6.6.109):

$$|\vec{z}|_{C([0,T];H_z)} \le C|u|_U. \tag{6.6.124}$$

Moreover, by once more using Lemma 6.6.4 we obtain with any $\epsilon > 0$ small enough

$$K\vec{\mathbf{v}} \in C([0,T];\mathcal{D}(A_v^{1/2-\epsilon})) \subset C([0,T];H_v),$$
$$|K\vec{\mathbf{v}}|_{C([0,T];H_v)} \le C|u|_U.$$

Finally, since

$$\vec{\mathbf{v}}(t) = K(\vec{\mathbf{v}})(t) + e^{-A_v(t)} \begin{pmatrix} 0 \\ \mathcal{M}_\gamma{}^{-1}\mathcal{B}u \end{pmatrix},$$

we also obtain by (6.6.121)

$$|\vec{\mathbf{v}}(t)|_{H_v} \le C|\vec{\mathbf{v}}|_{L_2(0,T;H_v)} + C\frac{1}{t^r}|u|_U \le C\left[1 + \frac{1}{t^r}\right]|u|_U. \tag{6.6.125}$$

Combining the inequalities in (6.6.124) and 6.6.125) yields the desired conclusion in the proposition. □

Remark 6.6.6. *In the case of rotational forces present in the model with* $\mathcal{D}(\mathcal{M}_\gamma{}^{1/2}) \subset H^{1/2}(\Omega)$*, there is no need to use the sharp regularity of hyperbolic traces presented in Lemma* 6.6.1. *Indeed, it suffices to apply Myatake's regularity result* [167].

Remark 6.6.7. *The proof of Proposition* 6.6.2 *is much simpler when the absorbing boundary conditions are added on the portion of the boundary* Γ_0. *This is to say that one would consider*

$$\frac{\partial}{\partial \nu} z = v_t - g z_t \text{ on } \Gamma_0 \text{ for some } g > 0.$$

Indeed, in this case one can give a soft proof based on functional analysis and a rather straightforward energy type of estimate (see [116, 70, 118]*).*

6.7 Comments and Open Problems

- From Lemmas 6.5.1 and 6.4.3 and Corollary 6.5.2 we know that the Riccati operator $P(t)$ and the decoupling function $r(t)$ have additional regularity properties, provided that either the observation R is smoothing or the dynamics $e^{At}B$ have a controlled singularity at the origin. These properties allow us to derive (rigorously) the DRE and the pointwise optimal synthesis.

  An outstanding open problem, in the area of the abstract DRE, is to obtain similar results for more general control problems, without additional regularity assumptions imposed on the observation R or on the dynamics $e^{At}B$. In particular, the issue of differentiability of Riccati operator and of optimal evolution is paramount.
- In the context of the structural acoustic model, an interesting open problem is an extension of the results in Theorem 6.3.1 to more general classes of structurally damped plates. This is to say, in (5.2.2) one would like to take different values of $\theta \geq 1/4$ and, of course, $\theta \leq 1/2$. For these values of θ and $\mathcal{M}_\gamma = 1$, the corresponding plates still represent the analytic dynamics. It is then plausible that the singular estimate of Proposition 6.6.2 is still valid. A particularly interesting case is that of the square root damping [185], when $\mathcal{M}_\gamma = I, \theta = 1/4$. A special case of a square root damping are thermoelastic plates. For thermoelastic models, a singular estimate and the resulting Riccati theory were established in [111]; for more general models, see [199, 142, 37].
- The plate models introduced in Chapter 5 do not apply directly to the thermoelastic walls. However, with appropriate adjustment of the notation [148, 111] one could consider thermoelasticity as well. If rotational forces are neglected, thermoelastic plates represent analytic dynamics. Therefore, the results presented in Theorem 6.3.1 are still valid for these models (at least for some subclasses of operators B that comply with Assumption 5.2.2) [111].
- In the case of a structural acoustic problem without the structural damping active on the wall, $\alpha = 0$, we saw in section 6.3.3 that the Riccati theory holds for the rather special structure of unbounded control operators, which are represented by derivatives of delta functions. It would be interesting to see whether similar results can be obtained for more general classes of control operators subject to Assumption 5.2.2.
- An extension of the results in section 6.2.3 to plate models that do not account for rotational forces ($\gamma = 0$) is an open problem. The main difficulty is a validation of Assumption 6.4.1 for the same class of control operators as considered in section 6.2.3. The results obtained in [201, 200] may be helpful in this direction.

- The results obtained for structural acoustic problems deal with the case when the observation R is bounded. It is possible to relax this assumption, provided that one calibrates the amount of the unboundedness of R with respect to the value of the parameter r_0 in Assumption 6.5.1 [140]. However, this was not done yet.
- One would also like to consider the functional cost (6.2.1) which penalizes the terminal state. This is to say, a term $|G\mathbf{y}(T)|_Z$ is added to the functional cost (6.2.1). If the operator G is assumed to have an appropriate degree of smoothness, then the analysis is straightforward. However, in general, this problem leads to singularities of the gain operator $B^*P(t)$ at the terminal point $t = T$. The technical details are substantial, and one should follow the parabolic analysis presented for this case in [132] (see also [65, 28, 140]).
- Extensions of Theorems 6.3.1 and 6.3.2 to the cases of more general functional cost, including nonpositive quadratic cost, are certainly possible. This would lead to nonstandard Riccati equations. To accomplish this goal one would need to combine the techniques discussed in this chapter with those developed in the context of mini-max problems and nonstandard Riccati equations [163, 140, 199] and references therein. This has not been done yet.
- A control system involving penalization of a pointwise pressure z_t at some specific points x_i in an acoustic chamber was considered in [11, 12]. This problem corresponds to the very unbounded observation operator R. It was shown in [12] that the corresponding control system displays properties of regularity. By this we we mean that there is a state space where uncontrolled dynamics evolves as a strongly continuous semigroup and both control-state maps and state-observation maps are bounded in appropriate topologies. This allows us to study control-theoretic properties of the system and to establish an existence of an optimal solution to an underlying control problem. A problem that is open in this context is that of optimal control synthesis and its representation via the Riccati operator. The difficulties are related to strong unboundedness of the observation R (along with the unboundedness of B), which in this case represents pointwise evaluation of the pressure.

Chapter 7

Feedback Noise Control in Structural Acoustic Models: Infinite Horizon Problems

7.1 Orientation

In this chapter we study an infinite horizon problem associated with two classes of structural acoustic problems discussed in Chapter 6.

The main goal here is to provide a meaningful optimal synthesis of the optimal control problem in terms of the Riccati operator—a solution to an algebraic Riccati equation (ARE)—and the decoupling function $r(t)$, which satisfies an appropriate differential (autonomous) equation with a forcing term determined by the disturbance. As in the finite horizon case, the main issue here is the regularity of the gain operators, which is the key factor in justifying the derivation of ARE and the equation representing $r(t)$. But for infinite horizon problems, one must in addition address vital questions related to stabilizability and controllability of the original system. Indeed, to claim an existence of optimal solutions for the controlled PDE system, one must know a priori that the PDE satisfies the necessary and sufficient condition referred to as the finite cost condition (FCC) [140]. This FCC, in turn, is guaranteed if the PDE system is stabilizable or exactly controllable. It is well known that the property of stabilizability or controllability (or lack thereof) for hyperbolic-like systems is topology sensitive. In consequence, there is a narrow range of functional spaces—control and state—one can choose from to satisfy the FCC. In particular, one usually must work within the finite energy spaces, where stabilizability (and in some cases, controllability) properties are established. However, the choice of finite energy spaces, for the sake of the FCC, will result in the control operator B being unbounded.

In the case of unbounded control operators, the infinite horizon problem displays a certain smoothing effect, which is not present in the analogous problem on the finite horizon. This is so because the crucial terms and operators that enter into the analysis are limits of the corresponding quantities that describe the finite horizon control problem. It turns out that these limits have nicer properties than the elements of their respective sequences. The upshot of this is that to obtain a meaningful, optimal theory for the infinite horizon problem, one cannot use a standard procedure applied in the context of bounded control operators, that being to approach the infinite horizon problem by a limit of finite horizon problems which then leads to an ARE defined via a limit process performed on a DRE. In fact, this observation was a motivation in [66, 129, 140], to study the problem by direct methods, without any appeal to the convergence analysis via a sequence of finite horizon problems.

By this direct approach, one is able to obtain a well-developed algebraic Riccati theory, some components of which include the existence and uniqueness of solutions to the ARE, differentiability of optimal evolution, and a good definition for the gain operators. Recall that in the case of hyperbolic dynamics with no smoothing assumptions imposed on the observation R, quantities such as the gain B^*P may not be defined at all (even on a dense set). Thus, it would be very difficult, if not impossible, to recover these desirable properties from a limiting process.

In addition to these difficulties, the infinite horizon theory with unbounded control operators furnishes several new phenomena that have no counterpart in that of control by bounded operators. The most prominent are the following: the gain operators, although they are always densely defined on the basic state space (in contrast with the finite horizon theory), are intrinsically unbounded within the hyperbolic framework. Even more striking, singularities of the gains B^*P are observed even when acting on the domains of the generator of the original dynamics, i.e., $D(A)$. An eloquent example illustrating this phenomenon is given in [204] (see also [198] for related examples). Since ARE are defined on $D(A)$, this leads to the question, in view of the singularity of the B^*P on $D(A)$, How do we define the quadratic term in the ARE equation? In fact, the theory presented originally in [66, 140] shows that this equation is satisfied with an appropriately defined extension of the gain operators. Later, in [198, 21], with motivation coming from the counterexample in [204], it was shown that the extension of the gain operator coincides with B_e^*P, where B_e^* is an appropriate extension of the adjoint to the control operator B^*. Nevertheless, the meaning and the structure of the ARE equation are different from the one presented in bounded control theory. This fact has obvious consequences on computational issues, particularly those related to the numerics and convergence of algorithms that approximate the AREs. Indeed, to obtain a reasonable convergence (in the hyperbolic case), one needs to introduce a carefully selected regularization, before the approximation [70]. Of course, the analytic case is much more regular where the analysis leads to bounded gains and an avoidance of the pathologies described above; see [129, 28, 159] and references therein. In particular, the ARE has a classical meaning.

An outline of this chapter is as follows. The infinite horizon problem to be considered in this chapter is described in section 7.2. The main results pertinent to structure acoustic interactions, with respect to the two canonical cases of hyperbolic-parabolic and hyperbolic-hyperbolic couplings, are formulated in section 7.3. Sections 7.4–7.6 are devoted to the proofs of these results. In the course of these proofs, we present two abstract theories dealing with a general theory for hyperbolic dynamics (section 7.4) and a theory subject to singular estimate of the dynamics $e^{At}B$ (section 7.6). Aside from their association with structural acoustic interactions, these theories are of interest on their own.

7.2 Optimal Control Problem

As in the previous chapter, we define the state space $H \equiv H_z \times H_v$, where H_z, H_v are given by (5.2.6) We introduce the state variable $\mathbf{y} \equiv [z, z_t, \mathbf{v}, \mathbf{v}_t]$. (Note that the space denoted by X in Chapter 5 now becomes H.)

Let R be a bounded operator from $H \to Z$, where Z is another Hilbert space (space of observations).

The following optimal control problem is studied.

Control problem. Minimize the functional

$$J(u, \mathbf{y}) \equiv \int_0^\infty [|Ry|_Z^2 + |u|_U^2]\, dt \tag{7.2.1}$$

for all $u \in L_2(0, \infty; U)$ and $\mathbf{y} \in L_2(0, \infty; H)$ which satisfy the coupled system of (5.2.1), (5.2.2), or equivalently (5.2.9), which can be written as

$$\mathbf{y}_t - A\mathbf{y} = Bu + F, \qquad \mathbf{y}(0) \in H, \tag{7.2.2}$$

where the operator $A : H \to H$, given by (5.2.7), was shown to generate a C_0 semigroup on H, and the operator $B : U \to [D(A^*)]'$ represents the unbounded control operator. The forcing term $F \in L_2(0, \infty; H)$ describes the effect of the deterministic disturbance.

Since we are dealing with the infinite horizon problem, the issue of stability and stabilizability and detectability is critical. In what follows, we assume that the semigroup corresponding to the state equation described by (5.2.9) is exponentially stable. This assumption corresponds to specifying that the damping parameters in the definition of the operator A are nonzero. Indeed, if we take, for example, the model for the wave (5.2.1) with $d_1 + d_2 > 0$, $d_3 = d_2$, and the model for the wall (5.2.2) with $\alpha + g > 0$, then the results of Chapter 5 apply and yield exponential stability for the corresponding semigroups. Motivated by this, we impose the following condition.

Assumption 7.2.1.

$$|e^{At}|_{\mathcal{L}(H)} \le Ce^{-\omega t}, \qquad t \ge 0, \omega > 0. \tag{7.2.3}$$

As we saw in section 5.4, Assumption 7.2.1 holds for various configurations of the damping placed on the plate or wave equation. To focus our attention, we list three canonical situations when a combination of internal and boundary damping provides the desirable stability, with reference to (5.2.1), (5.2.2), where we take $f \equiv 0$, $u \equiv 0$:

(i) *Internal-internal damping.* If $d_1 > 0, \alpha > 0$, then the system e^{At} is exponentially stable on H with the wave equation satisfying the Neumann boundary data ($d_2 = d_3 = 0$) and the plate or shell equation subject to zero boundary data (hinged, clamped, or free); see Theorem 5.4.3.

(ii) *Boundary-internal damping.* If $d_1 = 0, \alpha > 0$, then the system e^{At} is exponentially stable on H with the wave equation subject to the boundary damping with either $d_2 > 0, d_3 = 0$ or $d_2 = 0, d_3 > 0$, where in the latter case Γ_1 is subject to the convexity hypothesis (see section 5.3.1). The plate or shell component of the structure is subject to the zero boundary data (hinged, clamped, or free); see Theorem 5.4.1.

(iii) *Boundary-boundary damping.* If $d_1 = 0, \alpha = 0$, then the system e^{At} is exponentially stable on H with the wave equation subject to the boundary damping (see above) and the plate equation subject to the dissipation acting via either hinged boundary conditions, where the boundary moments are dissipative ($g_1 = 0, g_2 > 0$), or the free boundary conditions, where both the shears and the moments are dissipative (i.e., $g_2 > 0$ or $g_1 > 0$); see Theorems 5.4.4, 5.4.5, and 5.4.6. For $\gamma = 0$ we may take $g_1 = 0$ in the free case.

Specific and detailed formulations of the results mentioned above are given in the appropriate sections of Chapter 5.

The main goal of this chapter is to provide a feedback representation of the optimal control which, as is well known, involves a solution of an appropriate ARE.

As in the case of the finite horizon problem, the main challenge and difficulty of the problem is brought by the unboundedness of the operator B, as in the finite horizon case. The results presented below differentiate between parabolic-hyperbolic coupling and hyperbolic-hyperbolic coupling.

We will see that for systems made up of hyperbolic-parabolic coupling, the regularizing effect of analyticity enjoyed by one component of the structure propagates onto the entire structure and ultimately leads to boundedly defined gain operators and regular Riccati equations. This is not the case with hyperbolic-hyperbolic coupling, in which suitable extensions of the gain operators need to be constructed.

7.3 Formulation of the Results

Our main task is to show that the dynamics described by (5.2.9) along with the functional cost given in (7.2.1) admit an optimal synthesis in the form of an appropriate Riccati feedback operator and the decoupling function $r(t)$. The results obtained depend quantitatively on the type of the coupling. Thus we distinguish two cases: hyperbolic-parabolic coupling and hyperbolic-hyperbolic coupling.

7.3.1 Hyperbolic-parabolic coupling

The main result is formulated below.

Theorem 7.3.1 ($\alpha > 0$). *Consider the control problem governed by the dynamics described in* (5.2.9), *with* $\alpha > 0$, $\theta = 1/2$, *and the functional cost given in* (7.2.1). *The control operator* $\mathcal{B}$ *is subject to Assumption* 5.2.2 *and* $F \in L_2(0, \infty; H)$. *We also assume Assumption* 7.2.1. *Then, for any initial condition* $\mathbf{y}_0 \in H$, *there exists a unique optimal pair* $(u^0, \mathbf{y}^0) \in L_2(0, \infty; U \times H)$ *with the following properties:*

(i) *Regularity of the optimal pair.*

$$u^0 \in C([0, \infty); U), \quad \mathbf{y}^0 \in C([0, \infty); H).$$

(ii) *Regularity of the gains and optimal synthesis. There exist a self-adjoint positive operator* $P \in \mathcal{L}(H)$ *with the property*

$$B^* P \in \mathcal{L}(H)$$

and an element $r \in C(0,\infty;H)$ *with the property*

$$B^*r \in C([0,\infty);U) \cap L_2(0,\infty;U).$$

Moreover, one has the following representation for feedback control:

$$u^0(t) = -B^*P\mathbf{y}^0(t) - B^*r(t), \qquad t \geq 0.$$

(iii) *Feedback semigroup. The operator* $A_P \equiv A - BB^*P : H \to D(A^*)'$ *generates a strongly continuous semigroup on* H *which is, moreover, exponentially stable.*

(iv) *ARE. The operator* P *is a unique (within the class of self-adjoint positive operators subject to the regularity in part* (ii)*) solution of the following operator ARE: for all* $x, y \in D(A)$

$$(A^*Px, y)_H + (PAx, y)_H + (R^*Rx, y)_H = (B^*Px, B^*Py)_U. \quad (7.3.4)$$

(v) *Equation for* r. *The element* $r(t)$ *is the unique solution of the differential equation*

$$r_t(t) = -A^*r(t) + PBB^*r(t) - PF \text{ on } [D(A)]' \\ \text{with the terminal condition } \lim_{T\to\infty} r(T) = 0. \quad (7.3.5)$$

(vi) *Uniform bounds. The gain operators are bounded on the state space, i.e.,*

$$|B^*Px|_U + |B^*r|_{L_2(0,\infty,U)} + |B^*r|_{C(0,\infty,U)} \leq C_\alpha[|x|_H + |F|_{L_2(0,\infty,U)}],$$

where the constant C *above does not depend on* $\gamma \geq 0$ *(rotational inertia).*

Remark 7.3.1. *The results of Theorem* 7.3.1 *apply to all models introduced in section* 4.4 *with* $\alpha > 0, \theta = 1/2$. *In particular, one may take* $g = 0$ *in the plate model and* $d_1 + d_2 + d_3 > 0$ *for the wave equation. If* $d_1 = d_3 = 0$, *geometric Assumption* 5.4.1 *is needed.*

Remark 7.3.2. *We note that the results of parts* (i) *and* (ii), *Theorem* 7.3.1, *provide more regularity properties of the optimal solution than optimization alone predicts. Indeed, the synthesis is defined pointwise with all the gains represented by bounded operators. Solutions to control problems with unbounded control actions do not usually manifest this degree of regularity (see* [129, 28, 140]*). These regularity results are necessary to give a meaning to the classical Riccati equation.*

Remark 7.3.3. *The bounds in part* (vi) *of Theorem* 7.3.1 *are uniform with respect to the moments of inertia. This means that the additional regularizing effect of inertial forces does not have a predominant effect on the regularity of the gains.*

Remark 7.3.4. *The result of Theorem* 7.3.1, *in the case of shells with structural damping* ($\alpha > 0$), *was proved in* [118].

The statement of Theorem 7.3.1 provides a full pointwise feedback synthesis for the optimal control. It also gives an important regularity result for the Riccati operator. Indeed, one cannot typically expect to obtain bounded gains with the unbounded control operator B [66, 28, 129, 21]. In our case, the additional regularity properties of the gains result from the presence of the structural damping on the wall. The regularizing effect of parabolicity present in the wall model is partially propagated onto the entire structure. An interesting aspect of the problem is that the overall coupled model is not analytic (in which case the boundedness of gains is well understood; see [129, 28, 65]). The hidden regularity of B^*P allows us to show that the external noise gain operator is also pointwise defined. This, in turn, provides a meaningful optimal synthesis.

If we only consider a viscous passive damping acting on the wall (i.e., replace the term $\alpha \mathcal{A} v_t$ by αv_t), then the resulting gain operator will not be pointwise defined. The coupling on the interface is then hyperbolic-hyperbolic. This is a special case of a more general situation considered in the next section.

7.3.2 Hyperbolic-hyperbolic coupling: Abstract results

We next turn to a much less regular case, when the coupling on the interface is hyperbolic-hyperbolic, which corresponds to the case $\alpha = 0$.

Theorem 7.3.2 ($\alpha = 0$; see [21]). *Consider the control problem governed by the dynamics described in* (5.2.9), *with* $\alpha = 0$, *along with the functional cost given in* (7.2.1) *We assume that the generator* A *satisfies Assumption* 7.2.1 *and* $F \in L_2(0, \infty; H)$. *The control operator* $\mathcal{B}$ *is subject to Assumption* 5.2.2 *and, in addition, we assume that for some* $T > 0$, *the following holds.*

Assumption 7.3.1.

$$\int_0^T |B^* e^{At} \mathbf{y}|_U^2 dt \leq C |\mathbf{y}|_H^2; \mathbf{y} \in D(A^*).$$

Then, for any initial condition $\mathbf{y}_0 \in H$, *there exists a unique optimal pair* $(u^0, \mathbf{y}^0) \in L_2(0, \infty; U \times H)$ *with the following properties:*

(i) *Regularity of the optimal pair.*

$$u^0 \in L_2(0, \infty; U); \mathbf{y}^0 \in C(0, \infty; H).$$

(ii) *Feedback semigroup. There exists a self-adjoint, nonnegative operator* $P \in \mathcal{L}H$ *such that the operator* $A_P \equiv A - BB^*P$ *is a generator of a strongly continuous semigroup* $e^{A_P t}$ *on* H, *which is, moreover, exponentially stable.*

(iii) *Regularity of the gain and optimal synthesis. The operator* P *enjoys the following regularity:* $P \in \mathcal{L}(D(A_P); D(A^*)) \cap \mathcal{L}(D(A); D(A_P^*))$; *hence* $B^*P : D(B^*P) \subset H \to U$ *is densely defined on* H $(D(A_P) \subset D(B^*P))$ *and an element* $r(\cdot) \in C(0, \infty; H)$ *with the property such that* $Py^0(t) + r(t) \in D(B^*)$, *a.e. in* $t > 0$ *and*

$$u^0(t) = -B^*[Py^0(t) + r(t)] \qquad \textit{a.e. in } t > 0; \qquad y_0 \in H,$$

$$u^0(t) = -B^* P y^0(t) - B^* r(t] \textit{ for all } t > 0, \; y_0 \in D(A_P), \; F \in H^1(0, \infty, H).$$

(iv) *Riccati equation. The operator* P *is a unique (within the class of self-adjoint nonnegative operators subject to the regularity in part* (iii)*) solution of the* *following operator ARE:*

$$
\begin{aligned}
&(A^*Px, y)_H + (PAx, y)_H + (R^*Rx, y)_H \\
&\quad = (B^*Px, B^*Py,)_U \text{ for } x, y \in D(A_P), \\
&(A^*Px, y)_H + (PAx, y)_H + (R^*Rx, y)_H \\
&\quad = (B_e^*Px, B_e^*Py,)_U \text{ for } x, y \in D(A),
\end{aligned}
\tag{7.3.6}
$$

where B_e^* is a suitable extension of B^* constructed in [198, 21].

(v) *Equation for r. The element $r(t)$ is the unique solution to the differential equation*

$$
r_t(t) = -A_P^* r(t) - P(t)F; \text{ on } D(A)', \qquad \lim_{T \to \infty} r(T) = 0. \tag{7.3.7}
$$

Remark 7.3.5. *The main difference between the results in the parabolic-hyperbolic case given by Theorem* 7.3.1 *and these presented in Theorem* 7.3.2 *is the qualitative nature of the gain operator. Indeed, in the parabolic-hyperbolic case we have that the gain operator is actually a* bounded operator, *while in the hyperbolic-hyperbolic setup this is not the case. In fact, one can show by means of an example that the gain operator is* intrinsically *unbounded (see also* [66, 129] *in the context of stabilization). Also, the formulation of the Riccati equation on $D(A)$ requires a special extension of B^* (which depends on R; see* [21]*). This is because the range of P over $D(A)$ may not belong to $D(B^*)$.*

A similar comment applies to the regularity of the element $B^*r(t)$. Indeed, while in the case of Theorem 7.3.1 *the optimal synthesis holds pointwise with each entity B^*P and B^*r well defined on its own, this is not the case in the present, purely hyperbolic, context.*

Note that the formulation of Theorem 7.3.2 requires the trace regularity in Assumption 7.3.1. In general, such assumption may not hold (it holds automatically if $\alpha > 0$). However, in a special but important case when the control operator is given by the derivatives of the delta functions, and the structural dynamics (5.2.2) are given by Kirchhoff equation, this admissibility condition holds true. This particular situation is elaborated on below.

7.3.3 Hyperbolic-hyperbolic coupling: Kirchoff plate with point control

The goal of this section is to provide a prototype for a structural acoustic model complying with the requirements posed by Theorem 7.3.2. The main issues are, of course, Assumption 7.3.1 and stability property (7.2.1), without assuming an existence of structural damping (i.e., $\alpha = 0$). A model that meets these requirements is Kirchhoff plates with boundary or viscous damping and control operators represented by the derivatives of delta functions. This is a particular case of models already discussed in section 5.3. For the convenience of the reader we recall the main setup.

As before (see (6.3.6), (6.3.7)) we consider a two- or three-dimensional structurally acoustic chamber, where the flat elastic wall Γ_0 is modeled as a Kirchhoff equation. This is a hyperbolic equation, which accounts for rotational forces (see

$\gamma > 0$ in (7.3.9) below). Since we deal with the infinite horizon problem, finite cost condition and detectability conditions are needed. The finite cost condition amounts to the property of exact controllability (or uniform stabilizability) of the underlying dynamics on the state space H of regularity with $L_2(0,\infty)$ controls. However, such exact controllability (or stabilizability) on the space of regularity H given in (6.3.10) fails for the problem described by (6.3.6), (6.3.7). As noted, this is a general *pathology* of hyperbolic or Petrovsky-type PDE dynamics with point control, acting through δ or δ' (see, e.g., [129] and the introductions of [92, 93]). In addition, another feature of the problem prevents the requisite controllability or stabilizability results—the very hybrid nature of the coupling between the systems. It is known that in such cases exact controllability or stabilizability fails even with distributed controls acting on the wall [156]. To remedy the situation, we must modify the original conservative dynamics by adding damping terms to make it uniformly stable on H while preserving the same regularity in $C([0,T],H)$. This can be accomplished by replacing the original dynamics in (6.3.6), (6.3.7) with the following damped version:

$$\begin{aligned} z_{tt} = c^2\Delta z - d_1 z_t + f \text{ in } Q \equiv \Omega \times (0,\infty),\\ \frac{\partial}{\partial\nu} z = 0 \text{ on } \Sigma_1 \equiv \Gamma_1 \times (0,T),\\ \frac{\partial}{\partial\nu} z + d_3 z_t = v_t \text{ on } \Sigma_0 \equiv \Gamma_0 \times (0,\infty), \end{aligned} \tag{7.3.8}$$

$$\begin{aligned} v_{tt} - \gamma\Delta v_{tt} + \Delta^2 v + z_t + k_0\Delta v_t = \delta'(x_0)u(t) \text{ in } \Sigma_0,\\ v = 0, \Delta v = -k_1\frac{\partial v_t}{\partial\nu} \text{ on } \partial\Gamma_0 \times (0,\infty). \end{aligned} \tag{7.3.9}$$

Equations (7.3.8) and (7.3.9) are associated with the initial conditions

$$z(0) = z_0, \quad z_t(0) = z_1, \quad v(0) = v_0, \quad v_t(0) = v_1. \tag{7.3.10}$$

Here the nonnegative damping constants d_i, k_i satisfy $d_0 + d_1 > 0, k_0 + k_1 > 0$.

We show that the model described above fits into the abstract framework of Theorem 7.3.2. To accomplish this we need to show that regularity Assumption 7.3.1 and stability Assumption 7.2.1 are satisfied.

We note that the regularity of the corresponding mixed problem (7.3.8), (7.3.9) is the same as that of the undamped version; see [44] (see Remark 6.3.6). In particular, Theorem 6.3.3 applies and Assumption 7.3.1 is satisfied.

Regarding the stability Assumption 7.2.1, the following result, which is a specialization of Theorem 5.4.4, is critical.

Lemma 7.3.1. *Consider the system consisting of* (7.3.8) *with* $f = 0, d_1 + d_3 > 0$ *and of* (7.3.9) *with control* $u = 0$ *and* $k_0 + k_1 > 0$. *Moreover, if* $d_1 = 0$, *we assume the following geometric condition. There exists a point* $x_0 \in R^2$ *such that*

$$(x - x_0)\cdot\nu \le 0 \text{ on } \Gamma_1, \tag{7.3.11}$$

where ν *is an outward normal to the boundary and* Γ_1 *is locally convex* [144]. *Then, there exist positive constants* $C > 0, \omega > 0$ *such that*

$$|(z(t), z_t(t), v(t), v_t(t))|_H \le Ce^{-\omega t}|(z(0), z_t(0), v(0), v_t(0))|_H,$$

where the above inequality holds on a quotient subspace $H_0 \subset H$:

$$H_0 \equiv \{y = (y_1, y_2, y_3, y_4) \in H; \int_{\Gamma_0} (d_3 y_1 + y_3)dx + \int_{\Omega} (d_1 y_1 + y_2)dx = 0\},$$
$$H \equiv H^1(\Omega) \times L_2(\Omega) \times H^2(\Gamma_0) \cap H_0^1(\Gamma_0) \times H^1(\Gamma_0).$$

The result of Lemma 7.3.1 is given in [41, 42], subject to an additional damping on Γ_1. However, the arguments developed in [112] allow us to consider the damping active only on Γ_0, provided that suitable geometric conditions are assumed on Γ_1.

As we see from the above result, the conditions that guarantee the uniform decay rates are more demanding then the ones in the case $\alpha > 0$. Indeed, they require some additional damping either on the boundary or in the interior of the wall.

Remark 7.3.6. *The result of Lemma* 7.3.1 *is still valid if we replace the Neumann boundary data on Γ_1 by the Dirichlet data. This is to say we consider $z = 0$ on $\Gamma_1 \times (0, \infty)$, where Γ_1 is subject to the geometric hypothesis* (7.3.11) *(there is no need for convexity of Γ_1). The drawback of this configuration is that the theories in the literature* [41] *require that Γ_0 and Γ_1 be separated. This is due to potential singularities that may develop for solutions to mixed problems. In fact, the same constraints exist in the case of stabilization of the waves or plates* [99, 97]. *On the other hand, by using the corners theory due to Grisvard* [69], *one can still obtain sufficient regularity of elliptic solutions by imposing a corner at the junction of Γ_1 and Γ_0.*

Remark 7.3.7. *We note that stability result of Lemma* 7.3.1 *seems new even in the context of separate stabilization of waves and plates. Indeed, stabilization of the plate component requires only one boundary stabilizer (a bending moment), while stabilization of the wave component holds with a Neumann unobserved part of the boundary. Techniques used in* [41] *rely heavily on microlocal analysis.*

Recalling the definitions of $\mathcal{A}_N, \mathcal{A}, \mathcal{M}_\gamma, N$ from section 5.3, we further construct the Dirichlet map $D : L_2(\partial\Gamma_0) \to L_2(\Gamma_0)$ by

$$Dg = h \Leftrightarrow \Delta h = 0 \text{ on } \Gamma_0 \text{ and } h|_{\partial\Gamma_0} = 0.$$

By Green's formula we have $D^*\mathcal{A}^{1/2}v = -\frac{\partial}{\partial\nu}v$ on $\partial\Gamma_0$. With these quantities we can write out the dynamical system (7.3.8), (7.3.9) as the first-order system (7.2.2) with the operator A given by

$$-A \equiv \begin{pmatrix} 0 & -I & 0 & 0 \\ A_N & d_1 I + d_3(A_N + I)NN^*(A_N + I) & 0 & -(A_N + I)N(\cdot|_{\Gamma_0}) \\ 0 & 0 & 0 & -I \\ 0 & \mathcal{M}_\gamma{}^{-1}N^*(A_N + I) & \mathcal{M}_\gamma{}^{-1}\mathcal{A} & \mathcal{M}_\gamma{}^{-1}(k_0\mathcal{A}^{1/2} + k_1\mathcal{A}^{1/2}DD^*\mathcal{A}^{1/2}) \end{pmatrix}.$$

With regard to (7.2.2) with A defined as above and B given by (6.3.12) we can combine the results of Theorem 6.3.3 and Lemma 7.3.1 with Theorem 7.3.2 to obtain the following.

Theorem 7.3.3. *Consider the control problem governed by the functional cost given in* (7.2.1) *and the dynamics described by* (7.3.8), (7.3.9) *or, equivalently, by* *the abstract system* (7.2.2) *with the operator* B *given by* (6.3.12) *and the operator* A *given above. We assume that* $F \in L_2(0, \infty; H_0)$, $d_1 + d_3 > 0$ *and* $k_0 + k_1 > 0$. *Moreover, we assume the geometric assumption* (7.3.11) *if* $d_1 = 0$. *Then, the conclusions of Theorem* 7.3.2 *apply*

The remaining sections of this chapter are devoted to the proofs of the main theorems: Theorem 7.3.1 and Theorem 7.3.2.

7.4 Abstract Optimal Control Problem: General Theory

In this section we consider a rather general infinite horizon optimal control problem governed by a strongly continuous semigroup with a regular control operator. Our goal is to recall some of the results pertaining to solvability of the corresponding optimal control problem. These results are a starting point for analysis of a structural acoustic control problem and the proofs of Theorems 7.3.1 and 7.3.2.

7.4.1 Formulation of the abstract control problem

We repeat the formulation from section 6.4.

Let H, U, Z be given Hilbert spaces and let the following operators be given:

- A is a generator of a strongly continuous and exponentially stable semigroup on H with domain $D(A) \subset H \subset D(A^*)'$. We assume that e^{At} is exponentially stable. That is to say, there exist constants $C, \omega > 0$ such that

$$|e^{At}|_{\mathcal{L}(H)} \leq Ce^{-\omega t}, \qquad \omega > 0.$$

- The operator $B : U \to D(A^*)'$ is bounded and, moreover, satisfies the following inequality:

$$|R(\lambda, A)Bu|_H \leq C_\alpha |u|_U, \qquad \lambda = \alpha + i\beta, \quad \beta \in R, \ \lambda \in \rho(A). \tag{7.4.12}$$

- The operator $R : H \to Z$ is bounded.
- $F \in L_2(0, \infty; H)$ is a given element.

We consider the following dynamics:

$$\mathbf{y}_t = A\mathbf{y} + Bu + F \text{ on } D(A^*)', \qquad \mathbf{y}(0) = \mathbf{y}_0 \in H. \tag{7.4.13}$$

With (7.4.13) we associate functional cost

$$J(u, \mathbf{y}) \equiv \int_0^\infty [|Ry|_Z^2 + |u|_U^2] \, dt. \tag{7.4.14}$$

Optimal control problem. *Minimize the functional* $J(u, \mathbf{y})$ *for all* $u \in L_2(0, \infty; U)$ *and* $\mathbf{y} \in L_2(0, \infty; H)$ *which satisfy* (7.4.13).

We note that by virtue of exponential stability assumed on e^{At}, the finite cost condition is automatically satisfied. Hence, by the standard optimization argument [19, 14] one deduces the existence and uniqueness of an optimal solution to the optimal control problem. This is to say that for all $\mathbf{y}_0 \in H$, $F \in L_2(0,\infty;H)$, there exist unique optimal control $u^0 \in L_2(0,\infty;U)$ and optimal trajectory $\mathbf{y}^0 \in L_2(0,\infty;H)$.

Our main aim is to derive the optimal synthesis for the control problem along with a characterization via an appropriate Riccati equation.

To accomplish this, additional restrictions need to be imposed on control operator B. We begin with the case when B is admissible [183]. This is to say that the following condition holds true. There exists $T > 0$ such that

$$\int_0^T |B^* e^{A^* t}\mathbf{y}|^2 dt \le C_T |\mathbf{y}|_H^2, \qquad \mathbf{y} \in D(A^*), \tag{7.4.15}$$

where $(Bu, v)_H = (u, B^* v)_U$, $u \in U, v \in D(B^*) \supset D(A^*)$.

Remark 7.4.1. *Condition* 7.4.15 *automatically implies validity of condition* (7.4.12), *but not vice versa. In the case of time reversible dynamics we have an equivalence of* (7.4.15) *and* (7.4.12)*—see* [125].

7.4.2 ARE subject to condition (7.4.15)

Since the semigroup e^{At} is exponentially stable and the control operator B satisfies (7.4.15), we are now in a position to use general results pertaining to optimal synthesis of control problems with disturbance. To accomplish this we apply abstract results of [163, 21] (see also [66]). Indeed, from Theorem 3.1.1 and Corollaries 3.3.4 and 3.3.5 in [163] and Theorem 3.5.1 in [198] we obtain the theorem that follows.

Theorem 7.4.1 (see [21]). *Under Assumption* 7.2.1 *and condition* (7.4.15) *the following hold:*

1. *There exists a unique, positive, self-adjoint operator $P \in \mathcal{L}(H)$ such that the operator $A_P \equiv A - BB^*P$ generates a uniformly stable semigroup on H. $P \in \mathcal{L}(D(A_P).D(A^*)) \cap \mathcal{L}(D(A).D(A_P^*))$. Hence, the gain operator B^*P is bounded from $D(A_P) \to U$.*
2. *The optimal control $u^0 \in L_2(0,\infty;U)$ and optimal trajectory $\mathbf{y}^0 \in C([0,\infty); H)$ satisfy the following feedback relation:*

$$u^0(t) = -B^*[P\mathbf{y}^0(t) + r(t)] \in L_2(0,\infty;U), \tag{7.4.16}$$

where $P\mathbf{y}^0(t) + r(t) \in D(B^*)$ a.e. in t, and $r(\cdot) \in C([0,\infty);H)$ is a unique (weak) solution of the differential equation

$$r_t(t) + A_P^* r(t) = -PF \quad \text{on } D(A_P)' \text{ with } \lim_{T\to\infty} r(T) = 0. \tag{7.4.17}$$

For $\mathbf{y}^0 \in D(A_P), F \in H^1(0,\infty,H)$, the synthesis in (7.4.16) *is pointwise in $t \ge 0$ and $B^*Py^0 \in C([0,\infty),U)$, $B^*r \in C([0,\infty),U)$.*
3. *The operator P and the function $r(t)$ admit the following integral representation:*

$$P = \int_0^\infty e^{A^* t} R^* R e^{A_P^* t} dt; \quad r(t) = \int_t^\infty e^{A_P^*(s-t)} PF(s) ds. \tag{7.4.18}$$

4. *The operator P satisfies the following ARE:*

$$(A^*Px, y)_H + (PAx, y)_H + (R^*Rx, y)_H \\ = (B_e^*Px, B_e^*Py)_U \ \ \text{for } x, y \in D(A), \qquad (7.4.19)$$

where B_e^* denotes a suitable (nonunique) extension of the operator B^* which is densely defined on H (see Theorem 3.5.1 in [198, 21]).
The Riccati equation (7.4.19) holds for all $x, y \in D(A_P)$ with B_e^* replaced by the regular B^*

Remark 7.4.2. *The presence of the extension operator B_e^* is critical in the formulation of the ARE in (7.4.19). Indeed, it may happen that B^*P is not even well defined on $D(A)$. A negative example of this type is given in [204], when B is a δ-function, and it is further analyzed in [198]. In [198, 21] extensions B_e^*, as in (7.4.19), are explicitly constructed. On the opposite side, an example is given in [192] with boundary control, where in fact $B^*Px = 0$ for all $x \in D(A)$, so that no extension of B is needed and the ARE collapses to a (linear) Liapunov equation.*

We note that the results in [163] do not provide any information about the regularity of the gains B^*P and B^*r (unless $y^0 \in D(A_P), F \in H^1(0, \infty; H)$. All we know at this point is that $[P\mathbf{y}^0(t) + r(t)] \in \mathcal{D}(B^*)$, a.e. $t > 0$. For this reason, one cannot split the feedback formula (7.4.16) in 1 above, and the bracket after B^* is essential. To be more precise, there are results in [163] that allow for such decomposition, but these require an additional hypothesis of controllability (hypothesis H7 in [163]) which does not hold in our case (due to the presence of structural damping in the wall component).

In view of Theorem 7.4.1 and Remark 7.4.2, the main task is to establish the needed regularity of the gain operators. In fact, this is the most technical part of the proof, which will require additional information about the operator B.

7.5 ARE Subject to a Singular Estimate for $e^{At}B$

In this section we consider the same abstract problem as introduced in the previous section. However, we study this problem under an additional singular type estimate imposed on the $|e^{At}Bu|_H$. This estimate was introduced in the case of finite horizon problem and is recalled below.

7.5.1 Formulation of the results

As in the analogous finite horizon case, we operate under the following assumption.

Assumption 7.5.1. *There exists a constant $0 \le r_0 < 1$ such that for some $T > 0$*

$$|e^{At}B|_{\mathcal{L}(U;H)} \le \frac{C_T}{t^{r_0}} \quad \text{for } 0 < t \le T. \qquad (7.5.20)$$

To provide a meaningful and regular optimal synthesis, it is necessary to analyze the operators B^*P and $B^*r(t)$. The aim of this section is to provide a meaning for the gain operators. The main result in this direction is given next.

Theorem 7.5.1. *Under Assumptions 7.2.1 and 7.5.1, in addition to all conclusions of Theorem 7.4.1 we have*

(i) $B^*P \in \mathcal{L}(H \to U)$.

(ii) *For any* $\mathbf{y}_0 \in H$, $F \in L_2(0,\infty;H)$ *we have*

$$u^0 \in C([0,\infty);U) \cap L_2(0,\infty;U), \quad \mathbf{y}^0 \in C([0,\infty);H) \cap L_2(0,\infty;H),$$

$$B^*r \in L_2(0,\infty;U) \cap C([0,\infty);U).$$

(iii) *The function* $r(t)$ *satisfies the differential equation*

$$r_t(t) = -[A^* - PBB^*]r(t) - PF(t) \text{ on } [D(A)]'$$

with the terminal condition $\lim_{T\to\infty} r(T) = 0$.

(iv) $u^0(t) = -B^*P\mathbf{y}^0(t) - B^*r(t) \in C(0,\infty,U)$

(v) *ARE (7.4.19) is valid with* B_e^*P *replaced by* B^*P.

Remark 7.5.1. *The result of Theorem 7.5.1 is proved under the sole Assumption 7.5.1 without assuming (7.4.15). As expected, from Theorem 6.5.1, Assumption 7.5.1 alone should suffice to provide full optimal synthesis and a meaningful Riccati equation.*

Remark 7.5.2. *The condition on uniform stability of* e^{At} *can be replaced by a weaker condition, namely, that the optimal control problem satisfies the finite cost condition and detectability condition. Since the structural acoustic model under any of the situations presented above is exponentially stable, we choose the more restrictive framework to simplify the exposition and to connect our result to those already existing in the literature.*

7.5.2 Proof of Theorem 7.5.1

The proof of Theorem 7.5.1 follows through a sequence of propositions. To begin with we notice that the singular estimate in Assumption 7.5.1 together with the exponential stability of e^{At} postulated by Assumption 7.2.1 imply the infinite time version of singular estimate which is quantitatively expressed in the proposition below.

Proposition 7.5.1. *Assume that operators* A *and* B *are subject to Assumptions 7.5.1 and 7.2.1. Then*

$$|e^{At}Bu|_H \le \frac{Ce^{-\omega_0 t}}{t^{r_0}}|u|_U, \qquad 0 < r_0 < 1, \quad \omega_0 > 0, t > 0,$$

where r_0 *is the same as in Assumption 7.5.1.*

Proof. By duality, the statement of the proposition is equivalent to

$$|B^*e^{A^*t}|_{\mathcal{L}(\mathcal{H},\mathcal{U})} \le \frac{C}{t^{r_0}}e^{-\omega_0 t}, \qquad t > 0.$$

The inequality in Assumption 7.5.1 is equivalent to an existence of $T > 0$ such that

$$|B^*e^{A^*t}|_{\mathcal{L}(\mathcal{H},\mathcal{U})} \le \frac{C_T}{t^{r_0}}, \qquad 0 < t \le T.$$

By invoking exponential stability of e^{At}, we obtain for any $t = nT + \tau$, $\tau < T$,

$$\begin{aligned}
|B^* e^{A^* t}|_{\mathcal{L}(\mathcal{H},\mathcal{U})} &\leq |B^* e^{A^*(nT+\tau)}|_{\mathcal{L}(\mathcal{H},\mathcal{U})} \\
&\leq |B^* e^{A^*((n-1)T+\tau)} e^{-A^* T}|_{\mathcal{L}(\mathcal{H},\mathcal{U})} \\
\leq \frac{C_T}{T^{r_0}} e^{-\omega((n-1)T+\tau)} &\leq \frac{C_T}{T^{r_0}} e^{-\omega t} e^{\omega T},
\end{aligned} \tag{7.5.21}$$

which is the desired estimate for $t > T$. The estimate for $t \leq T$ follows from (7.5.1). □

Our proof proceeds through a similar sequence of steps as in the case of the finite horizon problem (see also [140]). We use the explicit formulas describing optimal solution and related control operators. Since the semigroup e^{At} is exponentially stable, the representation of the optimal quantities is explicit in terms of the solution operator L. To see this, we recall some definitions. The control-to-state operator L is defined the same way as in Chapter 6:

$$(Lu)(t) \equiv \int_0^t e^{A(t-s)} Bu(s) ds.$$

From the hypothesis imposed on the operator B, Proposition 7.5.1 and singular integral theory it follows that

$$L \in \mathcal{L}(L_2(0,\infty;U); L_2(0,\infty;H)) \cap \in \mathcal{L}(C([0,\infty);U); C([0,\infty);H)). \tag{7.5.22}$$

We also use the L^* adjoint of L given by

$$(L^* w)(t) = B^* \int_t^\infty e^{A^*(s-t)} w(s) ds,$$

and by duality, we obtain that

$$L^* \in \mathcal{L}(L_2(0,\infty;H); L_2(0,\infty;U)). \tag{7.5.23}$$

As before, we define the evolution (semigroup) operator by the formula

$$\Phi(t-s)x = [I + L_s L^* R^* R]^{-1} e^{A(\cdot - s)} x \in L_2(s,\infty;H) \text{ for } x \in H,$$

where

$$(L_s u)(t) \equiv \int_s^t e^{A(t-\tau)} Bu(\tau) d\tau.$$

The Riccati operator P is now defined (recall Assumption 7.2.1) by

$$P \equiv \int_0^\infty e^{A^*(t)} R^* R \Phi(t)\, dt. \tag{7.5.24}$$

Clearly, $P \in \mathcal{L}(H)$. Moreover, it is standard to show that P is self-adjoint and positive on H [66].

The adjoint variable p is defined by

$$p(t) \equiv \int_t^\infty e^{A^*(s-t)} R^* R \mathbf{y}^0(s) ds \in C([0,\infty);H).$$

Note $B^*p = -L^*R^*R\mathbf{y}^0 \in L_2(0,\infty;U)$ and optimality conditions read

$$u^0(t) = -B^*p(t). \tag{7.5.25}$$

Defining

$$r(t) \equiv p(t) - P\mathbf{y}^0(t) \tag{7.5.26}$$

we rewrite optimality condition (7.5.25) as

$$u^0(t) = -B^*[P\mathbf{y}^0(t) + r(t)]. \tag{7.5.27}$$

To study properties of $\Phi(t)$ and P we need more regularity results about control-to-state operator L.

Using Proposition 7.5.1 we can demonstrate the following properties (see the proof of Theorem 2.3.5.1 in [140]).

Proposition 7.5.2.

$$\begin{aligned}
&L_s : C([s,\infty);U) \to C([s,\infty);H),\\
&L_s^* : C([s,\infty);H) \to C([s,\infty);U),\\
&[I + L^*R^*RL_s]^{-1} \in \mathcal{L}(C([s,\infty);U)) \cap \mathcal{L}(L_2((s,\infty);U)),\\
&[I + L_sL^*R^*R]^{-1} \in \mathcal{L}(C([s,\infty);H)) \cap \mathcal{L}(L_2((s,\infty);H)),\\
&L_s \in \mathcal{L}(L_2(s,\infty;U) \to C([s,\infty);H)) \textit{ for } r_0 < 1/2,\\
&L_s \in \mathcal{L}(L_p(s,\infty;U) \to C([s,\infty);H)) \textit{ for } p > \frac{1}{1-r_0}, \quad r_0 + \frac{1}{p} < 1,\\
&L_s \in \mathcal{L}(L_p(s,\infty;U) \to L_{\bar{p}}(s,\infty;H)) \textit{ for } r_0 \geq 1/2,\ p > 1, \quad r_0 + \frac{1}{p} \geq 1,\\
&\bar{p} = \frac{p}{(r_0-1)p+1} > p.
\end{aligned}$$

From regularity properties in Proposition 7.5.2 and explicit representation for $u^0, \mathbf{y}^0$, given by the same formulas as in Lemma 6.4.1, we conclude the following additional regularity of the optimal pair.

Proposition 7.5.3.

$$\mathbf{y}^0 \in C([0,\infty);H), \qquad u^0 \in C([0,\infty);U).$$

The key point of the analysis is to show that $\Phi(t)$ is a strongly continuous semigroup on H. This is accomplished by the same arguments as those used in the finite horizon case, using critically Assumption 7.5.1 and Proposition 7.5.2 together with the bootstrap argument. This shows that the optimal trajectory $\Phi(t)x$ corresponding to optimal control problem with $F = 0$ and initial data x satisfies

$$\Phi(\cdot)x \in C([0,\infty);H) \ \text{ for } \ x \in H.$$

The aforementioned continuity of $\Phi(t)$ combined with the semigroup property satisfied by $\Phi(t)$ allows us to identify, by the standard semigroup argument, operator $\Phi(t)$ with a strongly continuous semigroup generated by A_P. (Note that the above property is not necessarily valid for nonautonomous problems (e.g., finite horizon).)

We introduce function $\hat{u}$, which is an optimal control $\hat{u}$ corresponding to $F = 0$. Hence

$$\hat{u} = -B^*P\Phi(t) = -B^*Pe^{A_Pt}x \in L_2(0, \infty; U), \tag{7.5.28}$$

and B^*P is bounded on $D(A_P)$. To see this last claim, we take $x \in D(A_P)$ and we compute

$$B^*Px = \int_0^\infty B^*{A^*}^{-1}\frac{d}{dt}e^{A^*(t)}R^*R\Phi(t)xdt$$

$$= -B^*{A^*}^{-1}R^*Rx - \int_0^\infty B^*{A^*}^{-1}e^{A^*(t)}R^*R\Phi(t)A_Pxdt. \tag{7.5.29}$$

Hence

$$|B^*Px|_U \leq C|B^*{A^*}^{-1}|_{\mathcal{L}(H;U)}|x|_H + C_\omega|B^*{A^*}^{-1}|_{\mathcal{L}(H;U)}|A_Px|, \tag{7.5.30}$$

which yields the desired boundedness on $\mathcal{D}(A_P)$.

Our next step is to establish exponential stability of this semigroup.

Proposition 7.5.4. *Under Assumptions 7.5.1 and 7.2.1 the semigroup e^{A_Pt} is exponentially stable on H. That is, there exist positive constants C, ω_P such that*

$$|e^{A_Pt}x|_H \leq Ce^{-\omega_Pt}|x|_H.$$

Proof. By a well-known linear stability criterion [175] it suffices to show that $e^{A_Pt}x \in L_2(0, \infty, H)$ for each initial datum $x \in H$. From variation of parameter formula we obtain

$$e^{A_Pt}x = e^{At}x + \int_0^t e^{A(t-s)}B\hat{u}(s)ds. \tag{7.5.31}$$

Applying the result of Proposition 7.5.1 yields

$$\int_0^\infty \left|\int_0^t e^{A(t-s)}B\hat{u}(s)ds\right|_H^2 dt$$

$$\leq C\int_0^\infty \left[\int_0^t |e^{A(t-s)}B|_{\mathcal{L}(U;H)}|\hat{u}(s)|_U ds\right]^2 dt$$

$$\leq C\int_0^\infty \left[\int_0^t \frac{e^{-\omega_0(t-s)}}{(t-s)^{r_0}}|\hat{u}(s)|_U ds\right]^2 dt. \tag{7.5.32}$$

By using Young's inequality and the fact that $r_0 < 1$, we obtain

$$\int_0^\infty \left|\int_0^t e^{A(t-s)}B\hat{u}(s)ds\right|_H^2 dt$$

$$\leq C\int_0^\infty \int_0^t \frac{e^{-\omega_0(t-s)}}{(t-s)^{r_0}}ds \int_0^t \frac{e^{-\omega_0(t-s)}}{(t-s)^{r_0}}|\hat{u}(s)|_U^2 dsdt$$

$$\leq C_{r_0}\int_0^\infty \int_0^t \frac{e^{-\omega_0(t-s)}}{(t-s)^{r_0}}|\hat{u}(s)|_U^2 dsdt$$

$$\leq C_{r_0}\int_0^\infty \int_s^\infty \frac{e^{-\omega_0(t-s)}}{(t-s)^{r_0}}|\hat{u}(s)|_U^2 dtds$$

$$\leq C_{r_0}\int_0^\infty |\hat{u}(t)|_U^2 dt \leq C|x|_H^2, \tag{7.5.33}$$

where in the last step we used the continuity implied by (7.5.28). Combining the results in (7.5.31) and (7.5.33) and recalling, again, Assumption 7.2.1 yields

$$\int_0^\infty |e^{A_P t}x|_H^2 dt \leq C|x|_H^2,$$

which completes the proof of exponential stability. □

Our next result deals with a fundamental question of the boundedness of the gain operator B^*P, improving on (7.5.30).

Proposition 7.5.5. *Assume Assumptions* (7.2.1) *and* 7.5.1. *Then* $B^*P \in \mathcal{L}(H;U)$.

Proof of Proposition 7.5.5. We exploit the explicit representation of the Riccati operator in terms of the evolution associated with feedback dynamics $e^{A_P t}$ which are given by (7.5.24):

$$|(B^*P)x|_U = \left| B^* \int_0^\infty e^{A^*\tau} R^* R e^{A_P \tau} x \, d\tau \right|_U$$

$$\leq \int_0^\infty |B^* e^{A^*\tau}|_{\mathcal{L}(H;U)} |R^*R|_{\mathcal{L}(H)} |e^{A_P\tau} x|_H \, d\tau. \tag{7.5.34}$$

By Propositions 7.5.1 and 7.5.4, duality, and the Cauchy–Schwarz inequality we obtain

$$\leq C \int_0^\infty \frac{e^{-\omega_0 \tau}}{\tau^{r_0}} |e^{A_F \tau} x|_H \, d\tau \leq C|x|_H, \tag{7.5.35}$$

which proves the desired inequality, since $r_0 < 1$. □

Our next step is to establish the singular estimate for the feedback generator, analogous to that in the finite horizon case.

Proposition 7.5.6. *Assume Assumptions* 7.2.1 *and* 7.5.1. *Then there exist positive constants* C, ω_p *such that*

$$|B^* e^{A_P^* t} x|_H \leq \frac{C}{t^{r_0}} e^{-\omega_P t} |x|_H \ \mathit{for}\ 0 \leq r_0 < 1, \tag{7.5.36}$$

where r_0 *is the same as in Assumption* 7.5.1.

Proof. By exactly the same argument as that used in the finite horizon case, we first prove that for some $T > 0$

$$|B^* e^{A_P^* t} x|_H \leq \frac{C_T}{t^{r_0}} |x|_H \text{ for } 0 < t < T. \tag{7.5.37}$$

By using the exponential stability of $e^{A_P t}$ and applying the argument identical to this in the proof of Proposition 7.5.1, we propagate the above estimate for all $t > 0$. The details, which are identical to those given in Proposition 7.5.1, are omitted. □

To justify the derivation of the Riccati equation we need to establish strong differentiability of $e^{A_P t}$ on $D(A)$ for $t > 0$.

Proposition 7.5.7. *Under Assumptions* 7.5.1 *and* 7.2.1 *we have the following estimate:*

$$|e^{A_P t} A_P x|_H \leq C \left[1 + \frac{1}{t^{r_0}} \right] |x|_{D(A)}, \qquad t > 0.$$

Proof. Let $x \in D(A)$. By Propositions 7.5.5 and 7.5.6,

$$\begin{aligned} |e^{A_P t} A_P x|_H = |e^{A_P t}[A - BB^* P]x|_H &\le |e^{A_P t} Ax|_H + |e^{A_P t} BB^* Px|_H \\ &\le C|e^{A_P t} Ax|_H + |e^{A_P t} B|_{\mathcal{L}(U;H)} |B^* Px|_H, \end{aligned} \tag{7.5.38}$$

$$|e^{A_P t} A_P x|_H \le C \left[|Ax|_H + \frac{e^{-\omega_P t}}{t^{r_0}} |x|_H \right], \tag{7.5.39}$$

which yields the desired conclusion. □

We are now ready to derive the ARE.

Proposition 7.5.8. *Under Assumptions* 7.5.1 *and* 7.2.1 *the following ARE is satisfied:*

$$\begin{aligned} &(A^* Px, y)_H + (PAx, y)_H + (R^* Rx, y)_H \\ &\quad = (B^* Px, B^* Py)_U, \qquad x, y \in D(A). \end{aligned} \tag{7.5.40}$$

Proof. With $x, y \in D(A)$ we compute the expression

$$\begin{aligned} (A^* Px + PAx, y)_H &= \int_0^\infty (A^* e^{A^* t} R^* R e^{A_P t} x, y)_H dt \\ &\quad + \int_0^\infty (e^{A^* t} R^* R e^{A_P t} Ax, y)_H dt \\ = \int_0^\infty \left(R^* R e^{A_P t} x, \frac{d}{dt} e^{At} y \right)_H dt &+ \int_0^\infty (e^{A^* t} R^* R e^{A_P t} Ax, y)_H dt \\ &= -(R^* Rx, y)_H - \int_0^\infty \left(R^* R \frac{d}{dt} e^{A_P t} x, e^{At} y \right)_H dt \\ &\quad + \int_0^\infty (e^{A^* t} R^* R e^{A_P t} Ax, y)_H dt \\ &= -(R^* Rx, y)_H - \int_0^\infty (R^* R e^{A_P t} A_P x, e^{At} y)_H dt \\ &\quad + \int_0^\infty (e^{A^* t} R^* R e^{A_P t} Ax, y)_H dt, \end{aligned} \tag{7.5.41}$$

where all the quantities are well defined by virtue of Proposition 7.5.7 and Assumption 7.5.1.

By the result of Proposition 7.5.5 we obtain

$$\begin{aligned} (A^* Px + PAx, y)_H = -(R^* Rx, y)_H &- \int_0^\infty (R^* R e^{A_P t}(A - BB^* P)x, e^{At} y)_H dt \\ &+ \int_0^\infty (e^{A^* t} R^* R e^{A_P t} Ax, y)_H dt \\ &= -(R^* Rx, y)_H - \int_0^\infty (R^* R e^{A_P t} Ax, e^{At} y)_H dt \\ + \int_0^\infty (e^{A^* t} R^* R e^{A_P t} Ax, y)_H dt &+ \int_0^\infty (R^* R e^{A_P t} BB^* Px, e^{At} y)_H dt \end{aligned}$$

$$= -(R^*Rx, y)_H + \int_0^\infty (R^*Re^{A_Pt}BB^*Px, e^{At}y)_H dt$$
$$= -(R^*Rx, y)_H + (B^*Px, B^*Py)_U, \tag{7.5.42}$$

where in the last step we evoked representation in (7.5.24). The proof of Proposition 7.5.8 is thus complete. □

Our final step is to establish properties of r. From optimal synthesis (7.5.25) and regularity of optimal control u^0 as well as the regularity of the gain B^*P we infer that

$$B^*r = -B^*P\mathbf{y}^0 - u^0 \in C([0,\infty);U) \cap L_2(0,\infty;U).$$

We are now ready to derive the differential equation satisfied by r.

Proposition 7.5.9. *Under Assumptions* 7.2.1 *and* 7.5.1 *we have*

$$r_t = -(A^* - PBB^*)r - PF \quad \text{on } D(A)' \quad \lim_{t\to\infty} r(t) = 0.$$

Proof. A starting point is the formula (7.5.26), which written explicitly gives

$$r(t) = \int_t^\infty e^{A^*(s-t)}R^*Ry^*(s)ds - \int_0^\infty e^{A^*s}R^*R\Phi(s)y^*(t)ds$$
$$= \int_t^\infty e^{A^*(s-t)}R^*Ry^*(s)ds - \int_t^\infty e^{A^*(s-t)}R^*R\Phi(s-t)y^*(t)ds, \tag{7.5.43}$$

where y^* denotes an optimal trajectory corresponding to a zero initial datum and we recall that r does not depend on the choice of the initial data. By applying the arguments identical to those in Proposition 6.5.3 we obtain the following structure for the derivative of $\Phi(s-t)y^*(t)$:

$$\frac{d}{dt}[\Phi(s-t)y^*(t)] = \Phi(s-t)[-BB^*r(t) + F(t)] \text{ in } Z_{r_0}(t,\infty;H), \tag{7.5.44}$$

where the spaces $Z_{r_0}(t,\infty;H)$ are as introduced in Chapter 5. Differentiating (7.5.43) with respect to time and accounting for regularity properties derived in the propositions stated above as well as in (7.5.44) we obtain

$$(r_t, y)_H = -(A^*r, y)_H - \left(\int_t^\infty e^{A^*(s-t)}R^*R\frac{d}{dt}[\Phi(s-t)y^*(t)]ds, y\right)_H$$
$$= -(A^*r, y)_H - \left(\int_t^\infty e^{A^*(s-t)}R^*R\Phi(s-t)[-BB^*r(t) + F(t)]ds, y\right)_H$$
$$= -(A^*r, y)_H - (P[-BB^*r(t) + F(t)], y)_H. \tag{7.5.45}$$

The proof is complete. □

Remark 7.5.3. *We note that the regularity of* B^*r *also can be derived from the explicit formulas defining* r. *This argument, quite useful in computations, is given below. Under Assumptions* (7.2.1) *and* 7.5.1 *the following computations are justified.*

From Proposition 7.5.9 *we have* $r(t) = \int_t^\infty e^{A_P^*(\tau-t)}PF(\tau)d\tau$. *Hence by Proposition* 7.5.6

$$|B^*r|^2_{L_2(0,\infty;U)} \leq \int_0^\infty \left(\int_t^\infty |B^*e^{A_P^*(\tau-t)}PF(\tau)|_U d\tau\right)^2 dt$$

$$\leq C\int_0^\infty \left(\int_t^\infty \frac{e^{-\omega_P(\tau-t)}}{(\tau-t)^{r_0}}|P|_{\mathcal{L}(H)}|F(\tau)|_H d\tau\right)^2 dt$$

$$\leq C\int_0^\infty \int_t^\infty \frac{e^{-\omega_P(\tau-t)}}{(\tau-t)^{r_0}}d\tau \int_t^\infty \frac{e^{-\omega_P(\tau-t)}}{(\tau-t)^{r_0}}|F(\tau)|^2_H d\tau dt$$

$$\leq C\int_0^\infty \frac{e^{-\omega_P s}}{s^{r_0}}ds \int_0^\infty \int_t^\infty \frac{e^{-\omega_P(\tau-t)}}{(\tau-t)^{r_0}}|F(\tau)|^2_H d\tau dt \leq C|F|_{L_2(0,\infty;H)}. \tag{7.5.46}$$

Completion of the proof of Theorem 7.5.1. Combining the results of Propositions 7.5.1–7.5.9 yields the desired conclusions in Theorem 7.5.1. □

7.6 Back to Structural Acoustic Problems: Proofs of Theorems 7.3.1 and 7.3.2

To prove the results of Theorems 7.3.1 and 7.3.2 we apply abstract results given in Theorem 7.4.1 and Theorem 7.5.1. To accomplish this, one needs to verify all the hypotheses imposed on abstract operators. Theorem 7.3.2 follows directly from Theorem 7.4.1. For Theorem 7.3.1 we need to verify Assumption 7.5.1. But these are the same as in the finite horizon case and were already verified in Proposition 6.6.2. Thus the proofs of Theorems 7.3.1 and 7.3.2 are completed. □

7.7 Comments and Open Problems

- It would be interesting to prove Theorem 7.3.3 with the damping on the wall component via the free boundary conditions. This case is technically more demanding than the case of hinged boundary conditions which satisfy the Lopatinski condition. The main issue is to prove regularity in (7.4.15). It is believed that the argument given in [44] and based on Fourier analysis should be applicable; however this has not yet been done. Similarly, the case of damping via clamped boundary conditions is an open problem.
- The model of the Kirchhoff plate considered in subsection 7.3.3 could be generalized to include a potential structural damping on the wall, represented by a positive function, say, $\alpha(x)$, possibly vanishing on some subsets of Γ_0. Then, the goal is to show that the boundary damping provides the exponential decay rates for the energy, which are uniform with respect to $\alpha(x)$. It is plausible that the techniques of [41] can be extended to obtain this kind of result.
- As in the case of finite horizon control problem, one could study the case of unbounded observations R. The results of this type should correlate the value of the parameter r_0 in Assumption 7.5.1 with the amount of the unboundedness generated by the operator R.

- The results of section 7.3 can be extended to cover mini-max theory and optimization with nonpositive cost [199]. This will lead to appropriate algebraic nonstandard Riccati equations. A study of a singular version of the control problem meets with a greater challenge. For related results in the context of single PDEs, see [38, 169, 170, 172, 171, 174, 173].
- As we recall, the hyperbolic theory presented in these notes hinges on the regularity assumption (7.4.15). There are, however, hyperbolic systems that do not comply with this hypothesis. A canonical example is the wave equation with the control via Neumann boundary conditions defined on the finite energy space $H^1 \times L_2$. It is possible to relinquish regularity hypothesis (7.4.15), but this comes with a price [133]. We no longer operate within a C_0 semigroup framework. Riccati equations and the optimal dynamics are formulated within the context of integrated semigroups.
- There is another old approach to Riccati equations and optimal control problem, which has been recently reviewed within the context of frequency domain and infinite-dimensional system theory. This approach is based on the concept of spectral factorization; see [187, 203] and references therein. This theory is at present not applicable to multidimensional PDEs, and in particular, interactive structures, due to a so-far unchecked assumption. Thus, we do not further pursue this topic.
- Some properties of structural acoustic models within the context of system theory were analyzed in [11, 12]. The main focus of analysis in [11] is the robustness of dynamic stabilizers for systems with delays. Since structural interactions with delays are beyond the scope of these lectures, we do not pursue the topic and we refer the reader to [179, 11, 12] and references therein.
- Control problems dealing with minimization of a pointwise pressure $(z_t(x_0))$ (an example of unbounded observation) are studied in [11, 12]. It was shown there that the corresponding control system has an admissible control and admissible observation operator defined on a suitable state space. In fact, this space is more regular than the classical state space. The wave component coincides with $H_z \equiv H^{3/2}(\Omega) \times H^{1/2}(\Omega)$, and the plate component H_v is the same as before. These results are technical and rely on microlocal analysis methods.

Bibliography

[1] F. ALABAU AND V. KOMORNIK, *Boundary observability, controllability, and stabilization of linear elastodynamic systems,* SIAM J. Control Optim. 37 (1999), pp. 521–542.

[2] G. AVALOS, *The exponential stability of a coupled hyperbolic/parabolic system arising in structural acoustics,* Abstract Appl. Anal. 1 (1996), pp. 203–219.

[3] G. AVALOS, *Sharp regularity estimates for solutions to wave equations and their traces with prescribed Neumann data,* Appl. Math. Optim. 35 (1997), pp. 203–221.

[4] G. AVALOS AND I. LASIECKA, *Differential Riccati equation for the active control of a problem in structural acoustics,* J. Optim. Theory Appl. 91 (1996), pp. 695–728.

[5] G. AVALOS AND I. LASIECKA, *Exponential stability of a thermoelastic system without mechanical dissipation,* Rend. Istit. Mat. Univ. Trieste 28 (1997), pp. 1–28.

[6] G. AVALOS AND I. LASIECKA, *Exponential stability of a thermoelastic system with free boundary conditions without mechanical dissipation,* SIAM J. Math. Anal. 29 (1998), pp. 155–182.

[7] G. AVALOS AND I. LASIECKA, *The strong stability of a semigroup arising from a coupled hyperbolic/parabolic system,* Semigroup Forum 57 (1998), pp. 278–292.

[8] G. AVALOS AND I. LASIECKA, *Uniform decay rates of solutions to a structural acoustic model with nonlinear dissipation,* App. Math. Comput. Sci. 8 (1998), pp. 287–312.

[9] G. AVALOS AND I. LASIECKA, *Uniform decays in nonlinear thermoelastic systems,* Optim. Control Theory Algorithms, 15 (1998), pp. 1–22.

[10] G. AVALOS, I. LASIECKA, AND R. REBARBER, *Uniform decay properties of a model in structural acoustic,* J. Math. Pure Appl. 79 (2000), pp. 1057–1072.

[11] G. AVALOS, I. LASIECKA, AND R. REBARBER, *Lack of time-delay robustness for stabilization of a structural acoustics model,* SIAM J. Control Optim. 37 (1999), pp. 1394–1418.

[12] G. Avalos, I. Lasiecka, and R. Rebarber, *Well-posedness of a structural acoustic control model with point observation of the pressure,* J. Differential Equations, to appear.

[13] G. Avalos, I. Lasiecka, and R. Triggiani, *Uniform stability in nonlinear thermoelastic plates with free boundary conditions,* Internat. Ser. Numer. Math. 133 (1999), pp. 13–32.

[14] A. V. Balakrishnan, *Applied Functional Analysis*, Springer-Verlag, Berlin, 1975.

[15] H. T. Banks, R. J. Silcox, and R. C. Smith, *The modeling and control of acoustic/structure interaction problems via piezoceramic actuators: 2-d numerical examples*, ASME J. Vibration Acoustics 2 (1993), pp. 343–390.

[16] H. T. Banks and R. Smith, *Active control of acoustic pressure fields using smart material technologies*, IMA Vol. Math. Appl. 68, Springer-Verlag, Berlin, 1995, pp. 1–33.

[17] H. T. Banks and R. C. Smith, *Well-posedness of a model for structural acoustic coupling in a cavity enclosed by a thin cylindrical shell*, J. Math. Anal. Appl. 191 (1995), pp. 1–25.

[18] H. T. Banks, R. C. Smith, and Y. Wang, *The modeling of piezoceramic patch interactions with shells, plates and beams*, Quart. Appl. Math. 53 (1995), pp. 353–381.

[19] V. Barbu, *Nonlinear Semigroups and Differential Equations in Banach Spaces*, Nordhoff, Leiden, The Netherlands, 1976.

[20] V. Barbu, *Analysis and Control of Nonlinear Infinite Dimensional Systems,* Academic Press, New York, 1993.

[21] V. Barbu, I. Lasiecka, and R. Triggiani, *Extended algebraic Riccati equations arising in hyperbolic dynamics with unbounded controls,* Nonlinear Anal. 40 (2000), pp. 105–129.

[22] C. Bardos, G. Lebeau, and J. Rauch, *Sharp sufficient conditions for the observation, control, and stabilization of waves from the boundary,* SIAM J. Control Optim. 30 (1992), pp. 1024–1065.

[23] J. Bartolomeo and R. Triggiani, *Uniform energy decay rates for Euler-Bernoulli equations with feedback operators in Dirichlet/Neumann boundary conditions,* SIAM J. Math. Anal. 22 (1991), pp. 46–71.

[24] J. Beale, *Spectral properties of an acoustic boundary condition,* Indiana Univ. Math. J. 9 (1976), pp. 895–917.

[25] A. Benabdallah, *Von Karman Model with Thermal Effects,* Prepublications de l'Equipe de Mathematiques de Besancon, Besancon, France, 1998.

[26] A. Benabdallah and I. Lasiecka, *Exponential decay rates for a full von Karman system of dynamic thermoelasticity,* J. Differential Equations 160 (2000), pp. 51–93.

[27] A. BENABDALLAH AND D. TENIOU, *Exponential stability of a von Karman model with thermal effects,* Electronic J. Differential Equations 7 (1998), pp. 1–13.

[28] A. BENSOUSSAN, G. DA PRATO, M. C. DELFOUR, AND S. K. MITTER, *Representation and Control of Infinite Dimensional Systems,* Vol. 2, Birkhäuser, Boston, Basel, Berlin, 1993.

[29] M. BERNADOU, *Methods d'elements finis pour les problems de cocques minces,* Masson, Paris, 1994.

[30] M. BERNADOU, P. G. CIARLET, AND B. MIARA, *Existence Theorems for Two-Dimensional Linear Shell Theories,* Tech. Report 1771, Unité De Recherche NRIA, Rocquencourt, 1992.

[31] M. BERNADOU AND P. G. CIARLET, *Sur l'éllipticité du modè linéar de coques de W. T. Koiter,* Computing Methods in Applied Sciences and Engineering, Lecture Notes in Economics and Mathematical Systems, Springer-Verlag, Berlin, 1976, pp. 89–136.

[32] M. BERNADOU AND B. LALANNE, *On the approximation of free vibration modes of a general thin shell application to turbine blades,* Proceedings of the Third European Conference on Mathematics in Industry, Kluwer Academic, Dordrecht, The Netherlands, 1990, pp. 257–264.

[33] M. BERNADOU AND J. T. ODEN, *An existence theorem for a class of nonlinear shallow shell problems,* J. Math. Pures Appl. 60 (1981), pp. 285–308.

[34] E. BRADLEY AND I. LASIECKA, *Local exponential stabilization for a nonlinearly perturbed von Karman plate,* Nonlinear Anal. 18 (1992), pp. 333–343.

[35] M. E. BRADLEY AND C. MCMILLAN, *Wellposedness for nonlinear spherical shells,* Nonlinear Analysis 34 (1998), pp. 405–427.

[36] H. BREZIS, *Operateurs maximaux monotones et semigroupes de contractions dans les espaces de hilbert,* North–Holland, Amsterdam, 1973.

[37] F. BUCCI, I. LASIECKA, AND R. TRIGGIANI, *Structural Acoustic Problems for Dynamics with Singular Estimate,* Preprint, 2001.

[38] F. BUCCI AND L. PANDOLFI, *The value function of the singular quadratic regulator problem with distributed control action,* SIAM J. Control Optim. 36 (1998), pp. 115–136.

[39] J. CAGNOL, I. LASIECKA, C. LEBIEDZIK, AND J. P. ZOLESIO, *Uniform Stability in Structural Acoustic Models with Curved Walls Modeled by Dynamic Shells,* Rapport de Recherche, Universitaire Léonard de Vinci, Paris, 2000.

[40] M. CAMURDAN, *Uniform stability of a coupled structural acoustic system by boundary dissipation,* Abstract Appl. Anal. 3 (1998), pp. 377–400.

[41] M. CAMURDAN AND G. JI, *Uniform feedback stabilization via boundary moments a three dimensional structural acoustic model,* Proceedings of the 37th IEEE CDC 3, 1998, pp. 2958–2964.

[42] M. Camurdan and G. Ji, *Decay rates for a 3-dimensional structural acoustic model with boundary feedback*, Contemp. Math. 268 (2000), pp. 41–61.

[43] M. Camurdan and R. Triggiani, *Control problems in noise reduction: The case of two coupled hyperbolic equations*, Proceedings of SPIE's 4th Annual Symposium on Smart Structures and Materials, Bellingham, WA, 1997.

[44] M. Camurdan and R. Triggiani, *Sharp regularity of a coupled system of a wave and Kirchhoff equation with point control, arising in noise reduction*, Differential Integral Equations 12 (1998), pp. 101–118.

[45] G. Chen, *Energy decay estimates and exact boundary value controllability for the wave equation in a bounded domain*, J. Math. Pures Appl. 58 (1979), pp. 248–274.

[46] G. Chen, S. G. Krantz, D. W. Ma, and C. E. Wayne, *The Euler Bernoulli beam equation with boundary energy dissipation*, Oper. Meth. Optim. Control Problems 108 (1987), pp. 67–96.

[47] G. Chen and D. Russell, *A mathematical model for linear elastic systems with structural damping*, Quart. Appl. Math. 39 (1982), pp. 433–454.

[48] S. Chen and R. Triggiani, *Proof of extensions of two conjectures on structural damping for elastic systems*, Pacific J. Math. 136 (1989), pp. 15–55.

[49] S. Chen and R. Triggiani, *Characterization of domains of fractional powers of certain operators arising in elastic systems and applications*, J. Differential Equations 64 (1990), pp. 26–42.

[50] Ph. Ciarlet, *Mathematical Elasticity, Theory of Shells*, North–Holland, Amsterdam, 2000.

[51] Ph. Ciarlet and P. Rabier, *Les Equations de von Karman*, Springer-Verlag, Berlin, 1982.

[52] E. F. Crawley and J. de Luis, *Use of piezoelectric actuators as elements of intelligent structures*, AIAA J. 25 (1987), pp. 1373–1385.

[53] E. F. Crawley, J. de Luis, N. W. Hagood, and E. H. Anderson, *Development of piezoelectric technology for applications in control of intelligent structures*, Proceedings of Applications in Control of Intelligent Structures, American Controls Conference, Atlanta, 1988, pp. 1890–1896.

[54] R. Curtain, *An Introduction to Infinite-Dimensional Linear Systems Theory*, Springer-Verlag, Berlin, 1991.

[55] C. Dafermos, *On the existence and asymptotic stability of solutions to the equations of linear thermoelasticity*, Arch. Rational Mech. Anal. 29 (1968), pp. 249–273.

[56] E. K. Dimitriadis, C. R. Fuller, and C. A. Rogers, *Piezoelectric actuators for distributed noise and vibration excitation of thin plates*, J. Vibration Acoustics 13 (1991), pp. 100–107.

[57] G. DUVAUT AND J. L. LIONS, *Les Inequations en Mecaniques et en Physiques,* Dunod, Paris, 1972.

[58] F. FAHROO AND C. WANG, *A new model for acoustic interaction and its exponential stability,* Quart. Appl. Math. 57 (1999), pp. 157–179.

[59] H. O. FATTORINI, *Second Order Linear Differential Equations,* North–Holland, Amsterdam, 1985.

[60] H. O. FATTORINI, *Infinite Dimensional Control Theory,* Cambridge University Press, Cambridge, UK, 1999.

[61] A. FAVINI AND I. LASIECKA, *Wellposedness and regularity of second order abstract equations arising in hyperbolic-like problems with nonlinear boundary conditions*, Osaka J. Math. 32 (1995), pp. 721–752.

[62] A. FAVINI, I. LASIECKA, M. A. HORN, AND D. TATARU, *Global existence, uniqueness and regularity of solutions to a von Kármán system with nonlinear boundary dissipation*, Differential Integral Equations 9 (1996), pp. 267–294.

[63] A. FAVINI, I. LASIECKA, M. A. HORN, AND D. TATARU, *Addendum to the paper: Global existence, uniqueness and regularity of solutions to a von Kármán system with nonlinear boundary dissipation*, Differential Integral Equations 10 (1997), pp. 197–200.

[64] A. FAVINI, I. LASIECKA, AND T. TANABE, *Abstract differential equations and nonlinear dispersive systems*, Differential Integral Equations 6 (1993), pp. 995–1008.

[65] F. FLANDOLI, *Riccati equation arising in a boundary control problem with distributed parameters*, SIAM J. Control Optim. 22 (1984), pp. 76–86.

[66] F. FLANDOLI, I. LASIECKA, AND R. TRIGGIANI, *Algebraic Riccati equations with non-smoothing observations arising in hyperbolic and Euler-Bernoulli boundary control problems,* Ann. Mat. Pura Appl. 153 (1988), pp. 307–382.

[67] P. GRISVARD, *Characterization de quelques espaces d'interpolation*, Arch. Rational Mech. Anal. 26 (1967), pp. 40–63.

[68] P. GRISVARD, *Elliptic Problems in Nonsmooth Domains*, Pitman, Boston, MA, 1985.

[69] P. GRISVARD, *Controlabilite exacte des solutions de l'equations des ondes en presence de singularities*, J. Math. Pures Appl. 68 (1989), pp. 215–259.

[70] E. HENDRICKSON AND I. LASIECKA, *Convergence of numerical algorithms for the approximations of Riccati equations arising in smart material acoustic structure interactions*, Comput. Optim. 8 (1997), pp. 73–101.

[71] K. HOFFMANN AND N. BOTKIN, *Homogenization of a model describing vibration of nonlinear thin plates excited by piezopatches,* Optimal Control of Partial Differential Equations, Birkhäuser, Boston, Basel, 1999.

[72] K. HOFFMANN AND N. BOTKIN, *Homogenization of von Kármán plates excited by piezoelectric patches,* ZAMM Z. Angew Math. Mech. 19 (2000), pp. 579–590.

[73] L. HORMANDER, *Linear Partial Differential Operators,* Springer-Verlag, New York, 1964.

[74] L. HORMANDER, *The Analysis of Linear Partial Differential Operators,* Vols. I–IV, Springer-Verlag, New York, 1984.

[75] M. A. HORN, *Exact controllability and uniform stabilization of the Kirchhoff plate equation with boundary feedback acting via bending moments,* J. Math. Anal. Appl. 167 (1992), pp. 557–581.

[76] M. A. HORN, *Exact controllability and stabilization of the Euler Bernoulli plates via bending moments only on the space of optimal regularity,* J. Math. Anal. Appl. 167 (1992), pp. 557–581.

[77] M. A. HORN, *Implications of sharp trace regularity results on boundary stabilization of the system of linear elasticity,* J. Math. Anal. Appl. 223 (1998), pp. 126–150.

[78] M. A. HORN, *Sharp trace regularity of the traces to solutions of dynamic elasticity,* J. Math. Systems Estimation Control 8 (1998), pp. 217–229.

[79] M. A. HORN, *Stabilization of the dynamic system of elasticity by nonlinear boundary feedback,* Internat. Ser. Numer. Math. 133 (1998), pp. 201–210.

[80] M. A. HORN, *Uniform stabilization of anisotropic elastodynamic systems with nonlinear boundary dissipation,* Contemp. Math. 268 (2000), pp. 177–191.

[81] M. A. HORN AND I. LASIECKA, *Asymptotic behavior with respect to thickness of boundary stabilizing feedback for the Kirchhoff plate,* J. Differential Equations 114 (1994), pp. 396–433.

[82] M. A. HORN AND I. LASIECKA, *Asymptotic behaviour with respect to thickness of boundary stabilizing feedback for the Kirchhoff plate,* J. Differential Equations 114 (1994), pp. 396–433.

[83] M. A. HORN AND I. LASIECKA, *Uniform decay of weak solutions to a von Karman plate with nonlinear dissipation,* Differential Integral Equations 7 (1994), pp. 885–908.

[84] M. A. HORN AND I. LASIECKA, *Global stabilization of a dynamic von Karman plate with nonlinear boundary feedback,* Appl. Math. Optim. 31 (1995), pp. 57–84.

[85] V. ISAKOV, *On uniqueness in a lateral Cauchy problem with multiple characteristics,* J. Differential Equations 97 (1997), pp. 134–147.

[86] V. ISAKOV, *Inverse Problems for Partial Differential Equations,* Springer-Verlag, Berlin, 1998.

[87] G. Ji, *Uniform Decay Rates and Asymptotic Analysis of the von Karman Plate with Nonlinear Dissipation in Boundary Moments,* Ph.D. thesis, University of Virginia, Charlottesville, VA, 1998.

[88] G. Ji and I. Lasiecka, *Nonlinear boundary feedback stabilization for a semilinear Kirchhoff plate with dissipation acting only via moments-limiting behaviour,* J. Math. Anal. Appl. 229 (1999), pp. 452–479.

[89] J. U. Kim, *On the energy decay of a linear thermoelastic bar and plate,* SIAM J. Math. Anal. 23 (1992), pp. 889–899.

[90] T. Von Karman, *Festigkeitprobleme in Maschinenbau,* Encyklopedie Math. Wissenschaften 4 (1910), pp. 314–385.

[91] M. A. Kazemi and M. V. Kiblanov, *Stability estimates for ill-posed Cauchy problems involving hyperbolic equations and inequalities,* Appl. Anal. 50 (1993), pp. 93–102.

[92] A. Khapalov, *Controllability of the wave equation with moving point control,* Appl. Math. Optim. 31 (1995), pp. 155–175.

[93] A. Khapalov, *Interior point control and observation for the wave equation,* Abstract Appl. Anal. 1 (1996), pp. 219–236.

[94] H. Koch, *Slow decay in linear thermoelasticity,* Quart. Appl. Math. 58 (2000), pp. 601–612.

[95] H. Koch and I. Lasiecka, *Hadamard Wellposedness of Finite Energy Solutions in Full von Karman Systems*, manuscript.

[96] H. Koch and A. Stachel, *Global existence of classical solutions to the dynamical von Karman equations,* Math. Methods Appl. Sci. 16 (1993), pp. 581–586.

[97] V. Komornik, *Exact Controllability and Stabilization*, Masson, Paris, 1994.

[98] J. Lagnese, *Boundary stabilization of linear elastodynamic systems,* SIAM J. Control Optim. 21 (1983), pp. 968–984.

[99] J. Lagnese, *Boundary Stabilization of Thin Plates*, SIAM, Philadelphia, 1989.

[100] J. Lagnese, *Modeling and stabilization of nonlinear plates,* Internat. Ser. Numer. Math. 100 (1991), pp. 247–264.

[101] J. Lagnese, *Uniform asymptotic energy estimates for solutions of the equations of dynamic plane elasticity with nonlinear dissipation at the boundary,* Nonlinear Anal. 16 (1991), pp. 35–54.

[102] J. Lagnese and J. L. Lions, *Modeling, Analysis and Control of Thin Plates,* Masson, Paris, 1988.

[103] J. LAGNESE AND G. LEUGERING, *Uniform stabilization of a nonlinear beam by nonlinear boundary feedback,* J. Differential Equations 91 (1991), pp. 355–388.

[104] I. LASIECKA, *Exponential decay rates for the solutions of Euler-Bernoulli equations with boundary dissipation occurring in the moments only,* J. Differential Equations 95 (1992), pp. 169–182.

[105] I. LASIECKA, *Existence and uniqueness of the solutions to second order abstract equations with nonlinear and nonmonotone boundary conditions*, Nonlinear Anal. 23 (1994), pp. 797–823.

[106] I. LASIECKA, *Intermediate solutions to full von Karman system of dynamic nonlinear elasticity,* Appl. Anal. 68 (1998), pp. 121–145.

[107] I. LASIECKA, *Uniform stabilizability of a full von Karman system with nonlinear boundary feedback,* SIAM J. Control Optim. 36 (1998), pp. 1376–1422.

[108] I. LASIECKA, *Mathematical control theory in structural acoustic problems,* Math. Models Methods Appl. Sci. 8 (1998), pp. 1119–1153.

[109] I. LASIECKA, *Boundary stabilization of a 3-dimensional structural acoustic model,* J. Math. Pures Appl. 78 (1999), pp. 203–232.

[110] I. LASIECKA, *Uniform decay rates for full von Karman system of dynamic elasticity with free boundary conditions and partial boundary dissipation,* Comm. Partial Differential Equations 24 (1999), pp. 1801–1849.

[111] I. LASIECKA, *Optimization problems for structural acoustic models with thermoelasticity and smart materials,* Discuss. Math. Differ. Incl. Control Optim. 20 (2000), pp. 113–140.

[112] I. LASIECKA AND C. LEBIEDZIK, *Uniform stability in structural acoustic systems with thermal effects and nonlinear boundary damping,* Control Cybernetics 28 (1999), pp. 557–581.

[113] I. LASIECKA AND C. LEBIEDZIK, *Boundary stabilizability of nonlinear structural acoustic models with thermal effects on the interface,* C.R. Acad Sci. Paris Sér. B 328 (2000), pp. 187–192.

[114] I. LASIECKA AND C. LEBIEDZIK, *Decay rates in nonlinear structural acoustic models with thermal effects and without a damping on the interface,* Appl. Math. Optim. 42 (2000), pp. 127–167.

[115] I. LASIECKA AND C. LEBIEDZIK, *Asymptotic behaviour of nonlinear structural acoustic interactions with thermal effects on the interface,* Nonlinear Anal., to appear.

[116] I. LASIECKA AND R. MARCHAND, *Control and stabilization in nonlinear structural acoustic problems,* Proceedings of SPIE's 4th Annual Symposium on Smart Structures and Materials, 1997.

[117] I. Lasiecka and R. Marchand, *Uniform decay rates for solutions to nonlinear shells with nonlinear dissipation*, Nonlinear Anal. 30 (1997), pp. 5409–5418.

[118] I. Lasiecka and R. Marchand, *Riccati equations arising in acoustic structure interactions with curved walls*, Dynamics Control 8 (1998), pp. 269–292.

[119] I. Lasiecka and R. Marchand, *Optimal error estimates for FEM approximations of dynamic nonlinear shallow shells*, Math. Model. Numer. Anal. 34 (2000), pp. 63–84.

[120] I. Lasiecka and D. Tataru, *Uniform boundary stabilization of semilinear wave equations with nonlinear boundary damping*, Differential Integral Equations 6 (1993), pp. 507–533.

[121] I. Lasiecka and R. Triggiani, *Factor Spaces and Implications on Kirchhoff Equations with Clamped Boundary Conditions*, Preprint, 2001.

[122] I. Lasiecka and R. Triggiani, *Regularity of hyperbolic equations under $L_2(0,T;L_2(\Gamma))$-Dirichlet boundary terms*, Appl. Math. Optim. 10 (1983), pp. 275–286.

[123] I. Lasiecka and R. Triggiani, *Riccati equations for hyperbolic partial differential equations with $L_2(0,T;L_2(\Gamma))$-Dirichlet boundary terms*, SIAM J. Control Optim. 24 (1986), pp. 884–925.

[124] I. Lasiecka and R. Triggiani, *Uniform exponential energy decays of the wave in a bounded region with L_2 feedback control in the Dirichlet boundary conditions*, J. Differential Equations 66 (1987), pp. 340–390.

[125] I. Lasiecka and R. Triggiani, *A lifting theorem for the time regularity of solutions to abstract equations with unbounded operators and applications to hyperbolic equations*, Proc. Amer. Math. Soc. 102 (1988), pp. 745–755.

[126] I. Lasiecka and R. Triggiani, *Regularity theory for a class of nonhomogenous Euler-Bernoulli equations: A cosine operator approach*, Bull. UMI 7 (1989), pp. 199–228.

[127] I. Lasiecka and R. Triggiani, *Sharp regularity theory for second order hyperbolic equations of Neumann type*, Ann. Mat. Pura Appl. 157 (1990), pp. 285–367.

[128] I. Lasiecka and R. Triggiani, *Exact controllability and uniform stabilization of Kirchhoff plates with boundary control only on $\Delta w|_\Gamma$ and homogenous boundary displacement*, J. Differential Equations 95 (1991), pp. 62–101.

[129] I. Lasiecka and R. Triggiani, *Differential and Algebraic Riccati Equations with Applications to Boundary/Point Control Problems, Continuous Theory and Approximation Theory*, Lecture Notes in Control and Inform. Sci. 164, Springer-Verlag, New York, 1991.

[130] I. Lasiecka and R. Triggiani, *Regularity theory of hyperbolic equations with non-homogenous Neumann boundary conditions 2: General boundary data*, J. Differential Equations 94 (1991), pp. 112–164.

[131] I. Lasiecka and R. Triggiani, *Uniform stabilization of the wave equation with Dirichlet and Neumann feedback control without geometrical conditions*, Appl. Math. Optim. 25 (1992), pp. 189–224.

[132] I. Lasiecka and R. Triggiani, *Riccati differential equations with unbounded coefficients and nonsmooth terminal condition—the case of analytic semigroups*, SIAM J. Math. Anal. 23 (1992), pp. 449–481.

[133] I. Lasiecka and R. Triggiani, *Algebraic Riccati equations arising from systems with unbounded input-solution operator*, Nonlinear Anal. 20 (1993), pp. 659–695.

[134] I. Lasiecka and R. Triggiani, *Sharp trace estimates of solutions to Kirchhoff and Euler Bernoulli equations*, Appl. Math. Optim. 28 (1993), pp. 277–306.

[135] I. Lasiecka and R. Triggiani, *Recent advances in regularity of second-order hyperbolic mixed problems, and applications*, in Dynamics Reported—Expositions in Dynamical Systems, Vol. 3, Springer-Verlag, Berlin, 1994, pp. 25–104.

[136] I. Lasiecka and R. Triggiani, *Analyticity and lack thereof, of thermoelastic semigroups*, ESAIM Proc. 4 (1998), pp. 199–222.

[137] I. Lasiecka and R. Triggiani, *Analyticity of thermo-elastic semigroups with free B.C.*, Ann. Scuola Normale Superiore 27 (1998), pp. 457–482.

[138] I. Lasiecka and R. Triggiani, *Two direct proofs on the analyticity of the S.C. semigroup arising in abstract thermo-elastic equations*, Adv. Differential Equations 3 (1998), pp. 387–416.

[139] I. Lasiecka and R. Triggiani, *A sharp trace result on a thermo-elastic plate equation with coupled hinged/Neumann boundary conditions*, Continuous Discrete Dynamical Systems 5 (1999), pp. 585–598.

[140] I. Lasiecka and R. Triggiani, *Control Theory for Partial Differential Equations*, Cambridge University Press, London, 2000.

[141] I. Lasiecka and R. Triggiani, *Feedback noise control in an acoustic chamber-mathematical theory*, Stability and Control Theory: Nonlinear Problems in Aviation and Aerospace (S. Sivasundaram, ed.), Vol. 11, Gordon and Breach, Amsterdam, 2000.

[142] I. Lasiecka and R. Triggiani, *Optimal control problems for systems with singular estimate*, Proceedings of the Conference on Control of PDE's, University of Ohio, Athens, 2000, Dekker, New York, to appear.

[143] I. Lasiecka, R. Triggiani, and P. Yao, *Inverse observability estimates for second order hyperbolic equations with variable coefficients,* J. Math. Anal. Appl. 235 (1999), pp. 13–57.

[144] I. Lasiecka, R. Triggiani, and X. Zhang, *Non conservative wave equation with purely Neumann boundary conditions: Global uniqueness and observability inequalities in one shot,* Contemp. Math. 268 (2000), pp. 227–327.

[145] G. Lebeau and E. Zuazua, *Sur la decroissance non uniforme de l'energie dans les systemes de la thermoelasticite lineaire,* C.R. Acad. Sci. Paris 324 (1997), pp. 409–415.

[146] C. Lebiedzik, *Boundary stabilizability of a nonlinear structural acoustic model with thermoelastic effects and clamped/hinged b.c.*, Control of Nonlinear Parameter Distributed Systems, Marcel Dekker, New York, 2000.

[147] C. Lebiedzik, *Boundary stabilizability of thermo-structural acoustic models with unaccounted structural damping*, Contemp. Math. 268 (2000), pp. 327–341.

[148] C. Lebiedzik, *Exponential stability in structural acoustic models with thermoelasticity*, Dynamics Continuous Discrete Impulsive Systems 7 (2000), pp. 369–385.

[149] H. C. Lester and C. R. Fuller, *Active control of propeller induced noise fields inside a flexible cylinder*, Proceedings of AIAA 10th Aeroacoustics Conference, Seattle, WA, 1986.

[150] J. Lions, *Quelques methods de resolution des problemes aux limits nonlinearies*, Dunod, Paris, 1969.

[151] J. L. Lions and E. Magenes, *Non-Homogenous Boundary Value Problems and Applications*, Springer-Verlag, New York, 1972.

[152] J. L. Lions, *Control optimale de systemes gouvernes par des equations aux derivees partielles*, Dunod, Paris, 1968.

[153] J. L. Lions, *Exact controllability, stabilization and perturbations for distributed systems*, SIAM Rev. 30 (1988), pp. 1–68.

[154] J. L. Lions, *Controllabilite exacte et stabilization des systems distribues*, Masson, Paris, 1988.

[155] W. Littman and B. Liu, *On the spectral properties and stabilization of acoustic flow,* SIAM J. Appl. Math. 59 (1998), pp. 17–34.

[156] W. Littman and L. Markus, *Stabilization of hybrid system of elasticity by feedback boundary damping*, Ann. Math. Pura Appli. 152 (1988), pp. 281–330.

[157] K. Liu and S. Zheng, *Semigroups Associated with Dissipative Systems,* Chapman & Hall/CRC, Boca Raton, FL, 1999.

[158] W. Liu, *Partial exact controllability and exponential stability in higher-dimensional linear thermoelasticity,* ESAIM Proc. 3 (1998), pp. 23–48.

[159] X. Liu and J. Jong, *Optimal Control Theory for Infinite Dimensional Systems*, Birkhäuser, New York, 1994.

[160] Z. Liu and M. Renardy, *A note on the equation of thermoelastic plate,* Appl. Math. Lett. 8 (1995), pp. 1–6.

[161] P. Martinez, *Precise decay rate estimates for time-dependent dissipative systems*, Israel J. Math. 119 (2000), pp. 291–324.

[162] Y. Mayer, *Etude d'un modele mathematiques du controle des structures spatiales deformables*, Res. Notes Math., Pitman, Boston, MA, 1985, pp. 234–242.

[163] C. McMillan and R. Triggiani, *Min-Max game theory and algebraic Riccati equations for boundary control problems with continuous input-solution map. Part* II: *The general case,* Appl. Math. Optim. 29 (1994), pp. 1–65.

[164] S. Micu and E. Zuazua, *Boundary controllability of a linear hybrid system arising in the control of noise,* SIAM J. Control Optim. 35 (1997), pp. 1614–1637.

[165] N. Morozov, *Non-linear vibrations of thin plates with allowance for rotational inertia,* Sov. Math 8 (1967), pp. 1137–1141.

[166] P. M. Morse and K. U. Ingard, *Theoretical Acoustics*, McGraw-Hill, New York, 1968.

[167] S. Myatake, *Mixed problems for hyperbolic equations,* J. Math. Kyoto Univ. 130 (1973), pp. 435–487.

[168] N. Ourada and R. Triggiani, *Uniform stabilization of the Euler Bernoulli equation with feedback operator only in the Neumann boundary conditions,* Differential Integral Equations 4 (1991), pp. 277–292.

[169] L. Pandolfi, *From singular to regular control systems,* Control of PDE's, Lecture Notes in Pure and Appl. Math. 165, Dekker, New York, 1994, pp. 153–165.

[170] L. Pandolfi, *Singular Regulator Problem for Holomorphic Semigroup Systems*, Internal Report 28, Politechnica di Torino, 1994.

[171] L. Pandolfi, *The Kalman-Yakubovich-Popov theorem: An overview and new results for hyperbolic control systems*, Nonlinear Anal. 30 (1997), pp. 735–745.

[172] L. Pandolfi, *Dissipativity and the Lur'e problem for parabolic boundary control systems*, SIAM J. Control Optim. 36 (1998), pp. 2061–2081.

[173] L. Pandolfi, *On the Regularity of the Solutions of the Linear Operator Inequality,* Internal Report 7, Politechnico di Torino 7 (1998).

[174] L. Pandolfi, *The Kalman-Yakubovich-Popov theorem for stabilizable hyperbolic boundary control systems*, Integral Equation and Oper. Theory 34 (1999), pp. 478–493.

[175] A. PAZY, *Semigroups of Linear Operators and Applications to Partial Differential Equations*, Springer-Verlag, Berlin, 1986.

[176] G. DA PRATO, I. LASIECKA, AND R. TRIGGIANI, *A direct study of Riccati equations arising in hyperbolic boundary control problems,* J. Differential Equations 64 (1986), pp. 26–47.

[177] J.-P. PUEL AND M. TUCSNAK, *Boundary stabilization for the von Kármán equations*, SIAM J. Control Optim. 33 (1995), pp. 255–273.

[178] J.-P. PUEL AND M. TUCSNAK, *Global existence for the full von Karman system,* Appl. Math. Optim. 34 (1996), pp. 139–161.

[179] R. REBARBER, *Conditions for equivalence of internal and external stability for distributed parameter systems,* IEEE Trans. on Automat. Control 38 (1993), pp. 994–998.

[180] J. E. MUÑOZ RIVERA, *Energy decay rate in linear thermoelasticity,* Funkcial. Ekvac. 35 (1992), pp. 19–30.

[181] J. E. MUÑOZ RIVERA AND R. RACKE, *Smoothing properties, decay, and global existence of solutions to nonlinear coupled systems of thermoelastic type,* SIAM J. Math. Anal. 26 (1995), pp. 1547–1563.

[182] A. RUIZ, *Unique continuation for weak solutions of the wave equation plus a potential,* J. Math. Pures Appl. 71 (1992), pp. 455–467.

[183] D. L. RUSSELL, *Controllability and stabilizability theory for linear partial differential equations: Recent progress and open questions,* SIAM Rev. 20 (1978), pp. 639–739.

[184] D. L. RUSSELL, *Mathematical models for the elastic beam and their control-theoretic properties,* Semigroups Theory and Applications, Pitman Research Notes 152, 1986, pp. 177–217.

[185] D. L. RUSSELL, *On the positive square root of the fourth order operator,* Quart. Appl. Math. 46 (1988), pp. 751–775.

[186] V. I. SEDENKO, *On uniqueness of the generalized solutions of initial boundary value problem for Marguerre-Vlasov nonlinear oscillations of the shallow shells*, Russian Izvestiya, North-Caucasus Region, Ser. Natural Sciences 1–2 (1994), pp. 1–12.

[187] O. STAFFANS, *Quadratic optimal control of stable systems through spectral factorization,* Math. Control Signals Systems 8 (1995), pp. 167–197.

[188] D. TATARU, *On the regularity of boundary traces for the wave equation,* Ann. Scuola Normale Superiore 6 (1998), pp. 185–206.

[189] D. TATARU, *A priori estimates of Carleman's type in domains with boundary,* J. Math. Pures Appl. 73 (1994), pp. 355–387.

[190] D. TATARU AND M. TUCSNAK, *On the Cauchy problem for the full von Karman system*, Nonlinear Differential Equations Appl. 4 (1997), pp. 325–340.

[191] H. TRIEBEL, *Theory of Function Spaces*, Birkhäuser, Boston, Basel, 1983.

[192] TRIGGIANI, *An optimal control problem with unbounded control operator and unbounded observation operator where the ARE is satisfied as a Lyapunov equation,* Appl. Math. Lett. 10 (1997), pp. 95–102.

[193] R. TRIGGIANI, *Lack of uniform stabilization of non-contractive semigroups under compact perturbations,* Proc. AMS 105 (1989), pp. 375–383.

[194] R. TRIGGIANI, *Lack of exact controllability for wave and plate equations with finitely many boundary controls,* Differential Integral Equations 4 (1991), pp. 683–705.

[195] R. TRIGGIANI, *Interior and boundary regularity of the wave equation with interior point control,* Differential Integral Equations 6 (1993), pp. 111–129.

[196] R. TRIGGIANI, *Regularity with interior point control, part* II*: Kirchhoff equations*, J. Differential Equations 103 (1993), pp. 394–420.

[197] R. TRIGGIANI, *Control problems in noise reduction: The case of two coupled hyperbolic equations,* Smart Structures and Materials, Mathematics, Modeling and Control, Proc. SPIE 3039, New York, 1997, pp. 382–392.

[198] R. TRIGGIANI, *The algebraic Riccati equations with unbounded coefficients; Hyperbolic case revisited,* Contemporary Mathematics: Optimization Methods in PDE's, Vol. 209, AMS, Providence, RI, 1997, pp. 315–339.

[199] R. TRIGGIANI, *Minimax Problems for Systems with Singular Estimate,* manuscript.

[200] M. TUCSNAK, *Control of plates vibrations by means of piezoelectric actuators,* Discrete Continuous Dynamical Systems 2 (1996), 281–293.

[201] M. TUCSNAK, *Regularity and exact controllability for a beam with piezoelectric actuator*, SIAM J. Control Optim. 34 (1996), pp. 922–930.

[202] J. VANCOSTENOBLE AND P. MARTINEZ, *Optimality of energy estimates for the wave equation with nonlinear boundary velocity feedbacks*, SIAM J. Control Optim. 39 (2000), pp. 776–797.

[203] G. WEISS AND M. WEISS, *Optimal control of weakly stable regular linear systems,* Math. Control Signals Systems 10 (1997), pp. 287–330.

[204] G. WEISS AND H. ZWART, *An example in lq optimal control*, Systems Control Lett. 8 (1998), pp. 339–349.

Index